Certified Information Systems Security Professional (CISSP) Exam Guide

Master CISSP with hands-on bonus content and practical coverage of all eight exam domains

Ted Jordan, CISSP

Ric Daza, PhD, CISSP

Hinne Hettema, PhD

Certified Information Systems Security Professional (CISSP) Exam Guide

Authors: Ted Jordan, Ric Daza, and Hinne Hettema

Reviewers: Prashant Mohan, Navya Lakshmana, Commander Saurabh Prakash Gupta, and Richard Carpenter

Publishing Product Manager: Anindya Sil

Editorial Director: Alex Mazonowicz

Development Editor: Shubhra Mayuri

Digital Editor: M Keerthi Nair

Senior Development Editor: Megan Carlisle

Presentation Designer: Shantanu Zagade

Editorial Board: Vijin Boricha, Megan Carlisle, Simon Cox, Ketan Giri, Saurabh Kadave, Alex Mazonowicz, Gandhali Raut, and Ankita Thakur

First Published: September 2024

Production Reference: 3231225

Published by Packt Publishing Ltd.
Grosvenor House
11 St Paul's Square
Birmingham
B3 1RB

ISBN: 978-1-80056-761-0

www.packtpub.com

Contributors

About the Authors

Ted Jordan, M.S., CISSP, Linux+, is a seasoned cybersecurity professional with over 30 years of experience. His career includes work with NASA, General Motors, Silicon Graphics, Sun Microsystems, Fakespace, and AM General. Ted has trained over 2,500 students to achieve their CISSP, Security+, and Linux+ certifications with The Training Camp and Learning Tree. He is also the author of five books on Linux and CISSP.

In his free time, Ted enjoys a good game of tennis or watching the complexities of carom three-cushion billiards. You can follow him at `linkedin.com/in/tedjordan` and `youtube.com/jordanteamlearn`.

Ricardo "Ric" Daza, PhD, is a cybersecurity mentor with the Tampa Bay Wave Accelerator, a committee member with West Florida ISACA, and a recipient of two NSA fellowships. He is also an adjunct cybersecurity professor and frequent speaker at regional and international conferences with a doctorate in Information Assurance and holds double CCIE[2] (R&S, Security), CISSP, CRISC, CISA, ISO 27001 Lead Auditor, PMP, and RHCE certifications.

Dr. Daza builds networks and develops cybersecurity solutions for foreign and domestic government agencies, as well as Fortune 500 companies in the financial, technology, defense, healthcare, and manufacturing sectors. He contributes to the cyber defense of organizations across the Americas. He specializes in an evidence-based approach to tackling process and technology challenges, including networking, risk management, security analysis, incident response, risk communication, vulnerability management, metrics and maturity programs, data science, programming, and more.

In addition to being a seasoned executive cybersecurity consultant, Dr. Daza was an exam content developer, crafting the tests like CISSP for ISC2, the largest cybersecurity certification body in the world.

Hinne Hettema, a PhD in theoretical chemistry and philosophy of science, focuses especially on the implementation of security practices and mentoring others to become proficient security professionals. Working in IT since the early 1990s and focusing on security since the early 2000s, he has held a variety of roles working as a consultant, as part of a team, or as a leader of a security team. With over two decades of experience in the security field, Hinne has also served as an adjunct senior research fellow in cybersecurity at the University of Queensland, Australia.

He has experience in developing, implementing and running security operations, incident response and security service definition and execution. He focuses current engagements primarily on how organizations can optimize current practices, develop improvements, and make sensible decisions about their future direction. To that end, he uses his skills in architecture, security posture management, data science, threat intelligence, risk assessment and situational awareness to ensure an optimal spend of the security dollar.

He also has extensive experience in incident handling and response, including OT and ICS environments. He is a confident public speaker and can present to various audiences, from the general public to boards to cybersecurity experts to people needing training in specific aspects of cybersecurity. Hinne has authored several books including *Agile Security Operation* by Packt.

About the Reviewers

Prashant Mohan is a seasoned Information Security professional with over 15 years of expertise across sectors like finance, healthcare, and services. He is known for his ability to assess security, improve processes, and mitigate risks, while also mentoring aspiring professionals and sharing his knowledge through webinars and forums. Prashant has authored two acclaimed books, including "Cirrus – 8000 Ft. view of CCSP Exam" and "The Memory Palace – A Quick Refresher For Your CISSP Exam". Additionally, his role as a technical editor for various publications underscores his commitment to enhancing industry knowledge.

Recognized as a thought leader in the realm of security architecture, Prashant Mohan leverages modern platforms to educate and empower others. His YouTube channel, "Lazy_Architect", (http://www.youtube.com/@Lazy_architect) serves as a hub for valuable tutorials, discussions, and insights into security architecture best practices.

In his leisure time, he is an avid explorer of cosmos delving into the mysteries of the Universe. When he is not doing any of these things, he relishes moments with his lovely wife and adoring daughter.

Navya Lakshmana, a cybersecurity professional with a decade of experience in information technology, earned her bachelor's degree in electronics and communication from Visvesvaraya Technological University (VTU) in Bengaluru, Karnataka, India. She is currently employed at Siemens Healthineers, a renowned healthcare service provider that creates advanced medical technology for everyone, everywhere, sustainably. Navya holds distinguished certifications, including CISSP, CCSP, GIAC Cloud Penetration Tester (GCPN), and GIAC Penetration Tester (GPEN).

Beyond her professional endeavors, Navya is dedicated to cybersecurity education. As the founder of CyberPlatter, a YouTube channel, she educates cybersecurity enthusiasts and professionals alike.

Commander Saurabh Prakash Gupta, CISSP, CCSP, CISM, GCIH, is a military veteran currently employed as a cybersecurity expert with Bosch Global Software Technologies in Bengaluru, India. Having started his journey as a marine engineer, he then developed expertise in the domains of information technology and information security over more than 20 years. He is currently leading the cybersecurity program for providing consulting and testing services to global customers in automotive, embedded, IoT, OT, cloud, and enterprise IT product domains. Previously, for the Indian Navy, he led the program for software induction and enterprise cybersecurity deployment at the Indian Navy headquarters. He loves traveling and is an avid reader.

Richard Carpenter, ACIIS CISSP, is an Information Security Professional specialized in the implementation and architecture of digital transformation and public cloud adoption for forward-thinking organizations.

With over 20 years of experience in the IT industry, Richard's expertise spans Information Risk Management, user training, operational policies, and security architecture. He began his career in infrastructure support, focusing on Identity and Access Management, and initially developed a passion for automation tools and supporting users with AS400-based technologies. Over time, he grew more interested in infrastructure design and support. A member of (ISC)² since 2016, Richard currently works as an Information Security Specialist in the media and advertising industry in the UK. He is frequently interviewed for industry media and podcasts. Richard's professional interests include multi-cloud security posture assessments, vulnerability management program design, managed service reviews, open-source software maturity programs, disaster recovery audits, and software readiness reviews.

Educated at the University of Portsmouth with a specialization in Electronic and Electrical Engineering, Richard holds CPD qualifications as a CISSP, AWS Solutions Architect, AWS Security Specialist, AWS Advanced Networking Specialist, and ISO 27001 Lead Implementer. He is also a member of the Chartered Institute of Information Security and is a passionate advocate for educating young people about online security and privacy awareness.

Table of Contents

3

4

8

Architecture Vulnerabilities and Cryptography   129

15

Designing and Conducting Security Testing   279

16

Planning for Security Operations   293

19

Business Continuity, Personnel, and Physical Security   375

20

Software Development Life Cycle Security   389

21

22

23

24

Preface

Information system security is critically important for enterprises as cybercrime continues to grow at a rapid pace. According to *Cybercrime Magazine*, cyber attackers inflicted damage totaling $6 trillion globally in 2021 and that is expected to grow to $10.5 trillion by 2025 (`https://packt.link/8qRsd`). As businesses move further with information systems to control various facilities such as water treatment facilities, automobiles, and nuclear plants, they need talented and certified professionals to help them secure these environments because cyberattacks could also be life-threatening.

This need for security has led to a high demand for knowledgeable and talented information system security engineers and architects who can help organizations design, build, and operate secure **Information Technology** (**IT**) environments. IT security certifications can help organizations identify and develop critical skills for implementing various cybersecurity initiatives. Certifications can also help individuals demonstrate their technical knowledge, skills, and abilities to potential employers to advance their careers.

The goal of this book is to help you pass the **Certified Information Systems Security Professional** (**CISSP**) certification exam by ISC2. The CISSP certification is the most sought-after global credential and represents the highest standard for information system security expertise. It confirms your ability to apply best practices to information system security architecture, design, and operations.

As you progress through this book, you'll engage with practical and straightforward explanations of cybersecurity concepts, designed to educate you on the challenges security professionals face in computing environments. The chapters in this book cover the domains of topics relevant to the CISSP exam, including developing a comprehensive information system security policy, conducting risk assessmentsfor IT deployments, implementing identity and access management solutions, securing data in system storage, and designing disaster recovery plans. Each chapter will guide you through scenarios that test your understanding of the CISSP domains, from architectural considerations to legal and compliance frameworks.

For additional practice questions and exams, acquire the *CISSP Certification Practice Exams and Tests* book. It includes over 1,000 practice questions critical to successfully passing the CISSP exam on the first try (ISBN: 1800561377).

By the end of this study guide, you'll possess a solid understanding of information system security principles and practices, as well as the confidence needed to apply this knowledge in your current role. You will also be well prepared to pass the CISSP exam the first time!

Who This Book Is For

This book is for those who are preparing to take and pass the CISSP exam. It is recommended that you have at least five years of experience in IT, with two of those years being focused on aspects such as IT security, application security, privacy, or data governance.

What This Book Covers

Chapter 1, Ethics, Security Concepts, and Governance Principles, introduces the most relevant information security concepts, which are the foundation of the entire book. We discuss the importance of ethics, fundamental security concepts, and the difference between due care and due diligence.

Chapter 2, Compliance, Regulation, and Investigations, discusses privacy regulations and country-specific legislation related to PII and PHI. We will review key jurisdictional differences in data privacy.

Chapter 3, Security Policies and Business Continuity, describes the common practices that organizations follow for defining security policies and deploying frameworks that prioritize business continuity.

Chapter 4, Risk Management, Threat Modeling, SCRM, and SETA, discusses the application of key risk management principles. This will include an in-depth look at threat modeling techniques and methodologies, along with **Supply Chain Risk Management (SCRM)** strategies. Additionally, you'll evaluate **Security Education, Training, and Awareness (SETA)** programs.

Chapter 5, Asset and Privacy Protection, delves into identifying and classifying information and assets, establishing appropriate handling requirements for them, and ensuring that resources are securely provisioned.

Chapter 6, Information and Asset Handling, further details asset security, focusing on the management of digital assets throughout their life cycle. It covers the usage and destruction phases of information, outlining the key requirements for effective oversight of digital assets.

Chapter 7, Secure Design Principles and Controls, guides you through the fundamental concepts of security models, helping you understand their role in protecting systems. Additionally, it covers the best practices for selecting appropriate security controls based on the specific requirements of a system.

Chapter 8, Architecture Vulnerabilities and Cryptography, discusses how you can assess and mitigate vulnerabilities in security architectures, select and implement cryptographic solutions as per your needs, and explore cryptanalytic attack methods to better recognize and defend against threats.

Chapter 9, Facilities and Physical Security, covers how to apply security principles in the design of buildings and other facilities, ensuring they are safeguarded against potential threats. The chapter will also cover the design and implementation of effective security controls tailored to different areas within a facility, including both restricted zones and general work areas. You will also learn how to incorporate utilities and HVAC systems into the overall security framework.

Chapter 10, Network Architecture Security, provides an overview of the key concepts of network architectures. We discuss network fundamentals, networking devices, and providing security channels around these architectures.

Chapter 11, Securing Communication Channels, discusses how organizations secure communications using various hardware and software solutions.

Chapter 12, Identity, Access Management, and Federation, discusses the implementation of security practices suited to an organization's environment, performing detailed accounting of user and system access, and securely managing the provisioning and deprovisioning of identities to minimize vulnerabilities.

Chapter 13, Identity Management Implementation, focuses on the implementation of effective authentication systems to verify user identities and control access. The chapter will also delve into authentication, authorization, and accounting, explaining how these systems work together to ensure that users are not only verified but also granted appropriate access and that their activities are properly logged.

Chapter 14, Designing and Conducting Security Assessments, discusses how you can develop effective methods to evaluate the security posture of systems and ensure they meet the required standards. The chapter covers how to conduct thorough security control testing, including how to execute and analyze tests to identify vulnerabilities and verify the effectiveness of implemented controls.

Chapter 15, Designing and Conducting Security Testing, reviews the most common ways to conduct audits of IT systems, covering the audit process, the methodologies, and the required adaptations for a cloud environment.

Chapter 16, Planning for Security Operations, discusses investigation procedures and how to comply with them so that all incidents are properly documented and reviewed. The chapter covers logging and monitoring activities that track and help you analyze system events for potential security issues.

Chapter 17, Security Operations, details how you can effectively execute the incident management process. The chapter covers the procedures for responding to and resolving security incidents, and also operating and maintaining both detective and preventive measures to continuously protect systems from threats.

Chapter 18, Disaster Recovery, discusses the specifics of preparing to withstand disasters and business disruptions so that businesses can continue the delivery of products and services within acceptable time frames.

Chapter 19, Business Continuity, Personnel, and Physical Security, teaches you how to actively participate in planning and conducting exercises to test and improve security measures. The chapter also covers physical security strategies, including measures to protect physical assets and facilities from threats and ensure that employees are safeguarded and trained to handle security-related situations effectively.

Chapter 20, Software Development Life Cycle Security, is dedicated to educating you on the **Secure Software Development Life Cycle (S-SDLC)**, including coverage of topics such as defining requirements, what methodology to use to apply the S-SDLC, threat modeling, and secure coding.

Chapter 21, Software Development Security Controls, details security controls identified and applied in the software development environment. We discuss the fundamentals of source code, compilation, and tools.

Chapter 22, Securing Software Development, describes maintaining secure software. We discuss tools that monitor code changes, risk analysis, and mitigation.

Chapter 23, Secure Coding Guidelines, Third-Party Software, and Databases, discusses the security impact of acquired software, whether it be commercial off-the-shelf, open source, or third-party. The chapter also covers security vulnerabilities at the source code level, the security of **Application Programming Interfaces** (**APIs**), including best practices for protecting these critical components, and secure coding practices to prevent common vulnerabilities and ensure robust software development.

Chapter 24, Accessing the Online Practice Resources presents all the necessary information and guidance on how you can access the online practice resources that come free with your copy of this book. These resources are designed to enhance your exam preparedness.

How to Get the Most Out of This Book

This book is crafted to equip you with the knowledge and skills necessary to excel in the CISSP exam through memorable explanations of major domain topics. It covers the eight core domains critical to the security expertise that candidates must be proficient in to pass the exam. For each domain, you'll work through content that reflects real-world IT security challenges. At certain points in the book, you'll assess your understanding by taking chapter-specific quizzes. This not only prepares you for the CISSP exam but also allows you to dive deeper into a topic as needed, based on your results.

Online Practice Resources

With this book, you will unlock unlimited access to our online exam-prep platform (*Figure 0.1*). This is your place to practice everything you learn in the book.

> How to Access These Materials
>
> To learn how to access the online resources, refer to *Chapter 24, Accessing the Online Practice Resources*, at the end of this book.

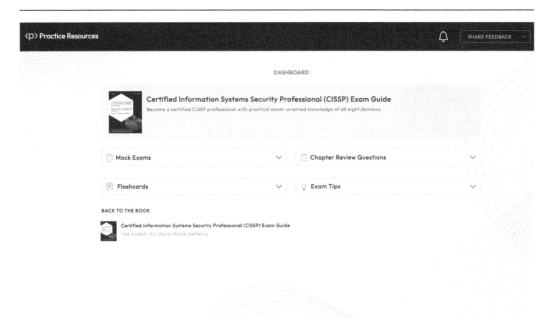

Figure 0.1: Online exam-prep platform on a desktop device

Sharpen your knowledge of CISSP concepts with multiple sets of mock exams, interactive flashcards, and exam tips, accessible from all modern web browsers.

Download the Color Images

We also provide a PDF file that has color images of the screenshots/diagrams used in this book. You can download it here: `https://packt.link/4ACMi`.

Conventions Used

There are several text conventions used throughout this book.

`Code in text`: Indicates code words in text, database table names, folder names, filenames, file extensions, pathnames, dummy URLs, user input, text on screen, and X (formerly Twitter) handles. Here is an example: "The `Unable to connect` response tells the tester that the connection is closed."

A block of code is set as follows:

```
rule TestExample
 {
    strings:
    $test_string = "foobar"
    condition:
    $text_string
 }
```

Bold: Indicates a new term or an important word and abbreviations. Here is an example: "These experts (all of whom hold a CISSP certification) gather every three years to review and revise the exam outline during the **job task analysis (JTA)** portion of the certification's life cycle."

> **Tips or important notes**
> Appear like this.

Get in Touch

Feedback from our readers is always welcome.

General feedback: If you have any questions about this book, please mention the book title in the subject of your message and email us at customercare@packt.com.

Errata: Although we have taken every care to ensure the accuracy of our content, mistakes do happen. If you have found a mistake in this book, we would be grateful if you could report this to us. Please visit www.packtpub.com/support/errata and complete the form. We ensure that all valid errata are promptly updated in the GitHub repository at https://packt.link/Zjmry.

Piracy: If you come across any illegal copies of our works in any form on the internet, we would be grateful if you could provide us with the location address or website name. Please contact us at copyright@packt.com with a link to the material.

If you are interested in becoming an author: If there is a topic that you have expertise in and you are interested in either writing or contributing to a book, please visit authors.packtpub.com.

Share Your Thoughts

Once you've read *Certified Information Systems Security Professional (CISSP) Exam Guide*, we'd love to hear your thoughts! Scan the QR code below to go straight to the Amazon review page for this book and share your feedback.

https://packt.link/r/1800567618

Your review is important to us and the tech community and will help us make sure we're delivering excellent quality content.

Free Benefits with Your Book

This book comes with free benefits to support your learning. Activate them now for instant access (see the *How to Unlock* section for instructions).

Here's a quick overview of what you can instantly unlock with your purchase:

PDF and ePub Copies   **Next-Gen Web-Based Reader**

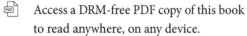

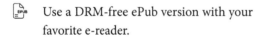

Access a DRM-free PDF copy of this book to read anywhere, on any device.

Use a DRM-free ePub version with your favorite e-reader.

Multi-device progress sync: Pick up where you left off, on any device.

Highlighting and notetaking: Capture ideas and turn reading into lasting knowledge.

Bookmarking: Save and revisit key sections whenever you need them.

Dark mode: Reduce eye strain by switching to dark or sepia themes

How to Unlock

Scan the QR code (or go to packtpub.com/unlock). Search for this book by name, confirm the edition, and then follow the steps on the page.

Note: Keep your invoice handy. Purchases made directly from Packt don't require one.

Becoming a CISSP

You have begun the journey to obtain the most prized cybersecurity certification in the world. The **Certified Information Systems Security Professional** (**CISSP**) is said to be *10 miles wide and an inch deep*. The eight domains of the CISSP cover a vast amount of information. However, despite the previous quote, you still need to understand the underlying concepts. This is because the exam does not just test your memory of concepts but also their application in scenarios to solve problems.

One of the reasons the CISSP is as broad and respected as it is is because it is built and maintained by experts from around the world and diverse industries. These experts (all of whom hold a CISSP certification) gather every three years to review and revise the exam outline during the **job task analysis** (**JTA**) portion of the certification's life cycle. During the JTA process, experts ensure that the knowledge embodied by the outline represents what a cybersecurity practitioner needs to know to perform their job effectively. This chapter will discuss why this is so critical. You'll review the CISSP exam itself, its structure, and the new CISSP **Computerized Adaptive Testing** (**CAT**) version of the exam. You'll also be provided with the best exam tips and tricks. Finally, you'll learn what it takes to become a CISSP.

Making the Most Out of This Book – Your Certification and Beyond

This book and its accompanying online resources are designed to be a complete preparation tool for your **CISSP Exam**.

The book is written in a way that you can apply everything you've learned here even after your certification. The online practice resources that come with this book (*Figure 1.1*) are designed to improve your test-taking skills. They are loaded with timed mock exams, interactive flashcards, and exam tips to help you work on your exam readiness from now till your test day.

> **Before You Proceed**
>
> To learn how to access these resources, head over to *Chapter 24, Accessing the Online Practice Resources*, at the end of the book.

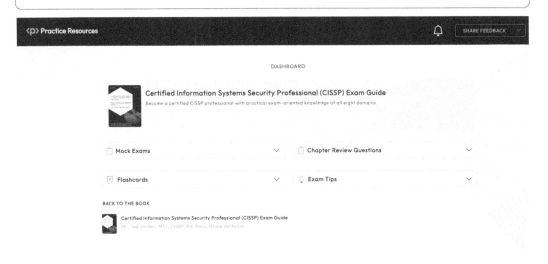

Figure 1.1: Dashboard interface of the online practice resources

Here are some tips on how to make the most out of this book so that you can clear your certification and retain your knowledge beyond your exam:

1. Read each section thoroughly.

2. **Make ample notes**: You can use your favorite online note-taking tool or use a physical notebook. The free online resources also give you access to an online version of this book. Click the BACK TO THE BOOK link from the Dashboard to access the book in **Packt Reader**. You can highlight specific sections of the book there.

3. **Chapter Review Questions**: At the end of this chapter, you'll find a link to review questions for this chapter. These are designed to test your knowledge of the chapter. Aim to score at least **75%** before moving on to the next chapter. You'll find detailed instructions on how to make the most of these questions at the end of this chapter in the *Exam Readiness Drill - Chapter Review Questions* section. That way, you're improving your exam-taking skills after each chapter, rather than at the end.

4. **Flashcards**: After you've gone through the book and scored **75%** more in each of the chapter review questions, start reviewing the online flashcards. They will help you memorize key concepts.

5. **Mock Exams**: Solve the mock exams that come with the book till your exam day. If you get some answers wrong, go back to the book and revisit the concepts you're weak in.

6. **Exam Tips**: Review these from time to time to improve your exam readiness even further.

In this section, we will cover the following topics:

- The need for CISSPs
- CISSP exam overview
- CISSP exam structure
- Exam tips and tricks
- Information about becoming a CISSP

The Need for CISSPs

One of the challenges facing the cybersecurity profession is satisfying the necessity for qualified cybersecurity practitioners to meet the demand. According to the Bureau of Labor Statistics, the rate of growth for jobs in information security is projected at 37% from 2012-2022 (`https://packt. link/FNAup`). That's much faster than the average for all other occupations. The **Human Resources (HR)** professionals who are on the front lines dealing with this challenge rarely possess the ability to quantify the expertise of a cybersecurity job candidate. Therefore, a respected, unbiased standard is necessary to help potential employers more easily determine qualified candidates from unqualified candidates. Enter ISC2 and their CISSP certification.

The **International Information System Security Certification Consortium (ISC2)** was established as a non-profit organization in 1989. Five years later, ISC2 launched its first certification, the CISSP, in 1994. At the time, the cybersecurity market was in desperate need of a baseline of cybersecurity knowledge to aid both the industry in standardizing the profession and those seeking to hire cybersecurity professionals. Since its founding, ISC2, through the CISSP and its other eight certifications, has established and maintained that standard.

In 2005, the United States **Department of Defense (DoD)** created the 8570 directive to assess and manage its cybersecurity workforce. The CISSP provides independent verification of a reliable baseline of knowledge and experience in cybersecurity of a practitioner. The CISSP tells the world that you know something about cybersecurity—not just something, but the right something about cybersecurity, as determined by industry experts who hold a CISSP certification. As per the 8570 directive and its current successor, the 8140 directive, many job roles in cybersecurity within the DoD require a CISSP certification to qualify.

In addition to helping HR professionals validate a baseline level of knowledge, the CISSP certification also validates experience. The CISSP certification requires not just a passing score but a minimum of five years of experience. ISC2 verifies this requisite experience before conferring the certification on any candidate who has achieved a passing score on the exam. You will learn more about this experience requirement in the *Information about Becoming a CISSP* section. This additional benefit of experience verification is of great value to employers.

The CISSP certification also comes with a 40-hour annual **Continuing Professional Education (CPE)** requirement to maintain the currency of your CISSP certification. See `https://packt.link/6EFMh` for more information. While ISC2 is a non-profit organization, they don't just track your CPE and maintain your currency for free; there is an annual maintenance fee of 125 USD per year. The bright side is that if you choose to pursue any of the other eight ISC2 certifications, you will pay only 85 USD per year, unlike other cybersecurity certification organizations.

CISSP Exam Overview

The CISSP exam outline is the most important tool when preparing for the certification. It is no exaggeration to say it is the roadmap of the test. This section will explain why it is so important to know it well. First and foremost, it is what ISC2 uses to build the test questions. The certification industry (organizations such as ISC2, ISACA, SANS, and CompTIA) calls exam questions **items**. The process of building test questions is called **item writing**, which for the CISSP exam and ISC2 is done by volunteer CISSPs in an **item writing workshop**.

If you search the web for item writing, you'll find many first-hand accounts from volunteers about their experiences of participating in an item writing workshop. There are some excellent ones on ISC2 where volunteers share their workshop experiences and details about the item writing process: `https://packt.link/SvggM`. ISC2 works very hard to protect the confidentiality and efficacy of their item bank (their database of exam questions). So, don't waste your time trying to find or use brain-dumps or allegedly real questions (most likely fake).

Your study time is much better spent understanding the material covered in the exam outline and how ISC2 uses it to build items. The exam outline is the product of another kind of volunteer workshop, known as a **JTA**. In this workshop, the volunteer CISSPs review the current outline and update it to more accurately reflect the knowledge and skills a CISSP should have today and over the next three-year cycle. Once this crucial step is complete, the existing items in the bank must be mapped to the new outline. This is also done by volunteer CISSPs in a workshop called an **item mapping workshop**.

The item mapping process is important for two reasons. First, categorizing items into the appropriate part of the outline is necessary to build every test with an exact balance of items from the appropriate part of the outline, as determined by the JTA. The weighting of the outline will be discussed in detail later. Second, item mapping is necessary to determine where and how big the holes are in the item bank. These holes are then assigned to subsequent item writing workshops to be filled with new items based on the new exam outline. See `https://packt.link/IqXal` to view the outline.

This aspect will be of particular interest to you as you prepare for the CISSP exam. Each item must map to a specific topic in the exam outline. No surprise items on topics not covered by the exam outline are allowed. So, the exam items are fixed by the exam outline—this is an unbreakable rule. That being said, the outline is divided into eight domains or areas of knowledge, which you will soon see can be quite broad.

Domains

A domain is a broad collection of related information. In this section, you will become more familiar with the exam outline. The top level of the outline represents the eight domains. The second level represents the subject areas within the domain that CISSP candidates need to be familiar with related to that domain. Many second-level subject areas have a third level to further clarify the knowledge that is to be tested in the exam at the level above it. Any concept under the umbrella of a domain is fair game as a potential exam item.

It is no coincidence that this book is laid out exactly like the CISSP exam's outline, as that is the information you need to know. Each domain in the exam outline will be covered by one or more chapters in this book. The goal is to introduce and explain each concept in the exam outline. Not only do you need to memorize this, but you also need to understand it as the exam tests your ability to correctly apply concepts to solve situations. It is not possible to capture every bit of potential information contained within a domain. This book will at least introduce every concept in the outline and delve deeper into those areas that are understood to have a high probability of showing up on your test.

CISSP CAT Examination Weightage

As mentioned earlier, each domain in the exam outline has a weight assigned. This means the Pearson VUE testing software must build your test with the exact percentage weights that are prescribed in the exam outline. So, if your test has 100 scored items, 16% or 16 items will be about concepts in *Domain 1, Security and Risk Management*.

While all ISC2 exam outlines provide domain-level weights, the CISSP exam outline provides weights for both linear testing and CAT. See `https://packt.link/UCB05` for more information. The following table shows the domain level (the top level) of the exam outline, along with its corresponding weights:

Domain	Weight
1. Security and Risk Management	16%
2. Asset Security	10%
3. Security Architecture and Engineering	13%
4. Communication and Network Security	13%
5. Identity and Access Management (IAM)	13%

Domain	Weight
6. Security Assessment and Testing	12%
7. Security Operations	13%
8. Software Development Security	10%

Table 1.1: CISSP CAT examination weights

The weights are the same for both versions (linear testing and CAT) of the test. ISC2 publishes item weight information for both linear testing and CAT in case you plan on taking a non-English version of the CISSP exam. All ISC2 exams besides the English CISSP exam are linear. See `https://packt.link/oNM7u` for the other languages available. While the domain weights are fairly evenly balanced, they do have a little difference among them. This may help you budget your time and help you decide where you want to focus your study efforts. This information, combined with the pre-assessment test in the next chapter, can provide insights into where and how to focus your time.

CISSP CAT Examination Information

In 2017, ISC2 began using CAT for all English CISSP exams worldwide. This version of the test covers the same material from the exam outline as the traditional test (linear testing). According to ISC2, "*CISSP CAT is a more precise and efficient evaluation of your competency*" (`https://packt.link/TxPI2`). Translation—it is a little less painful. If you know the material, the CAT exam can determine that in fewer items. You go from the linear test, which is 6 hours long and contains 250 items, to a 3-hour test with potentially as few as 100 items in the CAT exam.

Overall, the CAT exam is much nicer than the linear version. That being said, there are a few things about the CAT exam you should know so that you are not surprised. First, the CAT scoring algorithm is much more efficient. This means that you never really know when the test is going to end.

You know the absolute minimum (100 items) and the absolute maximum (3 hours), although it is unlikely that you will finish at either of those two extremes. The test ends as soon as the algorithm is confident you either know your stuff or you don't. If you don't know your stuff, the algorithm will not just let you run down the clock while exposing more items to you if it already knows you are not going to pass.

CISSP Exam Structure

The exam is made up of three types of items: multiple-choice questions, innovative questions, and scenario questions. The last two types of questions are legacy, meaning ISC2 will not be making any more questions of that type. The bulk of the questions are multiple-choice, and that is what this book will be focusing on. The other two types have been mentioned because you may see one or two in your exam.

"Innovative questions" is a fancy term for drag and drop. Imagine a graphic with four or five different boxes, where you have to drag the concept or term from one side of the screen to the other to match it up with an appropriate concept. If you know the material in this book, you should have no problem with this type of question. Another rare type of question is scenario questions. These questions have a long introduction scenario, followed by two to five questions based on that scenario.

As mentioned previously, today's CISSP exam is predominantly made up of **multiple-choice questions** (**MCQs**). These questions have a to-the-point question portion (known as the *item stem*) and they have four *options* (A, B, C, and D). Only one option is the *key* or the correct answer; there cannot be more than one correct answer. The other three options are called distractors; they are incorrect answers.

To pass the exam, you need 700 out of 1,000 points. These points are scaled, which means that not all the questions are worth the same. Additionally, 25 questions are worth zero points. These are known as pre-test questions. If a pre-test question performs well, it will be promoted to a scored item in a future exam. Obviously, ISC2 does not indicate which questions are pre-test and which are scored, so try your best on all the questions.

So, what makes one question worth more than another? The more cognitively difficult the question, the more points it is worth. This cognitive difficulty is based on **Bloom's Taxonomy**. See `https://packt.link/eLxTU` for more information on Bloom's Taxonomy. In short, Bloom explains that there are different levels of understanding regarding concepts, with the most basic being *Knowledge* and the highest being *Evaluation*. For the CISSP exam, you only need to learn *Knowledge, Application*, and *Analysis*, as shown in the following diagram:

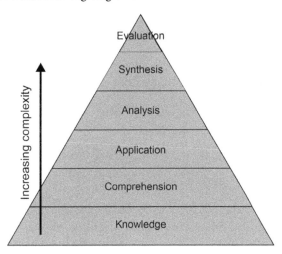

Figure 1.2: Bloom's Taxonomy

You can think of a knowledge-level question as pure memorization of a term or a concept you read. Application-level questions can be thought of as a deeper understanding of the underlying concept. Finally, the most challenging of cognitive levels is *Analysis*. It requires a deep understanding of multiple concepts; in particular, applying multiple concepts to solve a specific problem.

The idea of cognitive difficulty is best made clear with a few examples. Consider a concept from *Domain 4, Communication and Network Security*; specifically, *4.1*:

- At which layer of the Open System Interconnection (OSI) reference model does the Address Resolution Protocol (ARP) operate?

    A. 2 – Data Link

    B. 3 – Network

    C. 6 – Presentation

    D. 7 – Application

This is an example of a knowledge-level item. You only need to remember from reading or seeing an OSI model graphic that ARP is a layer 2 protocol. You need not know what it does, how it does it, about security issues with ARP, or how to fix them.

- What is the purpose of the Address Resolution Protocol (ARP)?

    A. To resolve a Fully Qualified Domain Name (FQDN)

    B. To request an Internet Protocol (IP) address for a host

    C. To resolve an Internet Protocol (IP) address to a Media Access Control (MAC) address

    D. To build a loop-free topology in Internet Protocol (IP) networks

This is an example of an application-level item. It requires a deeper understanding of what the ARP does, why it is needed, and where it fits into the OSI and **Transmission Control Protocol/Internet Protocol (TCP/IP)** models.

- Which attack leverages the Address Resolution Protocol (ARP)?

    A. Transmission Control Protocol (TCP) spoofing

    B. Distributed Denial of Service (DDoS)

    C. Man-in-the-Middle (MitM)

    D. Dynamic Host Configuration Protocol (DHCP) starvation

This is an example of an analysis-level item. Here, the exam is still just talking about ARP, but each question requires a progressively deeper understanding of the underlying ARP concept. For this item, you must understand what ARP is, how ARP works, and the cybersecurity attacks that use it. Notice that all the items are single sentences. Note that there is no correlation between the length of a question's portion (*item stem*) and its cognitive difficulty.

Information About Becoming a CISSP

What does it take to become a CISSP? Two things. First, you must demonstrate mastery of the knowledge encompassed in the CISSP exam outline, which this book and your diligent efforts will help you with. Second, you must meet the CISSP experience requirement. See https://packt.link/OkYeS for more details. Upon passing the exam, you must furnish ISC2 with proof of at least five years of *cumulative* paid work experience in at least two of the eight domains in the CISSP exam outline.

ISC2 is very specific regarding how much experience it takes to satisfy this requirement. By five years, they mean throughout your career, including full-time (35+ hours/week), part-time (20–34 hours/week), and internships. One year of experience equals 2,080 hours. So, a total of 10,400 hours is required.

At the time you pass the exam, you are not a full CISSP yet. A four-year college degree or a certification from an ISC2-approved list will satisfy one year of experience. If you do not currently meet the experience requirement yet, don't worry—you will be designated as an Associate CISSP and will be given six years to meet the job experience requirement.

Exam Tips and Tricks

This section will present some tried and tested exam tips and tricks to help you study for the CISSP exam, as well as some tips on how to approach the questions. First, consider the ISC2 website. There is a wealth of resources there, two of which you should be familiar with. The first is the ISC2 Community (https://packt.link/OT5sN), where you can explore the community you are trying to join. Be sure to check out the CISSP study group at https://packt.link/mEwXi.

The second resource is the ISC2 official acronym list, which is made available to you during the exam. However, you can preview it here: https://packt.link/AZxpN. Every acronym used anywhere in an ISC2 item bank is made public here. The item bank is a sneak peek into the concepts the items cover. Note that this covers all the acronyms for all nine of the tests that ISC2 offers (CISSP, CISSP-ISSAP, CISSP-ISSEP, CISSP-ISSMP, SSCP, CCSP, CAP, CSSLP, and HCISPP); they are not broken down by certification.

The goal of ISC2 is to ensure that exam candidates have a true command of the exam's material, thus avoiding *on-paper* CISSPs. These are people who have their certification, but once they are in a professional setting, they do not understand the CISSP **Common Body of Knowledge (CBK)**, which would threaten ISC2 and its CISSP certification's hard-fought reputation as the best in cybersecurity. To that end, this section will discuss your study strategy. Unlike other tests you may have taken in the past, the CISSP exam will require more than just memorization to pass. It is important to keep this in mind. With each concept you are exposed to as you prepare, ask yourself: *Why is this important? How does it work? What other concepts does it relate to?*

A good example of the axiom *understand, don't memorize* is aptly illustrated concerning security frameworks such as ISO 27001, NIST 800-53, and COBIT. While it is important to be familiar with these and other fundamental documents, these three frameworks all cover the same concepts—it is just that they are published by three different organizations: ISO, NIST, and ISACA, respectively. You do not want to spend your precious study time memorizing which framework says what or how it says it. Rather, focus your efforts on understanding the concepts contained within, why they are important, and who tends to use one framework over another and why.

Moving on to the test itself, take a look at some strategies to use during the test. Remember that the bulk of the exam questions will be in multiple-choice format, that is, the format where the question portion of the item is known as the *item stem*, and the four potential answers are known as *options*. Each part of these items is discussed next. First, keep in mind that the length of the item stem can mislead you into a false sense of security.

As mentioned earlier in this chapter, the length of the item stem is not representative of its underlying complexity. So, be sure you understand the nuance of what is being asked. It can be easy to quickly read a question, especially one with a short item stem, and assume you know what they are asking. The best way to avoid this pitfall is to read the question slowly and carefully. Your eyes can play tricks on you when you speed read. Missing or misreading just one word can change its meaning. Anxious test-takers tend to rush, afraid they will run out of time. If you know the material, then there will be plenty of time.

Now, take a look at the options portion of an item. Remember that there are only four options (A, B, C, and D). Only one of those options is the correct answer or the *key*. The other three options are aptly named *distractors*. ISC2 does not set out to trick you but to test how well you know the material. Sometimes, the difference between one answer option and another is one word or the sequence of a list. So, the wrong answer will look right to someone who only slightly knows the concept being tested. That is the mark of a good distractor: not to trick someone who understands the concept but to distinguish between the ones who do and do not know the concept well.

Sometimes, you can know the material too well. This can happen in a couple of ways. One way is that you work or have worked in the domain that is being tested so you have real-world experience. This can cause you to overthink the question. Keep in mind that every item on the CISSP exam must be backed up with a valid reference. Exam items are never based solely on an item writer's personal experience unless their personal experience is common practice. It would be unfair to expect any CISSP candidate to have knowledge that is not publicly available, such as from non-proprietary sources such as books, journals, and websites.

If you find yourself facing an item where, after reading the stem, you cannot find the right answer among the options, here are a few tips. First, look for the best answer from the given choices. Next, all else being equal, choose your answer while wearing your manager hat and not as a technical person. Remember that the CISSP is meant to be broad, not deep—a perspective prized among managers. Finally, if those two tips do not illuminate the best choice, try to understand the differences among and between all the options given. If all else fails, guess. In the CAT version of the CISSP exam, you cannot mark questions or go back to a question later, so never leave a question unanswered.

Summary

In this chapter, we discussed the CISSP certification and why it is so valuable in the cybersecurity industry. You also learned how it is built and maintained by CISSP-certified experts from around the world. You were introduced to the all-important CISSP exam outline provided by ISC2 and the foundation of how this book is organized and dug deeper into the CISSP exam's structure. You got some exam tips and tricks and learned about the experience requirements to fully become a CISSP.

The next chapter will give you a pre-assessment test to help you gauge your strengths and weaknesses in the exam outline.

II
Pre-Assessment Test

To successfully begin any journey, you need two things. First, you must know where you are starting from. Second, you must know where you are going (your destination). The second part is easy: your goal is to pass the CISSP exam. On every journey, it is usually desirable to take the most direct path. To that end, the purpose of this chapter is to help you determine where your knowledge might be stronger and where it might be weaker.

> **Take this test with a timer**
>
> You can also find the questions listed below on our online platform. From the *Dashboard*, go to **Practice Questions** and select **Pre-assessment Test** from the dropdown.
>
> Before you begin, ensure that you have unlocked the accompanying resources. If you have not, refer to *Chapter 24, Accessing the Online Practice Resources*, for detailed instructions.

This does not mean that if you pass all the pre-assessment questions from a given domain, it is safe to skip that domain entirely. If you plan to go through this book sequentially, chapter by chapter, then you can skip the questions in this chapter and jump right into *Chapter 1, Ethics, Security Concepts, and Governance Principles*. When you are done with the material in this book, you can use the questions in this chapter to help you prepare for the exam. However, if your particular study style is where you like to jump around using either the domain weights (see the previous chapter) or the results of this pre-assessment test as a guide, that strategy works as well. This pre-assessment test is weighted by domain, just like the actual test. So, get started by discovering your strengths and weaknesses.

Security and Risk Management – 16%

1. How many canons are there in the ISC2 Code of Ethics?

    A. 3

    B. 4

    C. 5

    D. 6

2. What is the purpose of conducting a Business Impact Analysis (BIA)?

   A. Enumerating vulnerabilities and prioritizing them for the business

   B. Reporting a breach and determining its impact on the business

   C. Determining and quantifying the cybersecurity risk to the business

   D. Identifying and evaluating the impact that unexpected events have on the business

3. Which of the following should be used to determine the risks associated with using a Cloud Provider (CP) for the backend of a mobile application?

   A. Control Objectives for Information and Related Technology (COBIT)

   B. Open Web Application Security Project (OWASP)

   C. Cloud Access Security Broker (CASB)

   D. Process for Attack Simulation and Threat Analysis (PASTA)

4. Which privacy regulation would an international company need to meet to be compliant in Canada?

   A. Health Insurance Portability and Accountability Act (HIPAA)

   B. Personal Information Protection and Electronic Documents Act (PIPEDA)

   C. General Data Protection Regulation (GDPR)

   D. Controlling the Assault of Non-Solicited Pornography and Marketing (CAN-SPAM)

Asset Security – 10%

5. Which of the following roles has technical control over an information asset dataset?

   A. Data creator

   B. Data custodian

   C. Data processor

   D. Data owner

6. Which classification type is BEST suited for information that, if compromised or accessed without authorization, could lead to criminal charges?

   A. Internal-only

   B. Confidential

   C. Restricted

   D. Public

7. How long should an organization retain its data?

    A. 1 to 3 years

    B. It depends on the kind of data being retained

    C. At least 7 years

    D. Destroy it as soon as it is no longer needed

Security Architecture and Engineering – 13%

8. Which class of fire extinguishers is **BEST** for electrical fires?

    A. Class A

    B. Class B

    C. Class C

    D. Class D

9. Which type of attack was MOST likely used if users visiting a website are seeing anti-virus warnings about malicious code?

    A. Cross-Site Scripting (XSS)

    B. Distributed Denial of Service (DDoS)

    C. Structured Query Language (SQL) injection

    D. Buffer overflow

10. Which method is BEST for protecting laptops?

    A. Full Disk Encryption (FDE)

    B. Advanced Encryption Standard (AES)

    C. Blowfish

    D. Multi-Factor Authentication (MFA)

Communication and Network Security – 13%

11. Which of the following is a well-known Transmission Control Protocol (TCP) port used by Simple Mail Transfer Protocol (SMTP)?

    A. 22

    B. 21

    C. 25

    D. 79

12. Which of the following wireless security protocols utilizes Simultaneous Authentication of Equals (SAE) for secure authentication?

    A. Wired Equivalent Privacy (WEP)

    B. Wi-Fi Protected Access 2 (WPA2)

    C. Wi-Fi Protected Access (WPA) Enterprise

    D. Wi-Fi Protected Access 3 (WPA3)

13. Which of the following communication protocols is vulnerable to a snooping attack?

    A. Secure Shell (SSH)

    B. Layer 2 Tunneling Protocol (L2TP) v2

    C. Point-to-Point Tunneling Protocol (PPTP)

    D. Internet Protocol Security (IPsec)

Identity and Access Management (IAM) – 13%

14. Which of the following biometric authentication methods is the fastest while also being accurate?

    A. Facial imaging

    B. Hand geometry

    C. Iris recognition

    D. Signature

15. Which of the following components of an access control system determines what a user is allowed to do?

    A. Authentication

    B. Authorization

    C. Identification

    D. Verification

16. Which of the following security management methodologies can make use of geo-location to grant access?

    A. Attribute-Based Access Control (ABAC)

    B. Mandatory Access Control (MAC)

    C. Discretionary Access Control (DAC)

    D. Role-Based Access Control (RBAC)

Security Assessment and Testing – 12%

17. Which application security testing approach is the MOST cost-effective and comprehensive?

    A. Dynamic Application Security Testing (DAST)

    B. Static Application Security Testing (SAST)

    C. Interactive Application Security Testing (IAST)

    D. Penetration Testing Execution Standard (PTES)

18. Which of the following information security metrics is the BEST Key Risk Indicator (KRI) for an e-commerce business?

    A. Mean Time To Contain (MTTC)

    B. Number of days to deactivate former employee credentials

    C. Number of systems with known vulnerabilities

    D. Percentage of business partners with effective cybersecurity policies

19. What is the PRIMARY limitation of a Common Vulnerability Scoring System (CVSS) score?

    A. It doesn't take into account the impact of a successful vulnerability

    B. It doesn't take into account the attack vector

    C. It doesn't take into account the damage to your company

    D. It doesn't take into account the attack's complexity

Security Operations – 13%

20. When performing disaster recovery planning, which of the following options is the MOST applicable to determine the data backup's frequency?

    A. Recovery Time Objective (RTO)

    B. Recovery Point Objective (RPO)

    C. Maximum Tolerable Downtime (MTD)

    D. Mean Time between Failures (MTBF)

21. Which internal control is BEST used to prevent a single user from having control of every aspect of a change?

    A. Separation of Duties (SoD)

    B. Two-Factor Authentication (2FA)

    C. Least privilege

    D. Job history verification

22. Which failure method for an inline Intrusion Prevention System (IPS) would BEST serve security in the event of a failure?

    A. Fail-safe

    B. Fail-open

    C. Fail-closed

    D. Failover

Software Development Security – 10%

23. What is the PRIMARY factor in determining whether Commercial-Off-the-Shelf (COTS) software should be acquired?

    A. Procurement

    B. Business needs

    C. Development timeline

    D. Maintenance requirements

24. What best practice should a software design team adhere to when designing secure code?

    A. Limit the runtime of all functions

    B. Write Don't Repeat Yourself (DRY) code

    C. Use open source libraries

    D. Validate input from all untrusted data sources

25. Which secure coding practice will assist in preventing the disclosure of sensitive information in error responses?

    A. Session management

    B. Communication security

    C. Database security

    D. Error handling and logging

Answer Key

1. Answer B.

   This is a simple knowledge question, but it can be hard if you have never seen the *Ethics* page on the official ISC2 website, specifically the *Code of Ethics Canons* section. See `https://packt.link/NRHh1`.

2. Answer D.

   The answer options can be wordy, but each describes a specific security work product. For example, A is a SAR. B is a breach report. C is a RAR. D is the best answer that matches the purpose of a BIA. See `https://packt.link/4iqjJ`.

3. Answer D.

   COBIT is a security framework. OWASP is a nonprofit foundation that works to improve the security of software, which might seem like the right answer but it is not the best option. A CASB might sound right if you don't know it is cloud-hosted software, on-premises software, or hardware that acts as an intermediary between users and cloud service providers. PASTA is the best option. See `https://packt.link/cGkUc`.

4. Answer B.

   You need to be familiar with all the options to know that only the PIPEDA is a Canadian regulation. The others only have jurisdiction in the US or EU. See `https://packt.link/mAwS5`.

5. Answer B.

    You need to be familiar with all of these roles to determine the correct one as the differences between them are very nuanced. See `https://packt.link/1fvfD`.

6. Answer C.

    Part of knowing data classification types is understanding, at least at a high level, examples of the kinds of data that can be found at each level. See `https://packt.link/4zufg`.

7. Answer B.

    The point of this item is that the data retention policy is situational. The correct response depends on the kind of data being retained, the industry, and its regulations. See `https://packt.link/oxjqz`.

8. Answer C.

    It's important to know the different classes of fire extinguishers and what types of fire they are made to extinguish. Electrical fires are most likely to occur in data centers, so a Class C extinguisher or a multipurpose one such as a C-D extinguisher is the best option. See `https://packt.link/ILrxX`.

9. Answer A.

    As one of the most prevalent issues in the OWASP Top 10, you should be familiar with the XSS attack. See `https://packt.link/zwEbu`.

10. Answer A.

    Read the question carefully and consider the BEST option for laptops. Remember that laptops can be physically stolen, so FDE is the best option, regardless of which encryption algorithm is used. See `https://packt.link/k8wUb`.

11. Answer C.

    Well-known port numbers are one of those things that are worth committing to memory, especially for protocol analysis. See `https://packt.link/6j8Wi`.

12. Answer D.

    It is important to be familiar with the various wireless security protocols, including their similarities, differences, strengths, weaknesses, and release timelines. See `https://packt.link/qeGbA`.

13. Answer B.

   While you do not need to know every RFC, you do need to know a few of the more important ones. You also need to be well-versed in VPN technology, in particular, the differences between the common formats. You also need to know which VPN protocols give you a network within a network and which protocols give you a secure network within a network. See https://packt.link/T2zND.

14. Answer C.

   Beware of the wording of this question. Note that it does not ask for the best (which would be very subjective) but the fastest and most accurate. See https://packt.link/J0rXH.

15. Answer B.

   Understanding the subtle differences between what these terms mean is core to your cybersecurity knowledge. See https://packt.link/dSAmn.

16. Answer A.

   It is important to know each of these security management methodologies, as well as the features of each one, to be able to answer this question. Only one of them can make use of the geo-location *attribute* as part of the authentication criteria. See https://packt.link/fbFwQ.

17. Answer B.

   For Domain 6, it is crucial to understand all of the security testing approaches for applications. What are the traits of each and when is the most appropriate time to use each one? See https://packt.link/dZQdI.

18. Answer C.

   KRIs are an important way to measure whether or not your cybersecurity program is improving or worsening. Considering the options given, which one is the best KRI to let you know about the health of your cybersecurity program, and which just provides noise? See https://packt.link/sx1Fo.

19. Answer C.

   One of the de facto metrics in cybersecurity that you should be intimately familiar with is the CVSS. See https://packt.link/DJTfg.

20. Answer B.

   These are important metrics regarding DR/BCP. You should understand each one in terms of what role they play in the DR/BCP process. See https://packt.link/306je.

21. Answer A.

    Each of the options listed is a tool used in security operations. Only two (A and C) are potentially correct. However, only A is suitable for limiting who can control every aspect of a change. See https://packt.link/0CDhP.

22. Answer C.

    You should be familiar with each of these failure methods and when it is appropriate to use each one. See https://packt.link/u2nXz.

23. Answer B

    COTS is a fairly common term in government contracting circles. It is important to know what it is, when it is appropriate to use it, and what the benefits of that choice are. See https://packt.link/ZvHZb.

24. Answer D.

    You should be familiar with how hackers exploit code vulnerabilities. It is also important to know the top coding practices for secure code development and what exploits each counteracts. See https://packt.link/JVOW4.

25. Answer D.

    The answer options are all secure coding practices, but only one helps prevent sensitive information from being disclosed in error responses. See https://packt.link/Tejyw.

Summary

This chapter represented a miniature version of the CISSP test and should have given you an idea of where your strengths and weaknesses are. The questions matched the domain weights specified in the exam outline, just at a smaller scale. Keep in mind that it would be impossible to truly test your knowledge of the exam outline without a pre-assessment test with a thousand questions. So, even if you got the answers right—or you guessed and got the answers right—you should still read the chapters that cover that domain.

The next chapter will start covering the material in *Domain 1*, including professional ethics, foundational security concepts, and governance principles.

Ethics, Security Concepts, and Governance Principles

Being a **Certified Information Systems Security Professional** (**CISSP**) carries several responsibilities, including adhering to professional ethics, applying security governance to organizations, understanding the requirements for investigations, enforcing security policies and procedures, applying risk management principles, and maintaining security awareness and training programs.

This chapter begins with the CISSP's understanding of professional ethics, which is a requirement of the **International Information System Security Certification Consortium** (**ISC2**). Next, you will learn about the basic concepts of security, such as data confidentiality, data integrity, and data availability.

Finally, a CISSP must be able to apply security governance principles such as aligning security functions to an organization's policies, strategies, and goals. By the end of this chapter, you will be able to answer questions on the following:

- The ISC2 Code of Ethics
- The definitions of **confidentiality**, **integrity**, and **availability** (**CIA**)
- The rivals of CIA
- The definitions of nonrepudiation and authenticity
- The alignment of security governance to an organization's mission

The ISC2 Code of Professional Ethics

The tools and techniques used by a CISSP are very similar to those that are employed by a criminal hacker. One key difference is that the CISSP follows the ISC2 Code of Ethics, which can be found at `https://packt.link/8vIti`. It is duplicated here for your convenience:

- **Code of Ethics Preamble**:

  - The safety and welfare of society and the common good, duty to our principles, and to each other, requires that we adhere, and be seen to adhere, to the highest ethical standards of behavior.

  - Therefore, strict adherence to this Code is a condition of certification.

- **Code of Ethics Canons**:

  - Protect society, the common good, necessary public trust and confidence, and the infrastructure.

  - Act honorably, honestly, justly, responsibly, and legally.

  - Provide diligent and competent service to principals.

  - Advance and protect the profession.

The ISC2 requires its certification holders to follow and support the ISC2 Code of Ethics. Violating these provisions could subject the individual to the revocation of their certification. The reporting of any ethics violations to ISC2 must be in writing.

Important Security Concepts

Information security professionals focus on three fundamental areas of protection:

- Confidentiality
- Integrity
- Availability

Without one of these areas, an organization is more vulnerable to harm or damage and is exposed to higher information security risks. The CISSP must plan and design good security governance using these three concepts.

Managing security is like the support of a three-legged stool, as shown in *Figure 1.1*. If one area of security falters, the entire organization is vulnerable to data theft, malware infections, and ransomware attacks. This places the entire firm at risk of going out of business, causing potentially thousands of people to lose their jobs.

Figure 1.1: The CIA three-legged stool

For example, hospitals such as St. Lawrence Health System in New York suffered major ransomware attacks because their staff clicked on links within emails, which is also known as phishing (you can read more about these incidents at https://packt.link/FyMjt). Ransomware is an availability attack where users cannot access critical files such as X-rays and prescriptions unless the attacker is paid a ransom fee. Once the attackers receive the ransom, in most cases, the hospital's data files are made available again. (Ransomware is discussed in *Chapter 11, Securing Communication Channels*.)

> **Note**
> If ransom gangsters gain the reputation of not restoring data files after they are paid, soon, victims will no longer pay the ransom fees.

This is an example of how one carefully crafted attack can cause harm to the entire organization, where an attacker intentionally causes downtime, degradation, and destruction—the adversaries of availability. There are also adversaries of confidentiality and integrity, known as unauthorized disclosure and alteration, respectively. Together, the adversaries of security are known as **Disclosure, Alteration, Destruction (DAD)**, as pictured in *Figure 1.2*:

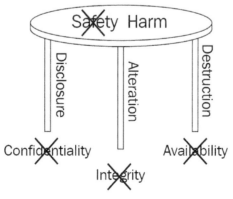

Figure 1.2: The CIA pyramid with DAD

The controls that are used to enforce CIA are discussed in the following sections.

Confidentiality Concepts

When someone submits a message so that only the intended individual can read it, it is considered confidential. Confidentiality controls protect data from being disclosed to unauthorized parties.

This is done via several controls, including the following:

- Encryption
- Passwords
- Access control lists
- Steganography
- Physical locks

Private records, financial reports, and tax identification numbers are generically called **objects**, and they are protected by a set of rules defined as the **access matrix monitor**. So, does the user, viewer, or reader, also known as the **subject**, have the privilege to access the data? How objects respond to subjects is referred to as the **access control model**, which is shown in *Figure 1.3*. There are several variations of access control models discussed in *Chapter 7, Secure Design Principles and Controls*. For example, some objects are defined as read-only. In this scenario, viewing these objects is allowed by subjects, but no changes or deletions can be made:

Figure 1.3: An access control model

Confidentiality can be compromised by several attack methods. Some of these attack methods are as follows:

- Social engineering
- Key loggers
- Brute-force password matching
- Dumpster diving
- Credential stuffing

Social engineering and dumpster diving are non-technical attacks. For example, tricking someone into revealing their password by pretending to be tech support is an example of social engineering. Going through trash to collect information about a target, such as tax identification numbers or credit card information, is referred to as dumpster diving.

Brute-force attacks and key loggers require technical knowledge. Key loggers installed on computers will send every keystroke to the attacker, revealing usernames and passwords. When a hacker attempts all possible passwords to break into a computer, they are conducting a brute-force attack. These, alongside other attacks, are discussed in *Chapter 8, Architecture Vulnerabilities and Cryptography*.

Authenticity and Nonrepudiation

One feature of encryption algorithms is their use in determining the authenticity of files, documents, music, and other objects. Authenticity ensures that the object originates from the source that is assigned to the file or data. Additionally, some encryption tools provide ways in which to assure you that email messages are truly from the sender; this is known as nonrepudiation. Authenticity and nonrepudiation controls use digital verification for software products, online stores, user identities, and more. Detailed workings of these features are discussed in *Chapter 8, Architecture Vulnerabilities and Cryptography*.

Integrity Concepts

After a user or subject downloads a file or object, they need to assure themselves of the following:

- That they have downloaded the entire file
- That the file is the correct version
- That the file has not been altered by malware

A file's integrity can be checked using several different methods, including hashing or version matching. These processes assure you that the object has not been altered. In other words, integrity checks assure you that there has been no data alteration.

Hashing is a technique in which a user processes a file through a hash generator with a type of serial number based on its contents. This number is called a **hash**. Once the file has been downloaded, you can ensure that the hash value matches by processing the file through a hashing algorithm. If the hash matches the value provided by the publisher, integrity is assured. If the hash does not match, the file has been altered. This could mean that, during download, the subject only retrieved a portion of the file, the file was modified and is a different version from the original, or the file has malware.

Once the subject or user understands why the hash does not match, they can determine whether to use the file as-is or obtain an object where the hash matches (for more details, please refer to *Chapter 8, Architecture Vulnerabilities and Cryptography*):

File HASH = F4A2416B378AB4FC

Figure 1.4: Integrity checking

Attackers can alter files for several reasons, including the following:

- To create a backdoor to a computer
- To introduce malware to elevate their privilege
- To set up a logic bomb to erase or encrypt all data

Generally, the malware installed on a system can perform multiple functions, one of which is creating a backdoor control so that the hacker can access the system remotely over the network. This allows the attacker to access the system from anywhere in the world through the internet. Often, when malware has been discovered, attackers will launch a logic bomb to erase or encrypt all of the user's files as punishment for attempting to remove the malware (an example of this is described at https://packt.link/q0LfZ). For more details, please refer to *Chapter 7, Secure Design Principles and Controls*.

Availability Concepts

The unavailability of data refers to a situation in which a user is ready to use their data but is unable to. Therefore, security professionals must ensure that systems are up and running so that they can be used when needed. Availability attacks include the following:

- Network data flooding
- A computer crashing
- A computer slowing down
- Encryption ransomware
- Earthquakes
- Severe weather

When an attacker floods a website with gigabytes of garbage data, a subject cannot get past the garbage to access the website. This is one type of **Denial-of-Service** (**DoS**) attack. For example, if an individual wants to sell their coffee table on Craigslist, during a DoS attack, the user will not be allowed to log in because of all the junk traffic targeted at the Craigslist web servers. This would be similar to 10,000 people waiting in line at Disneyland but never buying a ticket; because all these people are ahead of you and blocking you, you cannot enjoy the Disney experience. Please refer to *Figure 1.5* for a visualization of a network-based DoS attack (additional details of DoS attacks are covered in *Chapter 7, Secure Design Principles and Controls*, and *Chapter 10, Network Architecture Security*):

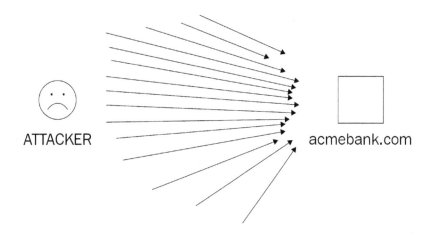

Figure 1.5: A network-based DoS attack

One reason attackers perform these attacks is to hold organizations at ransom. They cease the attack once they are paid for their extortion. Let us assume an organization has a website that generates revenues of $1,000,000 per month. Now, due to a DoS attack, customers cannot access the site, causing the firm's sales to reduce to $1,000 per month! Such a firm might be willing to pay a monthly $5,000 ransom to allow their website to function, earn income, and save the firm's reputation.

Chapter 11, Securing Communication Channels, details these and other types of DoS attacks along with how to mitigate such attacks so as not to fall prey to attackers, such as hacker gangs.

Availability can be assured by implementing the following:

- Using load balancing or clustering
- Implementing fault tolerance such as RAID5
- Keeping systems updated with security patches
- Installing a backup power generator
- Carrying out job rotations

Implementing recovery strategies is where most availability features are discussed. This book will dig deeper into patching, RAID, and redundancy through load balancing and clustering in *Chapter 18, Disaster Recovery*.

People Safety Concepts

The preamble to the ISC2 Code of Ethics states that the safety and welfare of society are critical tenets to holders of the CISSP certification. Safety is considered one of the *Golden Rule* concepts of the Code of Ethics. Holders of the CISSP certification should realize the importance of the following:

- People safety
- Management buy-in
- Everyone is responsible for security
- Training
- Following policies

So, if the CISSP candidate gets a scenario question such as the server room has caught fire, the candidate must understand that it is more important for the server room to *fail open*, allowing humans to escape the room, rather than to *fail closed* to protect the data and trap all the humans inside. In this scenario, failing closed would be unsafe (additional details on failing securely are covered in *Chapter 7, Secure Design Principles and Controls*).

Safety concepts also include physical security, such as the following:

- Fencing
- Lighting
- CCTV

- Escape plans
- Bollards

Note that additional safety-related details are mentioned in *Chapter 9, Facilities and Physical Security*.

Now that you have gone through CIA and public safety concepts, you can review the importance of not using a single safeguard but rather multiple controls to reduce the risk of exploits. This can be accomplished through defense in depth, as discussed in *Chapter 7, Secure Design Principles and Controls*.

Evaluating and Applying Security Governance Principles

The security governance process defines the policies and procedures of the firm along with the steps taken to arrive at security decisions. Security professionals must be careful to create policies that help the organization achieve its mission, goals, and objectives and not allow security features to become the mission itself. For example, perfect security means doors are closed, computers are powered off, and staff policies prevent users from doing their work. The organization is well secured, but this also means minimal revenue is generated.

Organizational Policies and Decision-Making

Security must be part of every organization's decision-making process. Policies that determine how these decisions are made must be put into place. Are decisions made by the management, the board, the committee, or the law? What effect will these policies have on security? Security advisors must be part of the policy definition process so that they can advise on the CIA aspects of the entire organization.

C-level positions, such as **Chief Executive Officer (CEO)**, **Chief Finance Officer (CFO)**, **Chief Operating Officer (COO)**, and **Chief Security Officer (CSO)**, define policies based on the mission and goals of the organization while balancing opportunity versus risk.

Standards are defined by management as quantifiable requirements to satisfy policies.

Procedures are created from standards made by middle managers such as security managers, HR managers, IT managers, and more. Procedures are practices performed to meet standards.

Systems administrators, engineers, HR specialists, technicians, and salespeople have the job of following the defined procedures, including making management aware of security incidents. Personnel must follow the policies dictated by management and sign agreements stating such. Please refer to *Figure 1.6* to view which governance decisions are made at which specific level of an organization:

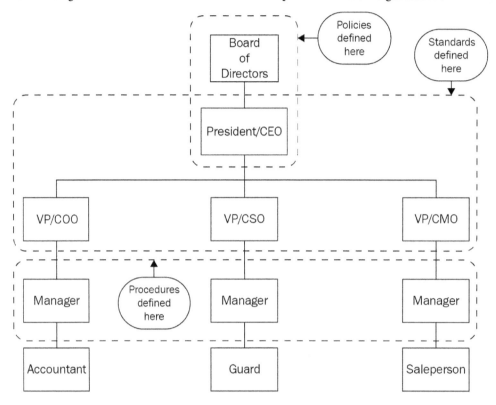

Figure 1.6: Decision levels in governance within an organization

As organizations mature, management considers mergers and acquisitions as methods to become more profitable. This can make an organization more competitive, but risks must be considered.

Mergers, Acquisitions, and Divestitures

Forward-thinking organizations form governance committees to address security concerns when creating their exit strategies, whether the firm plans to merge, acquire, or divest. Security specialists must plan for issues that come with an organization being purchased, bought out, partnered, or sold. If the company is sold or merged, which of the firm's policies will be used? Is the firm part of a regulated industry where government or industry standards affect how user data will be treated?

Governance committees have many security considerations to bear in mind during the restructuring of a firm. For example, if acquiring a new business, is their organization structured around Microsoft products when your organization is structured around Linux products? What new security vulnerabilities are created when these systems are interconnected?

Also, consider that the background check process of the new organization might not be as stringent as that of your organization. What security vulnerabilities are created using the new firm's employee background check process?

Other security considerations of corporate restructuring include, at the very least, the following:

- Differences in accounting methods
- The software products used
- Operational processes
- Cash flow methodologies
- Customer overlap
- Business reseller overlap
- Physical business locations
- Mixing of product portfolio
- The hardware products used
- Tax basis methods
- Corporate structure (for example, an LLC, S corporation, or C corporation)
- Management style
- Business relationships
- Regulatory requirements
- Conflicts of interest
- Investor conflict
- Transaction costs
- Industry commonality
- Stockholder interest
- The potential of synergy
- Supplier governance
- Market share

- Converting into a public or private company

- Valuation methods

- Forensic accounting

- Technology licenses

The governance committee not only weighs up the operational strategy for the new organization but the security framework and policy as well.

Essential Security Frameworks

Corporate boards and other organizational boards might advise the use of existing security frameworks because they are tested, approved, and trusted by auditors and regulators. Various industries have frameworks they are familiar with. For example, US Federal Government institutions use and follow **National Institute of Standards and Technology** (**NIST**) standards to reduce risk, including the *Special Publication (SP) 800-37: Guide for Applying the Risk Management Framework to Federal Information Systems* and *SP 800-53: Security and Privacy Controls for Federal Information Systems and Organizations*.

Many for-profit organizations use and follow **International Organization for Standardization** (**ISO**) standards such as *27001: Information Technology – Security Techniques – Information Security Management Systems – Requirements* and *27002: Code of Practice for Information Security Controls*.

US hospitals and medical institutions follow the **Health Insurance Portability and Accountability Act** (**HIPAA**) regulations. This is a federal law designed to protect patient health information.

Merchants that accept credit cards must follow **Payment Card Industry Data Security Standard** (**PCI DSS**) agreements. There are a dozen requirements that are all focused on protecting credit card information. For example, merchants cannot save the CVV code on credit cards. Additionally, merchants must perform an annual vulnerability scan (or if there are any major updates, a more frequent scan).

Many information technology organizations follow **Information Systems Audit and Control Association** (**ISACA**) standards, which focus on security related to IT, such as performance, malware, intrusion attacks, and more. This is called the **Control Objectives for Information and Related Technology** (**COBIT**) framework.

The **Information Technology Infrastructure Library** (**ITIL**) released detailed practices for information technology service management and asset management to align services with businesses. ITIL was introduced to reduce risk in IT service management.

The **Zachman Framework** looks at the who, what, when, where, why, and how of security events. Zachman values social collaborations to identify, define, represent, specify, configure, and initiate the security architecture. **Sherwood Applied Business Security Architecture** (**SABSA**), similar to Zachman, also uses a matrix to evaluate operational levels when defining enterprise security architecture and service management.

The **Deming cycle** reduces the possibility of risk through improved quality, waste reduction, and cost-cutting. Dr. Deming is known for saving the Japanese automotive industry. His framework around quality was not popular among US automakers because fuel efficiency and quality were not important at the time. During the gas crisis of the 1970s, Japanese vehicles rose to prominence because of their fuel efficiency and quality as a result of using the Deming framework.

The Open Group Architecture Framework (**TOGAF**) enforces security through its architecture development method, which focuses on specifying and meeting requirements through parameters such as architecture, opportunities, solutions, migration, implementation, and change management for the enterprise. It is popular with for-profit organizations worldwide.

Corporate and organizational boards choose and work with governance frameworks that are deemed the best for their organization. Frameworks are not one size fits all, so corporate management must fine-tune the framework to fit with their missions.

The Organizational Legal Liability Risk

Implementing a security framework shows due care. In other words, the organization understands there are operational, physical, and technical risks to itself and its customers. In the event of lawsuits, an organization needs to demonstrate that it performed some mitigation activities to reduce these risks, which is called **due diligence**. Not showing effective due diligence can present the firm as liable for damage by courts.

For example, consider a firm purchasing a building for 300 additional marketing staff. The policies dictate that the building should be structurally sound and have safe fire ratings. Due care is the process of meeting the requirements of safety. Due diligence is the act of ensuring the building is safe by reviewing earthquake analysis reports, reviewing flood tables, testing for asbestos, observing crime reports in the area, and more. Therefore, due diligence is the act of showing due care.

Summary

The important aspects of security revolve around three concepts: confidentiality, integrity, and availability. In this chapter, we learned that confidentiality allows users to protect data from being disclosed, integrity assures users that data has not been altered, and availability provides access to users when requested.

Authenticity ensures that the object originates from the source that is assigned to the file or data. Nonrepudiation ensures that messages are from the assigned sender, making them unable to deny sending the message.

A strong security governance principle uses multiple layers of protection, or defense in depth, because it provides the organization with a security backup if another control is defeated or fails.

Additionally, you learned about the importance of staff members following organizational policies. Failure to do so could result in data loss and human injury. Policies are defined by C-level executives and are created based on security governance and corporate goals.

To prevent companies from reinventing the wheel, there are security-governance frameworks that have been developed to help organizations create policies. These include ISACA, ISO, ITIL, NIST, HIPAA, and PCI-DSS. These frameworks are tested and trusted. Due care involves putting safety requirements in place, and due diligence is evidence that the mechanisms work. Finally, the CISSP must follow the ISC2 Code of Professional Ethics.

The next chapter details the contractual, legal, and regulatory requirements for keeping an organization and its data secure. Organizations that do not meet the requirements risk huge fines and lawsuits.

Further Reading

For more information regarding the topics covered in this chapter, please refer to the following resources:

- *White Paper: Defense in Depth, Todd McGuiness, SANS Institute, Security Basics*, 2001

- SELinux and Mandatory Access Control: `https://packt.link/6rezL`

- **Discretionary Access Control (DAC)**: `https://packt.link/Mi7Ki`

Exam Readiness Drill – Chapter Review Questions

Apart from a solid understanding of key concepts, being able to think quickly under time pressure is a skill that will help you ace your certification exam. That is why working on these skills early on in your learning journey is key.

Chapter review questions are designed to improve your test-taking skills progressively with each chapter you learn and review your understanding of key concepts in the chapter at the same time. You'll find these at the end of each chapter.

> **How to Access These Materials**
>
> To learn how to access these resources, head over to the chapter titled *Chapter 24, Accessing the Online Resources*.

To open the Chapter Review Questions for this chapter, perform the following steps:

1. Click the link – `https://packt.link/chapter01`.

    Alternatively, you can scan the following **QR code** (*Figure 1.7*):

Figure 1.7: QR code that opens Chapter Review Questions for logged-in users

2. Once you log in, you'll see a page similar to the one shown in *Figure 1.8*:

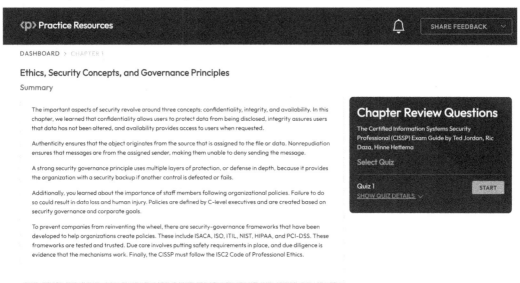

Figure 1.8: Chapter Review Questions for Chapter 1

3. Once ready, start the following practice drills, re-attempting the quiz multiple times.

Exam Readiness Drill

For the first three attempts, don't worry about the time limit.

ATTEMPT 1

The first time, aim for at least **40%**. Look at the answers you got wrong and read the relevant sections in the chapter again to fix your learning gaps.

ATTEMPT 2

The second time, aim for at least **60%**. Look at the answers you got wrong and read the relevant sections in the chapter again to fix any remaining learning gaps.

ATTEMPT 3

The third time, aim for at least **75%**. Once you score 75% or more, you start working on your timing.

> **Tip**
>
> You may take more than **three** attempts to reach 75%. That's okay. Just review the relevant sections in the chapter till you get there.

Working On Timing

Target: Your aim is to keep the score the same while trying to answer these questions as quickly as possible. Here's an example of how your next attempts should look like:

Attempt	Score	Time Taken
Attempt 5	77%	21 mins 30 seconds
Attempt 6	78%	18 mins 34 seconds
Attempt 7	76%	14 mins 44 seconds

Table 1.1: Sample timing practice drills on the online platform

> **Note**
>
> The time limits shown in the above table are just examples. Set your own time limits with each attempt based on the time limit of the quiz on the website.

With each new attempt, your score should stay above **75%** while your "time taken" to complete should "decrease". Repeat as many attempts as you want till you feel confident dealing with the time pressure.

2
Compliance, Regulation, and Investigations

Having a **Certified Information Systems Security Professional** (**CISSP**) certification carries the responsibility of complying with laws and regulations and understanding the different types of investigations. Organizations that comply with regulated standards do a better job of securing customer data. Standards also help organizations consistently compare results.

In the exam, you are tested on your understanding of contracts that allow merchants to accept credit cards, security regulations that protect hospital patient records, and other private data. You must also understand which investigatory process to follow when an organization's information systems are compromised.

By the end of this chapter, you will be able to answer questions on the following:

- The difference between contractual requirements, legal requirements, and industry standards
- Important user privacy principles
- The nature of cybercrimes and other threats
- How to protect copyrights with licenses and digital rights management solutions
- Protection of personal information through international laws, such as GDPR and OECD
- The difference between civil, criminal, and administrative investigations

Determining Compliance and Other Requirements

Organizations do not operate on internal missions, goals, and standards only; they also have external drivers that limit and dictate how they should work with employees, contractors, clients, suppliers, and affiliates. Organizations follow their imposed regulations and must comply with contracts, industry standards, and specific laws of their city, state, province, or country.

This section will cover compliance, which is when an organization follows specific directives or ordinances. There are specific tools, processes, and documentation to demonstrate adherence to these directives and ordinances that verify and validate compliance according to law, industry standards, and so on.

Important directives to understand in the CISSP exam are those that protect personal privacy, such as **personally identifiable information (PII)** and **protected health information (PHI)**. The general term for this category is *privacy*, and the individual ultimately responsible for these assets is titled the *chief privacy officer*.

> **Note**
>
> The role of the **chief privacy officer (CPO)** is different from the role of the **chief security officer (CSO)**. The CPO focuses on policies related to the protection of PII and PHI, whereas the CSO is responsible for policies for security at large, including physical, logical, and operational security. The **chief information security officer (CISO)** focuses on policies around cybersecurity.

Finally, auditing reviews are ways to ensure that compliance is met and that the tools and processes used to follow specific directives are validated for contracts, laws, industry standards, or regulations.

Contractual Requirements

Agreements between companies and individuals are known as contracts, and these contracts can be verbal or written. Contracts can specify in detail what assignment needs to be completed. The agreement can be as simple as the terms under which a plumber installs a water faucet, or as complex as maintaining a 10,000-user security operation center by a third-party organization.

> **Note**
>
> Although written and verbal contracts are both legal, written contracts are preferred. This is because during disputes, parties may disagree on the terms of the verbal agreement for various reasons, such as forgetfulness or there never being a true meeting of the minds. A written agreement explicitly spells out the terms, making it simpler to determine which party broke the agreement.

A contract contains terms and conditions as to which party should fulfill which parts of the agreement. For example, if an agreement is made to install a door, the installer agrees to follow the steps to properly install a working door and the customer agrees to make the payment. If any party falters in their part of the agreement, they can resolve issues informally between themselves or involve lawyers, arbitrators, or the courts to resolve the matter.

One contract agreement essential to CISSP professionals is the **Payment Card Industry Data Security Standard (PCI DSS)** agreement. The PCI DSS is designed to reduce credit card fraud. Businesses that accept credit cards are called *merchants*, as defined by the PCI Security Standards Council. Merchants must protect the data of their clients and follow a set of a dozen requirements to remain compliant with PCI DSS terms. These are shown in *Table 2.1*. Non-compliance could result in heavy fines for the merchant or even make them lose their ability to accept credit cards.

Goals	PCI DSS requirements
Build and maintain a secure network and systems	• Install and maintain a firewall configuration to protect cardholder data • Do not use vendor-supplied defaults for system passwords and other security parameters
Protect cardholder data	• Protect stored cardholder data • Encrypt transmission of cardholder data across open, public networks
Maintain a vulnerability management program	• Protect all systems against malware and regularly update antivirus software or programs • Develop and maintain secure systems and applications.
Implement strong access control measures	• Restrict access to cardholder data to those in a business who need to know • Identify and authenticate access to system components • Restrict physical access to cardholder data
Regularly monitor and test networks	• Track and monitor all access to network resources and cardholder data • Regularly test security systems and processes
Maintain an information security policy	• Maintain a policy that addresses information security for all personnel

Table 2.1: PCI DSS security best practices

The information above has been derived from the official documentation here: `https://packt.link/ExDZF`

PCI DSS standards categorize merchants into four levels, as shown in *Table 2.2*. The level of the merchant determines the requirements for **how** they handle private client data, such as their imprinted name, credit card number, billing address, and phone number.

Merchant level	Summary of criteria
Level 1	• Processes more than 6 million transactions annually • Annual reports on compliance • Quarterly network scans
Level 2	• Processes from 1 to 6 million transactions annually • Annual self-assessment questionnaire • Quarterly network scans
Level 3	• Processes from 20,000 to 1 million transactions annually • Self-assessment questionnaire • Quarterly network scans
Level 4	• Processes fewer than 20,000 annual transactions • Self-assessment questionnaire • Follows requirements issued by an acquiring bank

Table 2.2: PCI DSS merchant levels

One condition of PCI DSS that is consistent for all merchants is protecting the **card verification value** (**CVV**) code, which is the three or four-digit code on the back or front of the card. Merchants use the value to verify that the physical card is present during a transaction, reducing the risk of fraud. Merchants agree to discard and not save CVV codes.

> **Exam tip**
> The CISSP exam does not require knowledge of the four merchant levels.

Legal Requirements

External mandates, such as federal and local regulations, refer to the requirements a company or organization must adhere to to follow the law. Now, from a legal point of view, laws are created by governments and courts. Courts enforce *statutory laws* created by legislators and create new laws where vagueness exists. These *case laws* become precedents and end up becoming the standard unless legislators overturn them with a new statutory law. Depending on the law, legislators can overturn court precedents at the local, state, provincial, or national level.

There are legal standards that define whether organizations are liable for being non-compliant with safety or privacy laws. Organizations must put policies in place to ensure they can demonstrate due care and due diligence to protect themselves from being non-compliant. For example, consider a company that created its own encryption algorithm that gave it better results than using standard encryption algorithms.

Later, it was discovered that a breach occurred due to their inferior encryption algorithm, and privacy records were made available on the dark web by criminal hackers. As part of a judgment, a court may mandate that companies use standard encryption algorithms and make that a law. Organizations would then have to follow this directive to stay lawful.

One law of note for CISSP professionals is the **Federal Information Security Modernization Act** (**FISMA**), administered by the US Department of Homeland Security. FISMA defines a comprehensive framework designed to protect government information systems against threats. FISMA requires federal systems to meet **National Institute of Standards and Technology** (**NIST**) security standards by complying with the **special publication** (**SP**) *800-37: Risk Management Framework for Information Systems and Organizations: A System Life Cycle Approach for Security and Privacy*.

A critical component of FISMA is *NIST SP 800-53 rev 5: Security and Privacy Controls for Federal Information Systems and Organization*. This details the standards of security controls and safeguards for federal information systems to comply with FISMA. They are divided into 20 different families, as shown in *Table 2.3*.

Family	Family
Access Control	Personnel Security
Audit and Accountability	Physical and Environmental Protection
Awareness and Training	Planning
Assessment, Authorization, and Monitoring	PII Processing and Transparency
Configuration Management	Program Management
Contingency Planning	Risk Assessment
Identification and Authentication	Supply Chain Risk Management
Incident Response	System and Communications Protection
Maintenance	System and Information Integrity
Media Protection	System and Services Acquisition

Table 2.3: NIST SP 800-53 rev 5 security control families

Organizations that prove to be non-compliant risk losing federal funding and the right to do future business with the US government.

Industry Standards

For the best customer acceptance, organizations should comply with industry standards as well. For example, the **Institute of Electrical and Electronics Engineers (IEEE)** was initially chartered to continue the practice of providing education and furthering the advancement of innovation and technology through electricity and electronics. They developed standards that still exist to this day since their founding in 1884. One example is the color values that are shown on a resistor. Different colors have different meanings – for example, brown is one, red is two, and so on. These colors are combined to calculate the value of a resistor in ohms.

There is no legal force behind this, so it is not mandatory for companies to follow the industry standard. An organization should have a strong business reason not to follow the industry standard because their market is accustomed to it. For example, the IPv6 network standard has been available for over 25 years, but most network administrators still use IPv4 (`https://packt.link/YzMIM`).

If an organization ends up in court for a due diligence case, industry standards can be used as evidence to uphold its position. Regulators, and even contract agreements, refer to industry standards to affirm that a company is following appropriate methods for due diligence and care.

One series of industry standards important to CISSP security professionals is the **International Organization for Standardization (ISO)**. The ISO publishes industry standards for almost any type of organization, including automotive, manufacturing, and construction. Its standards do come with a fee, but it is recognized worldwide as being comprehensive and critical for organizations demonstrating that they apply due care and diligence.

The CISSP candidate should be familiar with *ISO 27001: Information Technology – Security Techniques – Information Security Management Systems – Requirements*, which provides requirements for **information security management systems (ISMSs)** and is the only certifiable framework among the ISO 27000 family. According to ISO 27001, risk is managed through 14 security groups, as shown in *Table 2.4*.

Information security policies	Operations security
Organization of Information Security	Communications Security
Human Resource Security	System Acquisition, Development, and Maintenance
Asset Management	Supplier Relationships
Access Control	Information Security Incident Management
Cryptography	Information Security Aspects of Business Continuity Management
Physical and Environmental Security	Compliance with Policies and Laws

Table 2.4: The ISO 27001 security groups

For example, *ISO 27002:Information Technology – Security Techniques – Code of Practice for Information Security Controls* delivers best-practice security control recommendations for those responsible for maintaining and securing information systems. It is not a certifiable framework because it must be interpreted and applied depending on the type and size of an organization. ISO 27002 provides a list of security controls that can be tailored to ISO 27001 for certification.

The Cloud Security Alliance is an organization that offers cloud security industry standards through its CSA STAR program, which is utilized for certifying **cloud service providers** (**CSPs**) such as Microsoft Azure, **Amazon Web Services** (**AWS**), and **Google Cloud Platform** (**GCP**). **STAR** is defined as **security, trust, assurance, and risk**, and it is a publicly available registry, documenting the privacy controls of cloud vendors that follow the rigorous standards outlined in the **Cloud Controls Matrix** (**CCM**).

The CCM is a cloud security control framework composed of 197 control objectives (such as a key policy, audit trail, and data recovery), structured into 17 domains that cover the important aspects of cloud security, as shown in *Table 2.5*. You can find more details about these domains at `https://packt.link/AU3D0`.

Audit and Assurance	Identity and Access Management
Application and Interface Security	Interoperability and Portability
Business Continuity Management and Op Resilience	Infrastructure and Virtualization Security
Change Control and Configuration Management	Logging and Monitoring
Cryptography, Encryption, and Key Management	Security Incident Management, E-Disc, and Cloud Forensics
Data Center Security	Supply Chain Management, Transparency, and Accountability
Data Security and Privacy	Threat and Vulnerability Management
Governance, Risk Management, and Compliance	Universal Endpoint Management
Human Resources Security	

Table 2.5: CCM cybersecurity domains

Laws, contracts, and industry standards are not the only tools used to secure organizations and their PII; regulations also play a strong part in protecting staff and customers.

Regulatory Requirements

Compliance regulations are created by local, state, and federal legislative bodies. Oversight is managed through related regulatory organizations called regulators. Penalties include fines and possibly imprisonment in cases of fraud or loss of life.

Regulations that the CISSP candidate must be familiar with start from the basics of the **General Data Protection Regulation** (**GDPR**), based in the European Union, which elevates data privacy to a *human right*. The GDPR addresses privacy standards – for example, is an organization allowed to share or sell its customers' private data, such as phone numbers and addresses? If an organization fails to deploy controls to protect PII, they face potential sanctions, bans on data processing, and fines of up to 20 million euros or 4% of worldwide gross sales, whichever is higher. *Table 2.6* lists what GDPR defines as personal data. (For details, visit `https://packt.link/GnazK`.)

First name, surname, and maiden name	Email address	Home address
Phone number	Photo	Date of birth
Bank account number	Credit card number	National identification number
Social insurance number	Passport number	Driver's license number
Vehicle license plate number	Employee number	IP address
Cookie ID	Location data (from a cell phone, for example)	MAC address
Handwriting	Login identifier	Patient identification number
Retinal scan	Fingerprints	Voice signature
Vehicle identification number	Birthday	Place of birth
Race	Religion	Weight
Employment information	Medical information	Financial information

Table 2.6: GDPR personal data identifiers

Another important regulation for security professionals to recognize is the **Health Insurance Portability and Accountability Act** (**HIPAA**). This US law states that healthcare providers must protect the PHI of their patients. Penalties for non-compliance can reach as high as $1,500,000 per year, per standard. Intentionally leaking PHI can lead to fines as high as $250,000 and up to 10 years in prison.

> **Note**
>
> The **Health Information Technology for Economic and Clinical Health (HITECH)** Act was implemented to extend the use of health information technology via **electronic health records (EHR)**. HITECH requires HIPAA-based organizations to report PHI breaches or face significant fines. For more information, refer to `https://packt.link/XArXo`.

Security professionals must be familiar with the **Gramm-Leach-Bliley Act (GLBA)** of 1999, which is a federal law allowing banks to safely merge with insurance providers under certain requirements. GLBA privacy requirements include providing privacy notices to clients and the customer's right to disallow the sharing of information. Financial institutions must create a written security policy and test the security policy. The security responsibility falls on the board of directors.

The **Sarbanes-Oxley (SOX)** Act of 2002 was created by the US Congress to mitigate fraud, committed generally by financial and accounting executives who alter accounting records. SOX made such manipulations of accounting records a criminal offense so that offenders could go to prison, instead of a civil case where offenders may only have to pay restitution.

US federal agencies must follow NIST security standards and guidelines to comply with FISMA to secure their IT environments. A segment of FISMA is targeted at government suppliers that have to follow the **Federal Risk and Authorization Management Program (FedRAMP)** program, ensuring they have security products and policies in place, which helps to keep federal organizations secure.

Privacy Requirements

The **Organization for Economic Co-operation and Development (OECD)**, a group that finds solutions to worldwide social challenges, developed privacy guidelines to foster transborder interoperability of personal data. They released eight privacy principles that address PII and are designed to protect an individual's liberty. The candidate should be familiar with these principles, as shown in *Table 2.7*. (For more details, visit `https://packt.link/0SgZT`.)

Privacy principle	Description
Collection Limitation	Private data must only be used for its collected purpose.
Data Quality	Private data must be kept accurate, complete, and up to date.
Purpose Specification	Private data must be used only for its specified purpose.
Use Limitation	Private data must not be disclosed without the consent of the data subject (the individual whom that data is about).
Security Safeguards	Private data must be reasonably protected.
Openness	Data owners (the party collecting PII) must be open about policies and practices around private data.

Privacy principle	Description
Individual Participation	Data subjects have the right to review their private data in a reasonable amount of time, as well as in a reasonable manner.
Accountability	A data controller (the party determining how the data is used) is accountable for complying with these eight principles.

Table 2.7: OECD privacy principle guidelines

Security professionals will notice that OECD principles are similar to GDPR requirements. One important difference is that the OECD's principles are simply international guidelines, but the GDPR requirements are legally binding. Also, the GDPR focuses on the automatic processing of personal data, whereas the OECD applies to private data involving dangers to liberty.

Finally, the GDPR enforces the right to be forgotten, where data subjects can request private data to be deleted. This can be difficult for a data owner because many of their supplier relationships that help maintain mailing lists or billing hold this private data.

Privacy laws and regulations require organizations to disclose a data breach to a supervisory authority. The GDPR states that breaches must be disclosed to the authorities within 72 hours. HIPAA requires breaches to be reported within 60 days to the authorities, giving data custodians more time to step through the incident management process compared to the GDPR. If organizations do not comply with the laws, they can face financial penalties.

Not all breaches are reported to data subjects. These reports are only required if personal information exceeds a specific level of risk to a subject's rights and freedoms. The GDPR requires breaches to be disclosed to the public *as soon as it is feasible to do so and without undue delay*.

This section detailed privacy laws and regulations and the specific tools used to demonstrate compliance with these directives and industry standards. The CISSP candidate must understand how regulations help secure personal privacy, such as PII and PHI.

Next, you will examine specific security threats that impact privacy and organizations.

Understanding Legal and Regulatory Issues

Information security professionals must not only be familiar with the risks within their organizations but also have a wider understanding of managing risk from outside the organization. Cybercrimes and data breaches could cause losses of private data, leaving the organization open to civil and criminal lawsuits.

Moreover, improper licensing of software could leave a company open to civil laws because they have unlicensed software. An information security professional must put mitigations in place so that this does not occur.

Finally, we will explain privacy terminology, such as the difference between the data subject and the data owner.

Cybercrimes and Data Breaches

Security operation centers (**SOCs**), data centers, and cloud service providers have to be critically concerned about protecting their data, especially their PII records. More and more regulations are being released, requiring organizations to protect the privacy records of their customers. If private data is lost or stolen, the company is held responsible and is liable to fines and civil lawsuits.

The following is a list of common cybercrimes and threats:

- **Malware**: In many territories, the distribution of malware is a crime.

- **Ransomware**: This is a type of malware where victims are extorted to pay a ransom so that they can recover their files. Most ransomware works by encrypting the victim's files with a public encryption key, and if the victim pays the ransom, the attacker provides the private key that decrypts the files so that the victim can recover their data.

- **Illegal use of systems**: Attackers gain unauthorized access to a victim's IT infrastructure and use it to mine bitcoins, launch **Distributed Denial-of-Service** (**DDoS**) attacks, and stash files such as copyrighted material, X-rated pictures, and illegal photos, exposing the victim to civil and criminal lawsuits.

- **Unauthorized access**: This refers to access to systems by an entity not authorized to do so. For example, an **advanced persistent threat** (**APT**) actor monitors the victim's activity, remaining undetected for as long as possible, and learns about the victim's activities.

- **Stealing of data**: In this case, attackers look to steal valuable data, such as financial records, intellectual property, PII, and competitive intelligence.

- **Fraud**: Fraud is the use of deception in order to induce another to part with something of value. One way this works is that the attacker pretends to be a relative or coworker and tricks the victim into giving them money. For example, the attacker may pretend to be a grandchild and contact a grandparent for emergency rent money by telling them a story about being evicted, or the attacker may pretend to work for an antivirus company.

   The victim will receive an email stating they were charged $300 for antivirus software. The victim will call the number in the email to request a refund for the product they did not purchase. The operator (attacker) apologizes but errs during the refund process, pretending to return $3,000, and tells the victim that they will lose their job if they do not return the $2,700 overcharge by sending cash in the mail or through gift cards (as they do not want their boss to see the charge reversal). Learn more about refund scams at `https://packt.link/3RDZl`.

- **Extortion**: This is a type of blackmail where the attacker extorts money by threatening to ruin the victim's reputation. For example, the attacker may claim to have embarrassing or compromising photos of the victim and threaten to distribute them to the victim's friends, or on the internet, unless they pay a fee to stop the distribution.

IT and networking infrastructures are vulnerable to criminals invading and attacking systems from afar, including offshore. Because of the internet, criminals no longer have to physically break into a building to commit malicious acts, and many countries' extradition policies make it difficult, and often impossible, to prosecute offenders.

Licensing and Intellectual Property Requirements

Intellectual property can be proprietary items such as software, agreements, trademarks, copyrights, licenses, and patents. When it comes to working with trademarks, organizations have to be prepared to protect them from duplication or other violations.

Inventions are protected through patents. Written works, such as poetry and even software coding, are protected by copyrights. We will go into depth on these intellectual property types in the following discussion.

Trademarks

Trademarks are used as a way for a company to brand its identity. For example, people can look at a symbol of colorful squares and recognize that the company behind it is Microsoft, or look at a symbol of a chameleon and recognize that it belongs to the SUSE Linux brand.

Common names cannot be trademarks, which is why companies create their own names, such as Kleenex tissues or Xerox copiers. Registered trademarks are registered with the **US Patent and Trademark Office** (**USPTO**) and are generally provided protection for up to 10 years, which can be renewed indefinitely.

Patents

Patents are applied for by organizations to protect their inventions. The patent has to be something new, useful, and non-obvious. Patents do not have to actually be built; rather, they exclude others from building and marketing the idea or concept.

Patent protection is provided for up to 20 years. Patents can also include utility patents, which include software patents, chemical patents, business method patents, and biological patents.

Copyrights

Copyrights protect the author of an original work and cover pictures, graphics, music, literature, sound recordings, sheet music, architectural drawings, film, and written works. Copyrights are different from trademarks and patents in that they protect the expression of an idea. Once something has been copyrighted, it is recommended to attach the © symbol to show that the work is owned by that specific author.

Copyrights protect items from unauthorized distribution, and even though it is not as strong as patent protection, copyright protection is provided for the life of the author, plus 70 years. Software source code and user interfaces can also be copywritten and protected from duplication by competitors.

Trade Secrets

Trade secrets refer to corporate information that is internally secret to an organization and vital for the firm's success. For example, the search algorithm used for Google is not publicly published and is privately held within Google. Employees must sign a non-disclosure agreement stating that they understand that this information must not be shared with anyone outside of Google. If they break the non-disclosure agreement, they can be sued by the organization. Some trade secrets include the recipes of Kentucky Fried Chicken or formulas for certain drugs.

Software Licensing

When it comes to licensing software, there are several types of agreements that can be managed. These agreements codify the terms, duration, and number of copies that govern the use of software. There are multiple types of licenses available to organizations. The majority fall into the following categories:

- **Site licenses**: The firm purchases the right to use the software for all employees of a physical/logical site or the entire organization. Oftentimes, there is a cap on how many versions can be used at a time, usually for a certain period of time.

- **Per-seat license**: An organization purchases a specific number of copies for its personnel, paying per individual or seat.

- **Per-CPU license**: This license allows organizations to run an application on more CPUs or cores. So, if the application uses an eight-core CPU system, it will cost more than if the software used just a dual-core CPU system. In addition, license fees can be based on the RAM installed.

- **Public domain license**: This software is commonly free and open source, meaning users can see the source code of the program (analogous to viewing the blueprints of a building). Developers can earn revenue by selling training, books, and tech support.

- **Shareware license**: Shareware involves some constraints on software use. For example, under a Creative Commons license, non-commercial use of software may be free, but business users have to pay a fee. An organization's software librarian is either managed by the CISO, tech support, or possibly the security office. No matter which department the software librarian falls under, software licenses are monitored for compliance.

Application allowlisting defines an allowed (and sometimes disallowed, called denylisting) list of software applications that employees and contractors can use. This process protects computers and networks from vulnerable applications. Unsupported software affects licensing and how systems operate. Therefore, having unsupported software is banned in most companies and organizations. Monitoring systems for disallowed or blacklisted software assures software librarians that they are conducting good license management and can properly support systems.

Digital Rights Management

One way intellectual property rights are managed is through **digital rights management** (**DRM**). DRM solutions are available for CDs, DVDs, within apps for smart devices, and over the cloud. Software can be limited to run in certain countries or regions only. For instance, some movies are released earlier in some countries than others. So, after a movie has completed a theater run, the title can be released and the DVD can be viewed in that country. DRM can attempt to determine what territory an individual may use the software in.

The CISSP candidate should know that DRM has the following seven features:

- **Verifications**: Verifying copyrighted material such as software, movies, or music with product keys, tracking and limiting the number of installations, or authenticating a user persistently online.

- **Encryption**: Encryption ensures that DRM restrictions cannot be bypassed.

- **Copy restriction**: Copy restriction mitigates copying copyrighted material, such as digital books or music.

- **Anti-tampering**: Anti-tampering prevents media from operating properly if the DRM recognizes tampering, such as unsigned software being used.

- **Regional lockout**: This feature disables software from running when a system recognizes forbidden locations.

- **Television**: This technology restricts content to what a customer has ordered from their cable provider.

- **Tracking**: DRM solutions use digital watermarks and metadata to determine whether copyrighted material is allowed to be used.

Many DRM solutions use an online agent to help provide digital rights protection. This agent is installed on every device in an organization that uses the software under protection and must be designed for the configuration management process.

Import and Export Controls

Security professionals must understand that some specific international laws can affect the security of a system. One notable area is laws around encryption. Certain countries are not allowed to use as high a grade of encryption as the United States. Most of these laws come from what is called the **Wassenaar Arrangement**, where participating countries, generally first-world countries with the capability of mass-producing weapons, agree to not distribute weapons and high-grade encryption technologies to countries that do not have these capabilities.

Security professionals must be aware of which 42 countries participate in such export-controlled agreements, such as the United States, Russia, Australia, and the European Union.

Transborder Data Flow

Businesses today are almost international by default. Websites such as eBay, Amazon, and Alibaba allow small businesses to create, ship, and manufacture items worldwide. The internet makes it simple for customers to communicate with stores worldwide.

However, CISSP candidates must have an understanding of some of the international legal issues. The main regulation that deals with transborder data flow is from the European Union and is called the GDPR, as discussed earlier in this chapter. Countries that do not abide by the GDPR cannot legally store the data of any EU citizen. Countries that follow the GDPR are listed in *Table 2.8*. The list includes EU member states, European Economic Area countries, and some other territories. Countries that do not participate in the GDPR include the United States and others not mentioned in *Table 2.8*.

Austria	Israel	Canada
Belgium	Hungary	Australia
Bulgaria	Ireland	The United Kingdom
Croatia	Italy	Iceland
Cyprus	Latvia	Lichtenstein
Czech Republic	Lithuania	Norway
Denmark	Luxembourg	Andorra
Estonia	Malta	Singapore
Finland	The Netherlands	Japan
France	Poland	Switzerland
Germany	Portugal	Argentina
Greece	Romania	
Uruguay	Slovakia	

Table 2.8: Countries protected by the GDPR (subject to change)

If a US organization wants to do business with EU citizens, they can voluntarily participate in a program called Privacy Shield by applying through the US Department of Commerce. Organizations that process data under Privacy Shield can manage EU customers' private data because the Privacy Shield program is compliant with the GDPR. Companies that apply must self-certify annually, be willing to submit to audits, and create policies that comply with the GDPR.

Contracts

A multinational company that is not headquartered in the EU and does not participate in the Privacy Shield program can agree, in a contract, that it will comply with the GDPR. These are termed standard contractual clauses. Therefore, organizations that work with EU organizations, and, therefore, normally work with EU private data, must add these clauses if they are not part of Privacy Shield. The standard contractual clauses have to be approved by the EU country. Once approved, the clauses can be used in any of the entity's agreements with EU businesses.

Privacy

The GDPR has some of the strictest privacy regulations, so it makes sense to define the different privacy terms based on their terminology, as follows:

- **PII**: This is any data that can be used to identify a person, such as their name and address. Separately, these items make it difficult to identify a person, but putting at least two of these together makes it highly likely to determine who an individual is. Elements considered PII under most privacy standards and regulations are listed in *Table 2.6*, and they include the tax ID number, work telephone number, mobile telephone number, MAC address, IP address, credit card number, bank account number, and photograph.

    HIPAA also includes additional PII privacy elements that must be secured for individuals, such as whether a person has a disease and the medications they take. This is known as **electronic protected health information (ePHI)**.

- **Data subject**: This is the person that the PII refers to.

- **Data owner**: Also known as the data controller. They collect the PII and are legally accountable for protecting it. In most cases, this will be some type of organization that has day-to-day control over the data.

- **Data processor**: They are the group or individual that has to process the data. For example, modifying, compiling, destroying, or creating a mailing list would be done by data processors. The data processor does have to comply with the law, and ultimately, the data owner is legally accountable for securing the data.

- **Data custodian**: They are in charge of managing the data – for example, making sure that it gets backed up properly. They are not concerned with the contents of the data, as it is their responsibility just to secure and protect it.

See *Figure 2.1*, which shows the different responsibilities and where legal liability applies.

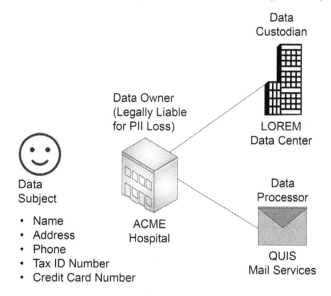

Figure 2.1: Specific data roles and liability

So, if a US-based organization hires a citizen of Belgium to work at their location in New York, would they need to comply with the GDPR? The short answer is *yes*. Since the company needs to document the name, address, and tax identification of the Belgian employee, they must comply with the GDPR.

Laws and regulations created for security are designed to protect the PII of consumers. In order for organizations to be compliant with PCI DSS, HIPAA, and other regulations, they add security controls that mitigate various security vulnerabilities. Mitigations include malware protection, ransomware protection, proper software licensing, and contracts between data owners and data custodians. Next, we will explore the different investigation types and where and when each type is used.

Understanding the Requirements for Investigation Types

Organizations conduct investigations depending on the type of incident. Investigation types important for the CISSP exam include the following:

- Administrative
- Criminal
- Civil
- Regulatory
- Industry standards

The CISSP candidate should understand when to use which investigation type. For example, an employee watching Netflix on company time warrants an administrative investigation because it goes against corporate policy. However, an employee viewing child pornography at work falls under a criminal investigation because this activity is against the law.

Administrative Investigations

The type of investigation depends on the type of incident. The CISSP launches an administrative investigation if there is an internal issue that goes against an organization's policy. Generally, these investigations are managed by human resources, and they would reach out to other departments or organizations if they need their expertise. Examples include contacting the chief security officer if they need expertise in matters concerning files downloaded by an attacker or working with the chief financial officer in matters dealing with fraud.

Administrative investigations are launched when an employee breaks some policy. Examples include consistently arriving late to work, using someone else's identification card to enter a SOC, or making personal phone calls using their corporate-owned mobile phone.

An administrative event can escalate to a criminal or civil event if we learn that an employee has broken the law. For example, if the employee uses illegal software on their company computer, they are violating the administrative policy by using software on the company's deny list. Considering the software was stolen, they are breaking intellectual property laws and triggering a criminal incident, which must be reported to law enforcement agencies.

Criminal Investigations

Criminal investigations launch when a user discovers a law is broken. The main job of the CISSP acting as a security engineer, security officer, or in any other role is to secure the crime scene – with yellow tape, for example, assuring no evidence gets tampered with. Law enforcement agencies deal with criminal matters every day and do their best in handling evidence and managing an investigation. Law enforcement agencies have all the tools to study and protect the evidence and can maintain the integrity and security required to go to court. The job of the CISSP is to preserve evidence and assist law enforcement agencies in any way possible so that a guilty person does not go free due to some technicality (such as tampering with evidence).

Therefore, the organization must cooperate with law enforcement agencies during a criminal investigation. Working with law enforcement agencies supersedes any corporate policies that are in place.

In a criminal case, the prosecution has to prove guilt to a level called *beyond a reasonable doubt*, so if there is even a chance that the individual is innocent, they are found *not* guilty. This contrasts with civil cases where the litigant must show the *weight of the evidence* to prove guilt.

Civil Investigations

Civil lawsuits require a much lower bar, where the defendant must prove by the preponderance of the evidence that they are innocent. So, if the plaintiff proves that the defendant is guilty, then the defendant is found guilty and is financially liable for the incident. In other words, the defendant gets no jail time, but they have to pay restitution.

As with other types of investigations, the evidence must be preserved and handled in a manner where it is not tampered with for justice to be properly distributed.

Civil lawsuits can trigger criminal investigations and vice versa. For example, if a hacker breaks into a bank via the internet and steals funds, this is a criminal act. A civil action results if the hacker damages systems or data as part of their attack. The victim could then sue for restitution.

Regulatory Investigations

Regulatory investigations start when there is a violation of a regulatory standard. For example, the **Occupational Safety and Health Administration (OSHA)** is a regulatory agency of the US Department of Labor and requires employers to comply with safety and health standards. A regulatory investigation starts when OSHA receives a complaint from an employee who reports unsafe conditions, such as not having masks in a dusty environment or not having helmets during building construction.

As with other investigation types, an employee needs to provide evidence for proper enforcement and mitigations. It is illegal to punish or fire workers who file complaints to OSHA. If the investigation proves that the complaints are valid, the organization must fix the unsafe issues and possibly pay a fine.

Users of medical services have the right to keep their health information private. Those rights are protected under HIPAA, and complaints are managed by the US Department of Health and Human Services Office for Civil Rights. Patients can file complaints if their health information was unwillingly shared, their health records were not updated appropriately, and so on. Patients should keep good records for evidence. Penalties could result in fines to the medical institution.

Businesses that process and store credit card transactions, or merchants, must comply with PCI DSS. The 12 requirements of PCI DSS that these merchants must follow include conducting annual vulnerability scans and penetration tests (or sooner if there has been a major update), regularly updating and patching systems, protecting systems with firewalls, configuring strong passwords, and encrypting transactions, as discussed earlier in this chapter.

Merchants can face fines from the payment processor, such as Visa or MasterCard, for non-compliance and data breaches. In serious cases, merchants can eventually lose the ability to accept credit cards.

As with other types of investigations, proper care and preservation of evidence are most important. Regulators generally have the force of law behind them with their own types of courts, the justice system, and arbitrators. Also, regulators do not need search warrants or subpoenas to collect evidence.

Industry Standards

Industry standards are generally compiled by organizations created for a single purpose – for example, the American Bar Association for lawyers, the Society of Automotive Engineers for vehicle engineers, or the International Information System Security Certification Consortium for cybersecurity professionals. The standards define how professionals should best perform their practice, such as how lawyers must maintain client confidentiality and how ISC2 professionals must follow a code of ethics.

As discussed in *Chapter 1, Ethics, Security Concepts, and Governance Principles*, as a CISSP, you practice cybersecurity at a higher ethical standard, and following the code of ethics is a condition of certification. Ethics complaints must be sent to the ISC2 Ethics Committee in writing, and ISC2 does its best to keep the complainant confidential during the investigation. If evidence shows that the CISSP does not meet the standard, they may lose their certification. For details on the ISC2 investigation process, refer to `https://packt.link/mlGIs`.

> **Note**
> The CISSP exam questions focus more on NIST, HIPAA, FISMA, and PCI DSS than OSHA.

In this section, you learned about five investigation types that can result from policy, legal, and regulatory violations. Policy violations are handled by human resources through administrative investigations. Violations that are serious enough can become civil or criminal investigations in which an offender can be sent to jail.

Summary

This chapter covered compliance, regulation, investigations, and the importance of protecting a user's private information, such as their name, address, phone, and tax identification number. If such information is stolen during a breach, it exposes the individual to identification theft, which could cost them thousands of dollars in losses and inconvenience.

Laws and regulations are in place to protect citizens by making companies and businesses responsible for these losses, and information security professionals must be aware of the various laws and requirements. Some requirements come from contractual agreements such as PCI DSS, which requires businesses not to save the CVV code on the back of credit cards.

Federal institutions must abide by FISMA, stating that information security mitigations should be put in place to protect privacy records. ISO 27001 is an industry standard that provides similar requirements for businesses.

Companies doing business overseas may have to consider the GDPR, and medical institutions must follow HIPAA requirements to protect an individual's health records.

The chapter concluded by detailing the different investigation types and privacy terminology. The next chapter will cover how standards, procedures, and guidelines are developed from security policies created by management.

Further Reading

- *File a Complaint*, Occupational Safety and Health Administration, US Department of Labor: https://packt.link/S5c4q

- *Security and Privacy Controls for Information Systems and Organizations*, NIST Special Publication 800-53 Revision 5: https://packt.link/tN5fT

- *Your Rights Under HIPAA*, US Department of Health and Human Services: https://packt.link/VH6uz

- *Italy fines Eli Gas e Luce 11.5 million neuroses for multiple GDPR violations*: https://packt.link/WZHG0

- *OECD Guidelines on the Protection of Privacy and Transborder Flows of Personal Data*: https://packt.link/KnWxe

- *GDPR Data Breach Reporting Requirements*: https://packt.link/AA2zy

Exam Readiness Drill – Chapter Review Questions

Apart from a solid understanding of key concepts, being able to think quickly under time pressure is a skill that will help you ace your certification exam. That is why working on these skills early on in your learning journey is key.

Chapter review questions are designed to improve your test-taking skills progressively with each chapter you learn and review your understanding of key concepts in the chapter at the same time. You'll find these at the end of each chapter.

> **How to Access These Materials**
>
> To learn how to access these resources, head over to the chapter titled *Chapter 24, Accessing the Online Resources.*

To open the Chapter Review Questions for this chapter, perform the following steps:

1. Click the link – https://packt.link/chapter02.

    Alternatively, you can scan the following **QR code** (*Figure 2.2*):

Figure 2.2: QR code that opens Chapter Review Questions for logged-in users

2. Once you log in, you'll see a page similar to the one shown in *Figure 2.3*:

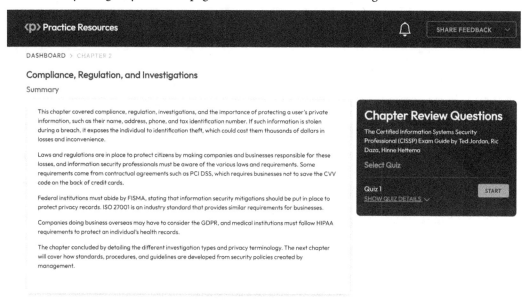

Figure 2.3: Chapter Review Questions for Chapter 2

3. Once ready, start the following practice drills, re-attempting the quiz multiple times.

Exam Readiness Drill

For the first three attempts, don't worry about the time limit.

ATTEMPT 1

The first time, aim for at least **40%**. Look at the answers you got wrong and read the relevant sections in the chapter again to fix your learning gaps.

ATTEMPT 2

The second time, aim for at least **60%**. Look at the answers you got wrong and read the relevant sections in the chapter again to fix any remaining learning gaps.

ATTEMPT 3

The third time, aim for at least **75%**. Once you score 75% or more, you start working on your timing.

Tip

You may take more than **three** attempts to reach 75%. That's okay. Just review the relevant sections in the chapter till you get there.

Working On Timing

Target: Your aim is to keep the score the same while trying to answer these questions as quickly as possible. Here's an example of how your next attempts should look like:

Attempt	Score	Time Taken
Attempt 5	77%	21 mins 30 seconds
Attempt 6	78%	18 mins 34 seconds
Attempt 7	76%	14 mins 44 seconds

Table 2.9: Sample timing practice drills on the online platform

---

Note

The time limits shown in the above table are just examples. Set your own time limits with each attempt based on the time limit of the quiz on the website.

With each new attempt, your score should stay above **75%** while your "time taken" to complete should "decrease". Repeat as many attempts as you want till you feel confident dealing with the time pressure.

3
Security Policies and Business Continuity

This chapter covers security policies and **business continuity** (**BC**). For security policies, you will be learning about what purpose they serve, the characteristics of a good policy, and the information security policy life cycle. You will examine how to identify, analyze, and prioritize BC requirements. Further, you will explore the **business impact analysis** (**BIA**) process and its role in the overall BC effort. Finally, you will look into the **business continuity plan** (**BCP**) construction process. By the end of this chapter, you will be able answer questions on the following:

- Developing, documenting, and implementing security policies, standards, procedures, and guidelines
- Identifying, analyzing, and prioritizing BC requirements
- BIA
- Developing and documenting the scope and the plan
- Contributing to and enforcing personnel security policies and procedures
- Candidate screening and hiring
- Employment agreements and policies
- Onboarding, transfers, and termination processes
- Vendor, consultant, and contractor agreements and controls
- Compliance policy requirements
- Privacy policy requirements

This chapter starts by discussing security policies, standards, procedures, and guidelines: what they are, their purpose, how they relate to each other, and how they can be built and implemented. These documents are all foundational to an effective cybersecurity program and are vital tools in the cybersecurity practitioner's toolbox; understanding how to make the most of these inexpensive and—often—free tools is invaluable knowledge. The following section discusses security policies.

Policies

What are security policies? Think of them as an organizational decree—a stated intent to achieve some ideal outcome or end state, similar to a law. These documents are meant to govern organizational behavior at a high level. Also similar to laws, they do not specify how their objectives are to be met—that is the purpose of a different document, which will be covered in the *Procedures* section.

A security policy serves several purposes. As mentioned previously, policies set high-level goals regarding their subject area. Organizations rely on security policies to decide how they should deploy limited resources and which success metrics will be applied. Security policies also provide a template for different related documents, such as processes and procedures. Procedure documents use policy directives as guiding principles to form more concrete instructions as to how policy objectives will be met within the organization.

These types of documents will be discussed in detail in the *Procedures* section of this chapter. Perhaps the most well-known purpose of a security policy is compliance. For both regulatory and standard compliance, security policies signal organizational commitment and maturity. Standards are yet another important security document that will be discussed later in this section. Compliance with a standard is a common aspect whereby compliance is audited by an independent third-party assessor. Security policies provide valuable evidentiary artifacts to the auditor in their quest to determine adherence to a standard.

Characteristics of a Successful Policy

The following are some of the characteristics of an effective security policy and the criteria to which they should adhere:

- **Affirmed**: First, a policy must be affirmed by management. A policy is of little value if no one follows it, especially management. This commitment must be consistently and continuously demonstrated by management in both words and deeds. Failure to ensure this first trait creates an organizational culture where following policies is not important.

- **Pertinent**: One of the common pitfalls in writing policies comes from the careless use of sample template language. This can lead to a lack of the next trait: pertinent policies. Effective policies cannot be one-size-fits-all. While it is good practice to leverage sample templates, care must be taken to ensure that every word of your policy is applicable to your organization's environment.

- **Pragmatic**: Most often, the work of developing security policies is delegated to **information technology (IT)** or information security folks. This can lead to policies representing the author's wish list. Instead, the author must strive for policy directives that are pragmatic, that is to say, based firmly on reality and not impractical to implement. It is a best practice to involve constituents in the policy development process to guarantee a pragmatic finished product.

- **Reachable**: A closely related trait to pragmatic policies is reachable policies. A sure way for policies to be ineffective is if they set impossible goals. Policy goals could be unattainable for a number of reasons. Ensure that your policy directives do not require the impossible. Remember— security is always competing with convenience. Always keep your constituents involved in the policy development process. They will let you know if you are asking for something that is unduly burdensome on their day-to-day work.

- **Adaptable**: An organization's information security department is often thought of as the department of *No*—the protector of the business process, perhaps, but rarely an enabler of business. Another important trait of a security policy is flexibility and adaptability. Policies must have the ability to adjust and change as the business changes. It is vital that policies are mindful of risk but also stay current with business innovation. This is one of the reasons why policies should be reviewed annually at a minimum.

- **Enforceable**: Another important policy trait is that of enforceability—that is to say, you need to have the ability to enforce the directives set forth in your policies by some means. In general, people believe that if there is a policy against a particular behavior but they can still perform that behavior, then the organization must not be serious about the policy. It is desirable to have some manner of control in place to minimally monitor, if not prevent, undesirable behaviors.

- **Comprehensive**: The modern-day organization is an ecosystem. IT is no longer a self-contained system behind the wall of one organization's firewall. IT now spans the cloud, customers, business partners, and employees. Your policies must also encompass all of those aspects of the IT system, taking into consideration a holistic view of threats to that ecosystem.

Information Security Policy Life Cycle

Security policies pass through four distinct phases in their life cycle. In this section, you will learn about these four phases and what happens in each one. To skip any one of these phases risks leaving out a critical step. These phases ensure that organizational responsibilities are adequately covered at each step in the process by the responsible party. Good and effective policies are not just created by IT but by many departments.

Development Phase

As with any important effort, it is vital to start with a plan. Every good plan begins with a clear view of the goal or objective of the policy. It is important to understand the business, regulatory, and contractual requirements and third-party relationships, as well as how they could be impacted during the planning step. Once you have a thorough impression of the task at hand, the next step is research. It is time to dig deeper into all these areas to avoid making critical mistakes in your policy. Remember to consider laws at the federal and state levels, and at the international level too if your organization operates globally.

You are now ready to begin writing your first draft. Keep in mind your audience and your research while writing. When you think your first draft is ready, it is time for the next step—stakeholder review. This includes both internal and external stakeholders. Some examples of internal stakeholders might be—but are not limited to—department heads, **human resources** (**HR**), in-house counsel, IT, and compliance. Some examples of external stakeholders might be industry experts, regulators, and compliance and legal counsel.

Keep track of and document the input from all stakeholders—the who, what, and when, and the outcome of each suggestion. This last tip can be invaluable as you move on to the next step and gather consensus toward the approval of your policy. Because a policy can affect the entire ecosystem of the organization, it is necessary to solicit consensus from all your stakeholders. The final step in the development stage of the policy life cycle is authorizing the policy—this is done by senior management. This authorizing step is perhaps the most important because it shows all those who read and sign the policy that it has the support and endorsement of senior management.

Publishing Phase

This next stage in the policy life cycle is critical to the success of your policies. It literally determines how well your policies are received, accepted, and followed within your organization. There are three steps within the publishing phase. The first step is getting the word out about your new policy or policies. How you communicate this information to your constituents matters greatly in the success of your policies. For example, leadership should play a critical role in the communication of new policies to the organization. If leaders do not participate, it will create the perception that leadership does not value the policies.

The next step in this publishing stage concerns making the policy or policies available to everyone. The goal here is to make policies available to your constituents in as painless and as quick a way as possible. The most common solution to this problem is to use an intranet site that every employee is introduced to during their new hire onboarding with HR and can access at any time in the future should they need a refresher. While you want to make policies widely available, you do not want to make them public, as policy documents can contain sensitive information that you may not want everyone to read. You may even want to restrict some policies to only certain individuals within the organization.

The final stage of the publishing phase in the policy life cycle is awareness of your policies. This is a perfect opportunity to instill security into the culture of your organization. Practically, this is done across the entire organization where new policies are concerned and during onboarding for new hires or individuals who are joining the organization. The logical place to align this effort is in the organization's **security education, training, and awareness (SETA)** program, which will be discussed further in the next chapter.

Adoption Phase

The next stage in the policy life cycle is the assimilation of your policies. In this key phase, you seek to monitor the performance of a policy with respect to the intended outcome. Whenever a policy deviates from its expected outcome, you have an opportunity to improve and reintroduce it. This phase comprises three distinct steps. The first step is the deployment of the policy into your environment.

Deployment goes beyond just announcing policies to the organization; rather, deployment is more about readying the organization to support and enforce policies. This entails implementing the administrative changes and technical controls that are necessary to ensure the success of a policy. Properly preparing the organization to enforce a policy can often require quite a bit of work and necessitates some capital outlay.

The next step is to closely monitor the performance of policies to ensure the desired effectiveness. During this step, it may become necessary to use temporary policy exemptions to overcome unforeseen circumstances. There are many ways to measure and determine the effectiveness of a policy, such as policy-specific metrics, interviews, surveys, number of incidents, and traditional audits.

The final step is the enforcement of policies. This step is vital to understand how effective a policy is in meeting its stated goal. If a policy is not or cannot be enforced, it may need to be redesigned. Worse still is a policy that is only intermittently enforced within the organization. If word gets out that a policy is not uniformly enforced, then soon, no one will adhere to the policy.

Evaluation Phase

The final stage in the policy life cycle is evaluation. There are two unique steps in this stage: collect observations and redeploy or withdraw a policy. The information security landscape is constantly evolving, as are organizational ecosystems and regulatory requirements. Because of this capricious environment, consistent vigilance over policies is mandatory to safeguard policy efficacy.

> **Note**
> Organizations with good policies create a solid foundation for their success. Organizations that scope and tailor existing frameworks to meet their needs reduce vulnerabilities because they are not reinventing the wheel; they are using standards tested and approved by other firms.

Minimally, policies should be reviewed annually. This review process should include collecting information about policies, their performance, and the organizational environment to verify that policies are still meeting objectives. This will likely require revisiting some of the research done in the first stage to be certain that policies are still current. If policies are found to be deficient but their subjects are still relevant to the organization, they should be updated and republished. However, if policies are no longer pertinent to the organization, they should be withdrawn.

You have explored what purpose policies serve, their role in compliance, some of the characteristics of a successful policy, and the four phases of an information security policy life cycle. Next, you will learn about standards and frameworks and how they are similar and different.

Standards

Cybersecurity standards can be described as a documented accumulation of best practices created by industry experts to protect organizations from cyber threats. Cybersecurity standards and frameworks are usually applicable to all organizations regardless of their size, industry, or sector, though some are a better fit than others. Now, what is the difference between a standard and a framework? A framework is always a standard, but a standard is not always a framework.

That should clear everything right up! The following unpacks the preceding statement. A framework is typically built from one or many standards. An example of a framework is the **National Institute of Standards and Technology Cybersecurity Framework (NIST CSF)**, which is built using several other frameworks, such as the **International Organization for Standardization (ISO)** *27001*, **Control Objectives for Information and Related Technologies (COBIT)**, the **Center for Internet Security Critical Security Controls (CIS CSC)**, and NIST **Special Publication (SP)** *800-53*. The *NIST SP 800-53* standard is the basis for several frameworks, such as the **Federal Information Security Management Act (FISMA)** and the **Federal Risk and Authorization Management Program (FedRAMP)**. So, that's an example of a framework made of other frameworks or standards.

What about standards that are not frameworks? There are thousands of standards that are not frameworks, such as every standard in the *ISO 27000* family, except *ISO 27001*, which is the framework. One other trait of frameworks is that their proper implementation within an organization can be independently certified by a certification body. This certification may be required by certain regulations such as FISMA and the **Defense Federal Acquisition Regulation Supplement (DFARS)**. These regulations will name a standard that an organization must implement and must be independently certified in before they can do business.

Some examples are the implementation and independent certification of the *NIST SP 800-53* standard for FISMA and federal information systems, and the *NIST SP 800-53* standard implementation to satisfy FedRAMP for federal cloud providers. Also, the **Cybersecurity Maturity Model Certification (CMMC)** is required by the **Department of Defense (DoD)** for defense and government contractors to satisfy DFARS and **Federal Acquisition Regulation (FAR)** requirements.

Procedures

Before you learn about procedures, you need to have an understanding of something that antecedes a procedure in the document hierarchy—a process. According to Merriam-Webster's dictionary, *"a process is a series of actions or operations conducing to an end."* In the case of information security, that end is to ensure compliance with a policy. According to Merriam-Webster, a procedure is also a particular way of accomplishing something or acting. In the case of information security, this refers to a specific, detailed series of actions that staff members must take in order to implement a process and comply with a process and its governing policy. The following figure represents the process of a defense contractor who needs to certify at CMMC **Maturity Level 1 (ML 1)**:

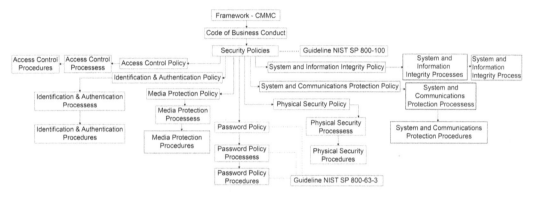

Figure 3.1: Hierarchy of documents

Guidelines

Guidelines are sometimes referred to as standards. As with standards, they are collections of best practices but are often too specific to a niche aspect of cybersecurity to be useful as a framework. The difference between standards and guidelines is that they are leveraged to inform specific aspects of cybersecurity programs. Take a look at an example use of guidelines in *Figure 3.1*. You may leverage the best practice in *NIST SP 800-100 Information Security Handbook: A Guide for Managers* as a guideline while researching and writing your security policies in a general sense. Through your research, determine a password policy appropriate to your organization. You can leverage a different set of best practices in *NIST SP 800-63-3 Digital Identity Guidelines* as a guideline to write your password policy, process, and procedures.

Identifying, Analyzing, and Prioritizing BC Requirements

Remember that it is not possible to plug every hole in a dam. Cybersecurity is an economic problem. Economics is the study of the allocation of scarce resources to satisfy unlimited demand. This applies to cybersecurity because the mitigation of risk requires resources. Resources typically refer to labor, capital, or time. Unlimited demand in this analogy is a risk, especially because you can never mitigate risk 100%. Even if you could, no organization has the resources to attempt to achieve 100% risk mitigation.

This section discusses contingency planning to protect an organization's ability to achieve its objectives. Coincidentally, this is the primary responsibility of all cybersecurity practitioners to their organization. In order to fulfill this responsibility, the cybersecurity practitioner must first understand how their organization creates value for its stakeholders. Cybersecurity practitioners cannot protect what they do not understand.

All organizations must create value for their stakeholders. Stakeholders are likely manifold, such as customers, business partners, shareholders/investors—even donors in the case of **not-for-profit** (**NFP**) entities. Organizations create value through their business processes. IT is merely the digitization of a business process. For example, consider a manufacturing company that makes widgets.

Suppose a salesperson for this hypothetical widget manufacturer makes a large sale of 10,000 units. The salesperson must report the sale to the company (that is, **accounts receivable** (**AR**), manufacturing, logistics, and perhaps countless other departments) to initiate and deliver value creation through their business processes. Now, imagine it is 1950. How is this business process carried out—through inter-office **memorandums** (**memos**), phone calls, or weekly meetings? Fast-forward to today—how is this same transaction initiated? Perhaps via an **enterprise resource planning** (**ERP**) app on the salesperson's smartphone or tablet? Therefore, it is the same or a similar business process (albeit easier), just digitized.

You have just been hired as this widget manufacturer's cybersecurity practitioner and are charged with protecting the company from rare but not unheard-of threats to their value creation—threats ranging from short-term internet or power outages, or minor hardware failures involving **hard disk drives** (**HDDs**), to more critical ones such as natural disasters (earthquakes, hurricanes, or tornados), fire, terrorist actions, or ransomware attacks. Where and how do you begin to inventory and prioritize the various components of their value creation, the business processes that underpin them, and the IT components and data that buttress those processes?

The first step is to secure management support for this important endeavor. The logical way to embody management support is through the use of a policy that defines high-level goals. This BCP policy should outline the steps and objectives to ensure BC. This policy should require the organization to perform a BIA. This crucial step is the foundation of BC efforts, which is discussed in the next section.

Business Impact Analysis (BIA)

The term BIA is used to refer to both a process and a final report that enables the BC effort to characterize system components, supported mission/business processes, and interdependencies. The first thing a BIA does is figure out an organization's value proposition, also referred to in US federal government circles as a *mission*. The next step—as you may remember from the widget example—is understanding the business processes that underpin the value proposition or mission. This is all in service of a much tougher task—triage.

Triage in the BC context relates back to the economics example—there are never enough resources to provide uninterruptible continuity to everything. So, each component's recovery criticality has to be prioritized or otherwise determined; this is also known as *impact* in US federal government circles. In order to do so, you must understand all the components that underpin each business process. Then, ask yourself what would happen to the business process and the organization if that component were to fail partially or completely. How long could the business survive without partial or full use of that component? That is, what will the *downtime* be? Also, it is important to understand how any specific component failure may cascade to other systems or processes that also depend on that component. This is known as **interdependency**.

In order to be able to triage properly, a thorough understanding of business processes, their value to the organization, interoperability with other business processes, and what data they rely upon is paramount. Keep in mind that components are not always digital. Components of a business process can be facilities, personnel, equipment, software, and data in digital form as well as archived physical records. Once all components have been inventoried, their correlation to business processes has been discerned, and all component interdependencies have been brought to light, you are ready for the next aspect of effective triage decision-making.

Using the information collated thus far through the BIA, you are set to prioritize BC resources. This task can seem straightforward until you realize how hard it is to achieve consensus on the value of a component to an organization. Part of information gathering includes several discussions with management, both internal and external, and department heads to identify a component's impact on the organization. Depending upon who you talk to, the response will likely be different.

During these discussions with stakeholders, your objective is to ascertain several key pieces of information, besides obviously reconfirming your understanding of organizational value creation, underlying business processes, and all requisite components. There are three key metrics common to BC planning that should be collected to facilitate triage. The first is **maximum tolerable downtime (MTD)**, which represents the total amount of time the organization is willing to accept for a business process failure (which includes partial or full impairment scenarios). This metric is invaluable in determining the appropriate recovery method.

The next metric is the **recovery time objective (RTO)**, which defines the maximum amount of time that a component can remain impaired or offline before there is an unacceptable impact on other components, supported business processes, and the MTD. RTO and MTD are often confused because both refer to a maximum. Generally, the RTO must confirm that the MTD is not exceeded, and the RTO must typically be shorter than the MTD. You can think of the MTD as the maximum outage for a business process and RTO as the maximum outage for a component that underpins one or more business processes. The following screenshot illustrates the relationships between these key metrics:

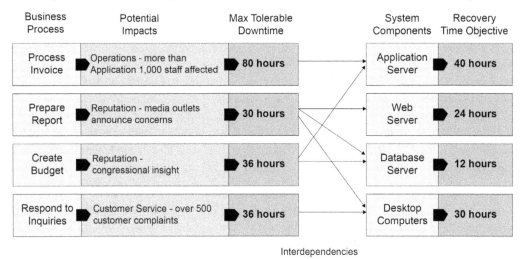

Figure 3.2: BIA process for an information system

The last of the three metrics is the **recovery point objective (RPO)**. It basically speaks to the status of a component, before the system is disrupted or there is an outage, until the point in time when normal operations are recovered. The RPO is dissimilar to the other two metrics, which are measured in time. The RPO involves a state of a process or, more specifically, the data that the process depends upon, often thought of as the normal state. A simpler way to think of the RPO is how much data loss (lost transactions/records since the last successful backup) a business process can tolerate during the process of recovery.

Exam Tip

Having a solid understanding of BIA terminology is key to passing the **Certified Information Systems Security Professional (CISSP)** exam. Knowledge of the RPO, the **work recovery time (WRT)**, the MTD, and their relationships will move you a step forward toward passing.

Now that you understand organizational value creation, the underlying business processes, and all requisite components, and have collected the corresponding MTD, RTO, and RPO, you are ready for the final step in the BIA process—setting priorities for business processes and requisite components. This cannot be achieved effectively without first collating the aforementioned information. There are many strategies to prioritize, such as business process criticality, outage impacts, tolerable downtime, and system components. The criteria used to prioritize vary by industry and even by organization, but a business process's contribution to the mission or value creation is paramount. This is usually ascertained by correlating, where possible, a business process and its components with its financial contribution when that is an applicable measure of value creation.

Developing and Documenting the Scope and the Plan

At this point in the BC planning journey, your organization already has a BC policy endorsed by management and a BIA has been compiled. The next step is to ascertain preventive controls—those that already exist and those that are lacking. It is usually cheaper to prevent an outage event rather than manage the aftermath. Many of the outage scenarios discovered during the BIA process may be mitigated or eliminated through preventive measures that deter, detect, and/or reduce impacts.

For example, consider fire as a potential outage event. Some preventative controls for fire may be smoke alarms, fire suppression systems (water or chemicals where appropriate), fire-resistant building materials, on-site backups (kept in fire-resistant storage), and off-site backups. Now that you know which preventative controls are in place to protect the business processes and their components from disruption, you can move to create contingency strategies based on the triage hierarchy determined through the BIA process.

The first thing to consider are backup and recovery methods and strategies that represent a means to restore system operations quickly and effectively after an outage. To best illustrate backup and recovery methods and strategies, consider the three different triage levels—low, medium, and high—and compare the different approaches you might take in each situation. First, take a look at the low-impact business processes associated with HR.

While this system is important for value creation, it is generally a supporting function and not part of the critical path of value creation. As such, HR—and the organization as a whole—can endure a longer MTD and RTO, even though they have components that have interdependencies with the finance department's business processes, specifically employee payroll. An appropriate backup and recovery method for HR might be a tape backup performed weekly as a full backup and nightly as an incremental backup. That tape could be kept in secure, fireproof storage on-site before eventually being moved to off-site cold storage.

Now, consider a medium-level triage business process and its requisite components. This business process has a shorter MTD and component-level RTO than HR is able to withstand. A simple tape backup, while desirable to the organization from a cost perspective, is not suitable from a time-to-perform-an-effective-restore standpoint. So, the organization has to close that restore-time window to meet or exceed MTD and RTO requirements, which will cost more money. Some technology options that can shorten the restore window are optical backup media (faster than tape to read and restore from) or a **local area network (LAN)**/**wide area network (WAN)** backup.

Finally, you will examine a high-level triage business process and its components. This level is reserved for mission-critical aspects of the organization. High-level triage is engaged for business processes and components with the shortest MTD and RPO (usually expressed in minutes or even seconds), so not surprisingly, it will draw upon the most expensive backup and recovery methods and strategies the organization can afford. Making use of **high-availability (HA)** technologies such as mirroring, clustering, active-passive standby, active-active standby, hot sites, or split-brain computing has the benefit of a recovery time ranging from a few seconds to sub-second recovery, usually imperceptible to the organization. However, they carry with them a price tag that can be **orders of magnitude (OoMs)** more expensive to implement and maintain.

Now that you have gone through a small sample of backup and recovery methods and understand what some of the contingency strategies at your disposal are, you will move on to the next step: developing a contingency plan. The plan covers the details about the roles and responsibilities and the procedures necessary in order to restore business processes after a disruption. One of the key aspects of a good plan is that it is flexible. Plans that are too detailed tend to break under the stressful situations in which these types of plans might be used. Keep in mind that these plans will be read under times of stress so they should be written in simple language that can be understood by anyone.

Your plan should be understood by anyone at any time, but during a crisis should not be the first time that somebody reads your plan if it can be avoided at all. That being said, that's usually what happens anyway, so keep that in mind as you're writing your document. One way to make sure your document is well written is to test it before you need it by running drills where people have to read the document for the first time and use it. Measure the performance of the plan and build that feedback into improvements in its writing for the future.

Training and drills are absolutely essential to writing and maintaining a good plan. Additionally, it is invaluable to make sure that everybody knows their job in the middle of a crisis, and isn't figuring out what that is on the day of a crisis. Dry runs and drills are not just essential for testing the plan; they are invaluable for making sure that your team knows their job and performs it properly. These tests should be run as frequently as an organization can allow, but at a minimum once a year, and the plan should be updated at that time as well.

Contributing to and Enforcing Personnel Security Policies and Procedures

It is said that people are the most valuable resource an organization has and, concurrently, its greatest risk. Cybersecurity comprises people, processes, and technology. Of these three, research demonstrates that people are most often the root cause of a failure of cybersecurity, hence the need for personnel security policies and procedures. This section will discuss the employee life cycle and the policy and procedure tools used to mitigate cybersecurity risks coming from personnel.

Candidate Screening and Hiring

No amount of processes or technology can withstand the sustained threat of a malicious trusted insider. People usually represent the weakest link in the people, processes, and technology triad. These considerations underscore the criticality of thoughtful screening and hiring processes. Organizations should conduct background and reference checks on potential employees to mitigate insider threat risks. These checks can be costly, and not all potential employees require the same level of scrutiny.

A personnel security policy should set the minimum standards for the organization regarding background checks or investigations while empowering information owners with the leeway to necessitate additional checks if they deem the information merits it. Employment law can be complicated; therefore, the development stage of personnel policy should involve outsiders such as legal counsel or employee representatives. There are a few rules you should be aware of when screening personnel. First, there are limits on the information you can gather when screening a candidate. Inquiries should be limited to the job the candidate is being considered for. Digging too deep could expose your organization to legal liability.

While each jurisdiction differs, it is wise to ask a candidate to consent to a background check prior to performing it. A candidate is within their rights to refuse a reasonable request for information, but you are allowed to not hire a candidate on that basis. Increasingly, organizations are leveraging social media for screening candidates. This can seem harmless on the surface, but social media profiles include information such as gender, race, and religious affiliation. The law expressly prohibits the use of this information for hiring. Use of this information could open your organization to discrimination charges.

Another area to be aware of during background checks is the use of educational, driving, and workers' compensation information. The **Family Educational Rights and Privacy Act (FERPA)** of 1974 requires schools to have written permission in order to release any information from a student's educational record. The federal **Driver's Privacy Protection Act (DPPA)** of 1994 necessitates permission from the individual before their personal **motor vehicle record (MVR)** may be sold or released to any third party. While workers' compensation cases are public records, the federal **Americans with Disabilities Act (ADA)** of 1990 states that you cannot discriminate against candidates using medical information or because an applicant has filed a workers' compensation claim. The law concerning criminal records varies from jurisdiction to jurisdiction, so it is better to consult legal counsel.

Major areas of candidate screening are financial history and credit checks. The **Federal Trade Commission** (**FTC**) says you can use credit reports when you hire new employees or during the evaluation of employees for promotion or job role changes, provided the **Fair Credit Reporting Act** (**FCRA**) is followed. Several sections of FCRA spell out employer responsibilities. Primarily, the employer is required to notify the employee of a negative employment decision stemming from their credit report. The **Fair and Accurate Credit Transactions Act** (**FACTA**) of 2003 added new language to the federal FCRA due to an exponential increase in identity theft. Additionally, employers are also prohibited from discriminating against someone who has filed for bankruptcy.

Employment Agreements and Policies

It is industry practice to have employees, contractors, and sub-contractors sign two basic agreements: a confidentiality agreement, also known as a **non-disclosure agreement** (**NDA**), and an **acceptable use policy** (**AUP**). NDAs are in place to prevent the unauthorized release of information and are a requisite even when an individual has access to information systems. Put simply, just because you have access to information, does not mean you have the right to share it at will. AUPs are concerned with the correct use of IT systems and cover things such as password management, internet access, remote access, and data classification standards. By signing an AUP, employees are also agreeing to the other policies mentioned previously.

NDAs are contracts between employees and employers where both sides agree that certain kinds of information are to remain confidential. The kinds of information that can be designated are practically limitless. Any kind of information can be considered confidential—intellectual property, customer lists, business processes, prototypes, engineering drawings, test results, tools, systems, and computer software. NDAs serve manifold purposes beyond the obvious; they can protect an organization from losing its patent rights. NDAs must clearly define the kind of information that may and may not be disclosed. That information must be classified and then labeled clearly and consistently. Most importantly, NDAs set forth how information must be handled when and if employment is terminated or when a contract or project concludes.

An AUP is an agreement between a company and its information systems users. By signing the policy, users acknowledge and agree to the rules in this and all affiliated policies regarding how to properly interact with information systems and handle information. This saves the company the trouble of having each information systems user sign individual policies initially and each time there is an update to said policy. Companies typically have the AUP reaffirmed by every employee annually; this recommits the policies in the minds of information systems users, reminding them of their agreement.

Onboarding, Transfers, and Termination Processes

The employment life cycle has several distinct stages with which a cybersecurity practitioner should be familiar. The stages vary from company to company, but this section will focus on three stages that are critical to personnel security—onboarding, transfers, and termination. Beginning with onboarding, this is the moment when the employee is added to the payroll and benefits systems. At this point, the employee provides a plethora of **personally identifiable information** (**PII**). It is the company's responsibility to classify and protect that newly onboarded data along with all other employee data. In the US, an employee must provide proof of identity, work authorization, and tax identification, all of which are also considered PII.

Another aspect of the onboarding stage is user provisioning, which is when the employee's user accounts, group memberships, and identification badges are created. Also, access rights and permissions, along with physical access devices such as traditional keys, **radio-frequency identification** (**RFID**) keycards, fobs, or smartcards, are assigned. Before granting access, the user should have reviewed and signed the terms and conditions of the AUP. The access rights and level of permissions granted to a user must match their role and responsibilities. The onus of determining the appropriate level of access belongs to the information owner, who must regulate what access should be granted and in which circumstances. Typically, supervisors request access for their subordinates; however, in some organizations, this could be handled by HR or IT. Employee transfers are similar to the user-provisioning aspect of onboarding. When an employee is promoted or changes job roles within the organization, their user provisioning may necessitate a change to their access rights.

In the final termination stage—widely considered the most dangerous phase—the employee departs the company. How the organization handles this depends on the nature of the departure—was it the result of resignation, firing, or retirement? In a nutshell, this stage reverses the previous two, which includes removing the employee from payroll, recovering information assets such as their smartphone, laptop, and tablet, removing access to devices, and deleting or disabling user accounts and access permissions. Studies show it can be dangerous to assume that termination of employment is amicable, even if the employee resigns for personal reasons. The terminated employee might attempt some act of revenge, create havoc, or steal information.

Vendor, Consultant, and Contractor Agreements and Controls

As mentioned in the *Employment Agreements and Policies* section, the risk from an employee's use of information systems and access to sensitive information is still present when using vendors, consultants, or contractors—collectively referred to as third parties. As with employee risk, third-party risk must be mitigated as well; the strategy and controls are predicated on the amount of access a third party requires to deliver its contractual obligations. You will explore a few different third-party scenarios and their requisite agreements and controls in the following paragraphs.

First, consider a scenario where your third-party business partner is rarely on-site (say, once or twice a month), but they have administrative access to information systems. Any third-party individual and the individual's organization that has access to information systems should be required to sign an NDA. Additionally, any third party with administrative access should also be required to sign your company's AUP, and their access to information systems should be monitored and logged. These third parties should be required to verify their identification upon arrival on-site and escorted while on-site to monitor their activities.

Now consider a slightly different third-party relationship, one where the business partner is on-site on a more permanent basis; that would be the case of a staff augmentation role using a contractor. This contractor also has administrative access as they perform an IT function. It would be wise to perform a background investigation, just as you would with a regular employee. This contractor should be required to verify their identification upon arrival on-site initially and be given a contractor badge identifying them as a non-employee with limited access. They should also be required to sign your company's AUP and NDA before starting their engagement.

Take a look at one final scenario. This one is different, not because of the amount of time spent on-site, but because of the sensitivity of the data the third party is accessing. In scenarios where your third-party business partner has access to sensitive information, the following two mitigations are required. The third party, or the third-party organization that has access to sensitive information, should be required to sign an NDA. These third parties should be required to verify their identification upon arrival on-site and escorted while on-site to monitor their activities.

Compliance Policy Requirements

Another key benefit of policies and procedures is their ability to satisfy compliance requirements as part of a larger compliance program. A compliance program is a collection of policies, processes, and procedures to facilitate compliance with laws and regulations and to protect the business's reputation. These practices are meant to prevent and detect violations of applicable laws and regulations. This is squarely in the realm of legal risk management and internal controls. Enforcing compliance helps an organization prevent and detect violations of laws and regulations, which protects the business from fines and lawsuits.

The compliance process is a continuous one. Most businesses establish a compliance program to govern their compliance responsibilities uniformly and faithfully over time. An example of compliance is during the preparation of a financial report that follows standard accounting principles such as **generally accepted accounting principles (GAAPs)**. Another example of compliance is when a data breach occurs, and the organization informs affected parties within a defined time period. Compliance is the state of being in accordance with relevant federal or regional authorities and their regulatory requirements. There are manifold industries with regulatory requirements that are best satisfied with policies and procedures.

Privacy Policy Requirements

The paradigm in the corporate world is that employees should not have any expectation of privacy regarding their actions done in company time, on company property, or with company resources. This lack of employee privacy encompasses electronic monitoring, video monitoring, and searches of their person and effects. Electronic monitoring includes phone calls, computer access, email, mobile (voice and data), internet access, and geolocation. Video monitoring includes company property, with the exception of cameras in restrooms or changing rooms, which is protected by law. Searches of their person and effects extend to searching an employee, an employee's work area, or an employee's property, including a car, as long as it is on company property at the time of the search. All searches of people and effects must be conducted in accordance with state and local regulations.

However, a company must disclose its monitoring activities to employees and obtain written acknowledgment of said policy. The court will only protect an organization's ability to monitor an employee's actions if it follows its own policy consistently. The primary takeaway is that organizations must clearly state their policies and uniformly enforce those policies. Minimally, privacy expectations must be clearly stated in the information security policy and acknowledged through a signed AUP in order to enjoy the court's protection to monitor employee activity.

Summary

This chapter covered security policies and BC. For security policies, you learned about their purpose, what makes a good policy, the information security policy life cycle, and personnel security policies and procedures. For BC, you looked at how to identify, analyze, and prioritize BC requirements. You reviewed the BIA process/report and its place in the overall BC effort. Lastly, you examined how to build a BCP. The next chapter will cover personnel policies, risk management, threat modeling, **supply chain risk management** (**SCRM**), as well as security awareness, education, and training programs.

Further Reading

For more information regarding the topics covered in this chapter, please refer to the following resources:

Develop, document, and implement security policy, standards, procedures, and guidelines:

- *Greene, Sari Stern. Security Policies and Procedures: Principles and Practices.*

Identify, analyze, and prioritize BC requirements:

- *Swanson, M., Bowen, P., Phillips, A.W., Gallup, D.*, and *Lynes, D. NIST Special Publication 800-34, Rev. 1, Contingency Planning Guide for Federal Information Systems. 149 p17* (2010).

- *Estall, Hilary. Business Continuity Management Systems: Implementation and Certification to ISO 22301. BCS, The Chartered Institute, 2012.*

Exam Readiness Drill – Chapter Review Questions

Apart from a solid understanding of key concepts, being able to think quickly under time pressure is a skill that will help you ace your certification exam. That is why working on these skills early on in your learning journey is key.

Chapter review questions are designed to improve your test-taking skills progressively with each chapter you learn and review your understanding of key concepts in the chapter at the same time. You'll find these at the end of each chapter.

> **How to Access These Materials**
>
> To learn how to access these resources, head over to the chapter titled *Chapter 24, Accessing the Online Resources*.

To open the Chapter Review Questions for this chapter, perform the following steps:

1. Click the link – https://packt.link/chapter03.

    Alternatively, you can scan the following **QR code** (*Figure 3.3*):

Figure 3.3: QR code that opens Chapter Review Questions for logged-in users

2. Once you log in, you'll see a page similar to the one shown in *Figure 3.4*:

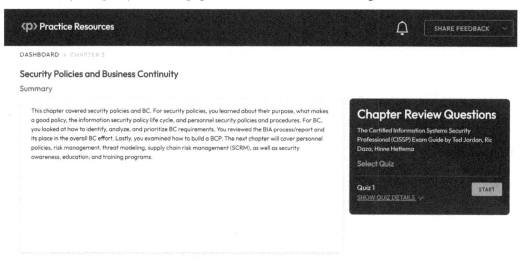

Figure 3.4: Chapter Review Questions for Chapter 3

3. Once ready, start the following practice drills, re-attempting the quiz multiple times.

Exam Readiness Drill

For the first three attempts, don't worry about the time limit.

ATTEMPT 1

The first time, aim for at least **40%**. Look at the answers you got wrong and read the relevant sections in the chapter again to fix your learning gaps.

ATTEMPT 2

The second time, aim for at least **60%**. Look at the answers you got wrong and read the relevant sections in the chapter again to fix any remaining learning gaps.

ATTEMPT 3

The third time, aim for at least **75%**. Once you score 75% or more, you start working on your timing.

> **Tip**
>
> You may take more than **three** attempts to reach 75%. That's okay. Just review the relevant sections in the chapter till you get there.

Working On Timing

Target: Your aim is to keep the score the same while trying to answer these questions as quickly as possible. Here's an example of how your next attempts should look like:

Attempt	Score	Time Taken
Attempt 5	77%	21 mins 30 seconds
Attempt 6	78%	18 mins 34 seconds
Attempt 7	76%	14 mins 44 seconds

Table 3.1: Sample timing practice drills on the online platform

> **Note**
>
> The time limits shown in the above table are just examples. Set your own time limits with each attempt based on the time limit of the quiz on the website.

With each new attempt, your score should stay above **75%** while your "time taken" to complete should "decrease". Repeat as many attempts as you want till you feel confident dealing with the time pressure.

4

Risk Management, Threat Modeling, SCRM, and SETA

This chapter is about risk management, threat modeling, **Supply Chain Risk Management** (SCRM), and **Security, Education, Training and Awareness**, commonly referred to as **SETA**. For risk management, you'll review how to apply risk management concepts. You will look into threat modeling concepts and methodologies and SCRM concepts. Finally, you will examine SETA programs.

By the end of this chapter, you will be able to answer questions on:

- Understanding and apply risk management concepts
- Understanding and apply threat modeling concepts and methodologies
- Applying Supply Chain Risk Management (SCRM) concepts
- Establishing and maintain a security education, training and awareness program

Will we start by reviewing and analyzing risk. Not all risks require the same risk response. If this is the case, how do security engineers evaluate risks and apply appropriate mitigations? They use accepted practices through a risk management framework.

Risk Management

A discussion of risk management concepts must first begin with defining the term *risk management*. The term *risk* is defined as the possibility that something unpleasant or unwelcome will happen. Put simply, risk management encompasses any of the myriad activities performed to control risk. This can seem like a very simple statement but, as you will see in this section, there is quite a bit covered under risk management.

At a bare minimum, risk management requires the identification of assets, determining the threats to said assets, the application of risk treatment, and the monitoring of treatment effectiveness. Entire books have been dedicated to this subject, and this is but a mere overview of risk management to provide you with a high-level understanding of the concepts. It is highly recommended that you explore these concepts in more detail to round out your cybersecurity knowledge of this foundational topic more thoroughly. The following section details how to identify threats and vulnerabilities – a key aspect of risk management.

Identify Threats and Vulnerabilities

A threat is typically considered as an intent by someone to cause harm to another. However, in cybersecurity, a threat refers to any situation or event with the possibility of negatively impacting the confidentiality, integrity, or availability of a system or asset. The term *vulnerability*, in the context of cybersecurity, describes the same areas as *threat*. The key difference is that vulnerability refers to the exposure of weakness, while threat delineates the ability to take advantage of a given vulnerability.

Now that you understand what threats and vulnerabilities are, you can begin exploring how to identify them. There are two popular approaches to identifying threats; the first is from the threat perspective and the second is from the asset perspective. From the threat perspective, consider the two example threats of a hurricane and a **distributed denial of service attack (DDoS)**. The next step is to consider which assets are affected by these threats. Assets that might be affected by a hurricane would be the ones that are adversely affected by high wind, flood water, or loss of electricity. Conversely, the assets affected by a DDoS attack are the ones that rely on unimpaired network connectivity for full functionality.

Now consider the process from the asset perspective with the example of an organization's customer list as the asset. Some of the threats to an organization's customer list might be theft, accidental deletion, or failure of the system hosting the list. Regardless of the approach taken, once the likely threats or affected assets are enumerated, the next step is to calculate exposure.

In the context of cybersecurity, the term *exposure* refers to the metric or rating assigned to the combination of the threat to an asset via a given vulnerability with consideration of its probability of occurrence (see *Figure 4.1*). This crucial step in risk management is necessary in order to prioritize the mitigation of risk. As is often the case, there are not enough resources to mitigate every risk. Without some metric to prioritize exposure (of which dollar value is most popular and universally understood), the triage work of risk assessment and analysis is vastly more difficult. Determining the exposure for a threat/vulnerability pair is actually part of the risk assessment process, which will be discussed in detail in the next section:

Figure 4.1: Threat and vulnerability relationship

So far, this chapter has covered what threats are and what vulnerabilities are and has made you familiar with the term *exposure*. You also briefly learned how threats and vulnerabilities are identified and about the importance of the concept of exposure in order to effectively prioritize risk. The next sections are going to talk about risk assessments and how risk is analyzed.

Risk Assessment/Analysis

In this section, you will be reviewing risk assessment and analysis, which is a foundational component of risk management. Risk assessment and analysis is concerned with the allocation of scarce resources as efficiently as possible to mitigate risk as effectively as possible. In order to begin the risk assessment and analysis process, you must first compile the asset inventory and then identify the threats and vulnerabilities of the assets. It is upon that inventory that you perform the risk assessment and analysis process to determine the applicable risk response and select the appropriate countermeasures. Refer to *Figure 4.2* to see where a risk assessment fits into the overall risk management process:

Figure 4.2: Risk management process

This section will dive deeper into the concept of exposure that was discussed previously. The term *exposure* is known by many other names as well, such as risk rating, likelihood, probability, or impact, just to name a few. The common denominator is the need to have a way to rank threat/vulnerability pairs for a given asset in order to facilitate the risk assessment process. Therefore, it doesn't matter what term you use or which way you choose to express it (dollar amount, high, medium, low, or some ordinal scale) as long as you are consistent with the method that you use to express it across all of your asset inventory. Additionally, it should be limited to a single threat to an asset, via one particular vulnerability, and the probability of its occurrence, which should be expressed in some form that facilitates comparison with other threats.

This section covered what a risk assessment is and how it relates to threats and vulnerabilities. Further, the concept of exposure, some of its other names, and its role in the risk assessment process were elaborated upon. The next section will take you through risk response (also known as risk treatment), which is the logical next step to risk assessment in the risk management effort.

Risk Response

As mentioned previously, risk response is also known as risk treatment, countermeasure, or risk mitigation. Once the risk assessment process is complete, you have the ability to prioritize which risks to mitigate first and which to defer, based on the asset's value to the organization, and begin the process of determining your risk response. The risk assessment process should conclude with producing a risk register. A risk register is a comprehensive listing of all of your assets (with an appropriate value assigned), all of the threats to each of those assets, the vulnerabilities that may be exploited by those threats, and the exposure level of each.

The first step in determining a risk response is taking a complete inventory of the risk response already in place – if there is one. Just because a threat and a vulnerability have been identified, it does not mean they have not already been mitigated. It also doesn't mean that the mitigations currently in place are still the most effective mitigations for the threat/vulnerability possible today. There may be nothing to do, or it may be a good time to revisit what the best mitigation technique is for the given threat/vulnerability in question.

Now that you have an idea of the first step involved in formulating a risk response plan, you can examine the four methods of responding to a risk: risk acceptance, risk mitigation, risk transference, and risk avoidance. There is a common misconception that every risk must be mitigated, but the reality is, as you'll review in the following four options, it is not your only choice:

- **Risk acceptance**: Risk acceptance is the most commonly used option out of the four, and people may not always be aware that they're using it. Whether you are aware of the risk or not, risk acceptance means that you have chosen to do nothing about it. People often do not realize that they are accepting a risk even if they don't know that the risk is present. Perhaps a better term for this option would be *risk ignorance*. A considerably smaller percentage of people are aware of the risk and consciously decide to do nothing about it. Mere awareness of the risk may potentially offer some mitigating effect.

- **Risk mitigation**: Risk mitigation is the option with which cybersecurity professionals are most familiar. It involves knowledge of the risk and actively taking steps to mitigate the risk through the implementation of countermeasures or controls.

- **Risk transfer**: Risk transference usually entails the purchase of insurance against the risk event. However, this option can be a little bit misleading. The term *transfer* can lead one to believe that once the insurance is purchased, the risk is no longer present. If a risk incident occurs, insurance can provide financial compensation, but it does not ameliorate the negative effects of the risk incident itself, such as the damage caused to an organization's brand.

- **Risk avoidance**: This option is often confused with risk acceptance. To clarify, risk avoidance refers to an organization terminating any activity that creates the risk in the first place. An example of risk avoidance is when a company chooses technology that does not carry risks with it compared to another technology option that does.

Assume for a moment that there is already a mitigation strategy present for a given threat/vulnerability. How do you determine whether it is effective enough to leave in place or requires an update? The answer is that it depends on the organization's risk appetite, and whether or not the control is mitigating risk to the organization's risk tolerance level.

Simply put, risk appetite is the level of risk an organization is willing to accept in the pursuit of its organizational objectives. Conversely, an organization's risk tolerance represents the limit to which they're willing to accept risk. The chief function of the risk response effort is to determine which of the four risk options to choose. If it is to mitigate a given risk, then that risk should be mitigated to a level somewhere between appetite and tolerance.

An additional factor to consider is whether or not the current mitigation technique can be monitored easily. Monitoring and measurement of mitigations/controls are covered in the *Monitoring and Measurement* section.

Exam Tip

One memory tool to help remember risk responses is **MATA**, for **Mitigation, Avoidance, Transfer, and Acceptance**. Risks remaining post-mitigation must always be accepted.

Countermeasure Selection and Implementation

This section will take you through countermeasure selection and implementation involved in the risk response technique of risk mitigation. There are a few factors to take into consideration when selecting countermeasures/controls. Not every control is a piece of technology; the applicable types of controls are discussed in the next section. In addition to choosing the type of control, there is also cost versus the asset value of a given control.

The prevailing wisdom for quite some time has been that you don't spend more on a countermeasure than the value of the asset you are protecting. While this can seem straightforward, it is much more complicated upon closer inspection. Take a look at the first aspect – the cost of a control or countermeasure. There are at least two components that make up the cost of a control: the cost of implementation and the cost of maintenance.

There are usually three aspects of the cost of implementation: *time, capital,* and *labor*. For example, if the control was a technology countermeasure such as a firewall, there would be a cost to acquiring the firewall, the labor to set it up, and finally, the time it takes to implement it. If the control was a policy instead, it would involve the time it takes to research and implement it, the labor by the people writing it, and the capital paid for any external assistance that contributed to the research, such as lawyers.

Similarly, there are three aspects of the cost of maintenance of a control: *time*, *capital*, and *labor*. Consider the firewall example again; there is the time it takes to maintain the firewall solution, the capital cost for software licensing and ongoing maintenance, and finally, the labor by staff to maintain the hardware, software, and configuration of the firewall. The same can also be applied to policy control: the time and labor by staff to manage and maintain the policy to keep it current, and the capital, time, and labor it takes to update, re-issue, and re-train staff on policy updates.

Therefore, the two factors of cost – implementation and maintenance – always have to be compared with the control's effectiveness at mitigating the risk to the desired level.

The comparative aspect of this assessment criteria is asset value. Asset value is complicated, primarily because there are many ways to assign value to an asset. The simplest of these is the replacement cost if the asset were to be destroyed somehow. However, the replacement cost could be considered an oversimplification of asset valuation because it doesn't take into consideration the impact of the loss of a particular asset on other dependent aspects of the organization. These impacts include loss of productivity, loss of data, damage to availability, damage to organizational reputation, and so on.

Additionally, there are a myriad of ways in which an asset can suffer damage/impairment, all of which have different costs to restore the asset. Consider the classic ransomware attack; this is where the attacker compromises a system, discovers the data that is most precious to the organization, and encrypts it. At this point, the underlying system has sustained no damage; the data is technically there and in place but inaccessible. The tricky part about this scenario is the longer the attack goes on, the more expensive it becomes for the organization, and therefore, the more expensive the asset will be deemed.

Fortunately, the most cost-effective countermeasure for ransomware is an effective backup and restore system. Unfortunately, this countermeasure is rarely implemented and tested to ensure it is operating properly. This a very common oversight that leaves many organizations unprepared for a ransomware attack. Also, attackers know more about an organization's **single points of failure** (**SPOF**) than the organization's cybersecurity staff do, so they can cripple an organization's business operations by encrypting only pieces of data. This is why the asset valuation portion of a risk assessment is so critical to the risk management process. Without it, cybersecurity professionals cannot effectively deploy the already limited resources to mitigate risk.

Types of Controls

According to ISO 27005:9.2, controls may provide one or more of the following types of protection:

- **Correction**: Rectifying or mitigating damage following a security event
- **Elimination**: Removing the source of the risk
- **Prevention**: Stopping security incidents from taking place
- **Impact minimization**: Reducing the potential consequences of a security incident

- **Deterrence**: Discouraging potential attackers from attacking

- **Detection**: Identifying security incidents or vulnerabilities as they occur

- **Recovery**: Restoring systems and data following a security incident

- **Monitoring**: Continuous observation and analysis to identify anomalies or suspicious behavior

- **Awareness**: Educating personnel to recognize and respond to security threats effectively.

This section will discuss the applicable types of controls. Instead of the numerous control types that ISO describes, this section will primarily focus on three: *preventive, detective,* and *corrective.*

Preventive

Preventive type controls are controls designed to prevent a given incident. Deterrent controls can be considered a type of preventive control because they are also intended to stop an attack. However, sometimes a distinction is made that they dissuade, rather than block, an attack. These controls can take many forms, such as firewalls configured to prevent certain traffic from passing, or security policies such as an **acceptable use policy** (**AUP**), which are written to prevent undesirable activity. While both controls in the previous examples are preventive, it would probably be more accurate to consider the policy a deterrent kind of preventive control since people can choose not to follow the policy. Conversely, a firewall control cannot be ignored once configured.

Detective

Detective controls are controls that detect undesirable behavior while it's in progress. A couple of examples of detective controls would be an **intrusion detection system** (**IDS**) that detects suspicious traffic and a camera pointed at the entrance to a building or room. The previous two examples are intended to detect undesirable activity; the camera can also be considered a preventive or deterrent control because the mere presence of the camera could deter people from engaging in activities that could harm the organization.

Corrective

Corrective controls are controls that are used after an incident has occurred to minimize the damage from the incident. Some examples of corrective controls would be a disaster recovery plan or a backup and restore capability. Both these examples don't come into effect until after the incident has occurred. A disaster recovery plan is only used after a disaster such as a hurricane; it involves letting everyone know what they should do to minimize the damaging effects. A backup and restore capability is the chief control in ransomware attacks. It does not detect or prevent a ransomware attack but an efficient restoration minimizes the damage caused by ransomware.

Control Assessments

Assessing controls is squarely within the purview of audit. Audit is concerned with confirming that *"what should be done"* or *"what has been claimed to have been done"* has actually been done and done properly. Without auditing, there would be little, if any, accountability for stakeholders and it would be difficult to establish trust with third-party business partners. Control assessments prevent ineffective controls from being deployed by raising the questions, *"Does this work the way we expected?"* and *"Does the control actually mitigate risk as expected?"*.

There are several reasons to assess controls. As the Russian proverb says *"Doveryai, no proveryai"* or *"Trust but verify."* Control assessments are about creating trust through the verification of controls. If security is a chain, then it's only as strong as its weakest link. Audit and verification of security controls highlight the importance of cybersecurity trust among business partners. This verification provides the impetus behind certifications such as ISO 27001, FISMA, and FedRAMP. These certifications, whether achieved voluntarily or mandated by regulation, provide clients and business partners assurance that best practice frameworks were implemented and independently verified through audit.

While establishing trust is an important reason to assess controls, there are still others, such as determining the effectiveness of controls. Effectiveness does not just pertain to whether the controls are working but also whether they are actually performing the job they were implemented to do. More specifically, effectiveness pertains to whether they are mitigating the risk to the intended level according to the security plan that prescribed them. A secondary benefit of this type of control assessment is to determine the overall effectiveness of a security plan. This kind of regular control assessment leads to opportunities for improvement of the efficiency of controls, which can sometimes lead to the replacement of controls in favor of new technologies that can do a better job. The next section discusses risk management, which entails monitoring and measurement of risk.

Monitoring and Measurement

According to NIST SP800–37, *"the purpose of the Monitor step is to maintain an ongoing situational awareness about the security and privacy posture of the information system and the organization in support of risk management decisions."* Monitoring and measurement is a logical extension of control assessment and a critical step in the risk management process. According to a saying often attributed to Peter Drucker, *"What gets measured gets managed."* If you want to manage your cyber risk, you must measure it and monitor it. Deploying cybersecurity controls, indeed cybersecurity programs, without the intention to assess, monitor, and measure those controls indicates that you don't care whether they succeed or not.

Most, if not all, of the reputable security frameworks that can be certified require proof that controls are being assessed, monitored, and measured to obtain certification. The monitoring of a security control or program is fairly straightforward to understand but the measurement is more complex. The goal of monitoring is to ensure that the amount of risk an individual control or an entire security program is mitigating has not changed from one moment to the next. On the other hand, there's the goal of measurement improvement; you cannot gauge whether a control or program has improved if you didn't take a measurement at some point in time in the past and then take a measurement at a future time for comparison. This process is referred to as establishing a baseline. The goal of a good security program is to mitigate the desired amount of risk, maintain risk mitigation, and finally, improve its efficiency over time.

Reporting

Reporting is an important concept in the risk management process. Cybersecurity cannot operate in the dark; it must have allies in the business. Stakeholders have a vested interest in understanding the cybersecurity risk in the organization with which they are associated and how it is being managed. Reporting is an opportunity to communicate to stakeholders about the risk management plan, its successes, its failures, and what it needs to succeed in the future. Risk management reporting must be tailored to match the intended audience, must include information that is most important to them, and must consider the audience's knowledge level about the technology and risk concepts being discussed in the report.

The first concept in reporting is to clearly understand who your stakeholders are. The next step is to understand what their top concerns are regarding the risk management of the organization. There can be more than one group of stakeholders; for example, investors could be one group, management or senior management could be another, and potential business partners could be yet another stakeholder group – all with different priorities. Each of these groups will likely have different knowledge levels and different priorities and will desire information at different intervals and in different formats. Success in risk management reporting depends on understanding these nuances among your audience.

Continuous Improvement and Risk Maturity Modeling

The topic of continuous improvement was briefly alluded to in the *Monitoring and Measurement* section. Programs (such as security programs or risk management programs) that are maturing are said to be improving over time; this is a highly desirable trait for any program. Most security frameworks, such as ISO 27001 and NIST, have a requirement to undergo continuous improvement. This section will discuss continuous improvement and what is meant by risk maturity modeling.

In the world of risk management, you can think of risk maturity as a spectrum with basic cyber hygiene on one end and proactive, advanced, or progressive activities on the other. In fact, the United States **Department of Defense (DoD)** has a new certification – **Cybersecurity Maturity Model Certification (CMMC)**. Given the sensitivity of the information being dealt with, a DoD contract requires a given level of maturity (levels 1 to 5, independently audited and assessed by a third-party assessor to be valid) according to the CMMC in order for that contractor to be awarded that contract.

This new trend in cybersecurity is no surprise as CMMC was built from ISO 27001 and NIST SP800–137. Requiring continuous improvement and increasingly mature risk processes within an organization serves two purposes. First, it ensures that the organization is not doing only the bare minimum checklist-style security but, instead, is striving to improve that security over time. Second, many frameworks that are certifiable by independent third-party assessment contain a maturity component because they are not permanent certifications but rather last a finite amount of time – usually three years – before the organization must be re-certified. Cybersecurity is a process, not a destination. Since threats are not static, cybersecurity cannot afford to be either; it must continually improve. The next section will move the discussion to risk frameworks.

Risk Frameworks

You explored security frameworks in the previous chapter. Just a reminder here: frameworks are collections of best practices. Risk frameworks are no different; they too are just collections of best practices. As examples of security frameworks, you've seen ISO 27001, NIST SP 800-53, COBIT, and others. For risk frameworks, you will look at NIST SP 800–37r2, ISO 31000, ISACA's Risk IT, and the **Committee of Sponsoring Organizations (COSO)**.

One key difference when discussing risk frameworks is that they do not just relate to cybersecurity risk but enterprise risk management as well. Specifically, people who use a risk management framework are charged with protecting value creation through the management of risk, aiding a company to achieve its objectives, and improving performance. These tasks are collectively called **enterprise risk management (ERM)**. Managing risk at an enterprise level is an integral part of governance and compliance.

> **Note**
> The CISSP exam questions focus more on NIST 800-37r2 and ISO 31000 when it comes to risk frameworks.

Taking a closer look at the managing risk aspect of ERM, every endeavor carries with it a level of risk – that is to say, a negative uncertainty. Conversely, every endeavor also carries a level of opportunity – that is, a positive uncertainty. Part of ERM's job is to determine an organization's appetite and tolerance for uncertainty in pursuit of the achievement of its goals. Risk frameworks provide a defined structure for decision-making to maximize positive uncertainty and minimize negative uncertainty in the pursuit of value creation.

In order to positively affect an organization's value creation, most risk management frameworks prescribe the following steps. First, an organization must clearly understand the uncertainties associated with its current endeavors. Next, the organization must understand its appetite and tolerance for those uncertainties. Then, similar to cybersecurity, the organization must understand what resources it has at its disposal to manage those uncertainties. Finally, it must do this with a clear understanding of the culture and human resource element in which it operates. This section covered a lot of ground in understanding risk management concepts, and yet we have just scratched the surface. The next section will take you through understanding and applying threat modeling concepts and methodologies.

Threat Modeling Concepts and Methodologies

In this section, you will gain an understanding of the concepts and methodologies involved in threat modeling. The first logical step is to understand what threat modeling is. As plainly as it can be expressed, threat modeling is a process for refining application, system, or business logic security by first identifying goals and weaknesses and then identifying countermeasures to mitigate the risks to the subject of the modeling exercise. First, some questions about the modeling subject, be that an application, a system, or a business process flow need to be considered.

One of the first considerations is the goal or objective of the subject of the modeling exercise. After all, the purpose of threat modeling is to ensure that the subject of the modeling exercise achieves its intended objective. You not only need to understand what the goal is but also how the objective is achieved. One of the more popular ways to achieve that understanding is through diagramming the application, system, or process being modeled. Diagramming is not only useful to understand the subject but also to ensure consensus among multiple parties as to the operation of an application, a system, or a process.

Diagramming also helps us with our next series of questions – understanding what can break the subject of our threat modeling. What often happens during the diagramming process is that you might start to think of what might break the process or where its weaknesses may be. This is especially true of cybersecurity professionals who are well-versed in breaking applications or systems as they are trained to spot such weaknesses. It is usually helpful to understand the dependencies at each stage in the flow of the application or system and what can disrupt those dependencies. In addition to knowing the dependencies, it is important to also know the assumptions being made at every stage in the flow. Keep an eye out for nested dependencies and assumptions. What does that mean? Often, a system or application will depend upon another system or application that has its own dependencies and assumptions, and those must be understood as well.

Once you are clear about all the dependencies and assumptions in each stage of the flow, you can move on to the next aspect of understanding what can go wrong. The next step in this process is understanding what is already in place (countermeasures) to prevent scenarios that could break this application or process. The most important aspect is to verify that the countermeasures are adequate given the holistic view of the application or process. At this point, you are now ready to move on to the next question: What should be done to ensure that your application or system is protected from the things you believe could go wrong?

Appropriate countermeasures should be prescribed to protect the application or system. Given the knowledge you now have about this system or application, its dependencies and assumptions, and its current countermeasures, you must make the all-important determination – what, if anything, must change in the current defense posture to shore up the defense agains threats? This part can seem pretty straightforward to a cybersecurity professional but, as mentioned in previous sections, cost must be considered when applied to the value of the asset being protected.

Supply Chain Risk Management

Up to three-quarters of a modern globalized (outsourced) organization's costs rests in its supply chain. This means that threats to the supply chain can have an exponential effect on ERM. The globalized nature of the supply chain also means much of the risk mitigation is out of an organization's direct control. Cyber threats are now thought by leading nations to be one of the greatest risks to business in the modern global, interconnected economy. This section will discuss procurement risk, assessing third-party risk, minimum security requirements, and service-level requirements.

Risks Associated with Hardware, Software, and Services

This section will discuss the risks associated with hardware, software, and the services associated with them – primarily, the risks found throughout their life cycle. A few of the risks associated with the IT life cycle are shadow IT, purchasing **consumer off-the-shelf** (**COTS**) products, risks associated with outsourcing, moving to the cloud, and emerging threats. Some of these threats and some best practice recommendations will be discussed in this section.

This section starts with a brief discussion about shadow IT – specifically, what it is, and what the risks associated with it are. Shadow IT is a colloquial umbrella term that refers to IT projects that are managed outside of the IT department. Consider the following example.

Suppose the manager of a software development team has asked IT to provide a server for their team. After six months, they still have not obtained a server. This manager has their own corporate credit card and the authority to spend money on it. Tired of waiting, they go to **Amazon Web Services** (**AWS**), and five minutes later, the team has a server. This is now a separate system from the company's main IT system, and this is known as shadow IT. The obvious risks are that the IT and security departments have no idea this virtual server exists, that it houses critical software development projects, or whether it is even secured. The best countermeasure against shadow IT is to have a policy that clearly states that shadow IT behavior will not be tolerated.

In terms of the acquisition of hardware, software, and services, it's generally considered a best practice to buy from **original equipment manufacturers** (**OEMs**) and their authorized resellers. While this practice does not completely ameliorate the risk, you can certainly minimize the risks of the more common threats in acquisition that can happen with open source products and inauthentic products. The next section will discuss third-party risk, how to assess it, and how to monitor it.

Third-Party Assessment and Monitoring

Third-party risk refers to threats originating from vendors with which the business has a relationship. In order to manage this risk, it is critical to know how the business relates to each of its vendors. Details such as dependencies, how IT systems are linked and secured, and what kind of information is being shared should be known.

One of the common misconceptions is that an assessment is the same as an audit. Assessments such as third-party assessments are designed to help organizations make better decisions, whereas audits are designed to confirm compliance with a standard or framework. Another common misconception is that third-party assessments are performed by a third party. While that may or may not be the case, that is not what we're talking about here. A third-party assessment can also be performed by in-house staff and refers more to what is being assessed than who is doing the assessment.

Third-party assessments generally focus on having a deeper understanding of the risks in the relationship, how they affect the organization's compliance posture, and gaining a better understanding of the data used in that relationship. Some questions the assessment would likely seek to answer about the data are as follows: what data is being shared, how it is being shared, where it is being stored, who has access to it, and whether it is protected through encryption.

A typical organization can have hundreds of third-party vendor relationships. It is not possible or practical to perform an assessment on all of them. Therefore, priority is given to those vendor relationships that have the highest level of access to data and systems within the organization. The higher the vendor's level of access, the more rigorous and in-depth the assessment that is performed. The monitoring aspect of a third-party assessment typically focuses on ensuring that nothing has changed since the full assessment; you can even think of it as a periodic mini-assessment.

Minimum Security Requirements

After the last section, you may be asking yourself "Doesn't monitoring and assessing third parties become expensive for organizations?" The answer is yes, it can cost millions of dollars a year for organizations to keep an eye on their third-party vendors through assessments and continuous monitoring. One of the ways organizations seek to keep third-party assessment costs down is through the use of minimum security requirements. A current example of the use of a minimum security requirement is the DoD CMMC standard – specifically, maturity level one (**ML 1**).

A minimum security requirement represents a standard vendors must meet if they wish to do business with the organization. In the example mentioned previously, CMMC ML 1 represents a minimum security requirement to do business with the federal government. The federal government goes a step further by requiring an independent third party to perform the assessment and verify compliance with the required maturity level. If the required maturity level is not verified, the contract will not be awarded to the vendor. The CMMC standard completely takes the burden of third-party assessment and monitoring and minimum security requirements off of the government's plate.

Service-level requirements

Service-Level Requirements (SLRs) are another critical piece of supply chain risk management. An SLR typically takes the form of a document that collects, from a client's perspective, the requirements for a successful engagement. An SLR is a precursor to another important document, the **service-level agreement (SLA)**. The SLR acts as a repository for the client's specifications of the IT services that the customer will be purchasing. Requirements such as capturing customers' tolerance for interruptions, how those interruptions will be handled, and what penalties will be exacted for exceeding contractual tolerances are contained in the SLR. These requirements are then implemented in the engagement contract under the SLA section of the contract. Supply chain risk management is so important that it is quickly becoming a specialization within cybersecurity. The next step in our journey is a discussion of SETA.

SETA Programs

It is often said that cybersecurity is made up of three things – people, processes, and technology. Of those three things, the people aspect is the weakest link. You can have all the policies, processes, and technology but if your people are working against you, then you have no chance. This section will talk about SETA programs, their importance, methods and techniques, and how to gauge their effectiveness. The objective is to create an educated workforce that is less vulnerable to cybersecurity attacks. One of the first things to understand about SETA is the most effective order of its implementation, which is *Awareness*, then *Training*, and finally, *Education*.

As NIST SP800-16 says "*Awareness is not training. The purpose of awareness presentations is simply to focus attention on security. Awareness presentations are intended to allow individuals to recognize IT security concerns and respond accordingly.*" Awareness is the crucial first step in the process designed to level set the audience for a further conversation about cybersecurity. Examples of awareness topics can be phishing, viruses, password protection and selection, and protecting your laptop.

Training involves imparting knowledge and skills to enable concepts taught by awareness. An example of training could be if the awareness topic was phishing, then the corresponding training might teach how to identify a phishing email. The distinction between training and education is a little more subjective. An obvious example of education is a degree program with an established curriculum. A certification is not traditionally considered education, but that line continues to blur.

Awareness and Training

With the goal of turning the liability of people into an asset for cybersecurity, we'll be discussing some of the methods and techniques to achieve that end through SETA. One of the first challenges to overcome in a SETA program is getting the audience's attention. Any method to grab the audience's attention that involves them in the cybersecurity problem is a good one. An example of training to prevent an attack such as social engineering is when an organization launches a phishing campaign against its own people. The results of this campaign can be leveraged in many ways to achieve the SETA objectives. For instance, the results of the campaign can be collected and presented at the beginning of the training to illustrate how vulnerable the organization is to phishing.

Another example of using a phishing training program might be establishing a special email address (phishing@yourorganization.com) for employees to forward suspected phishing emails to. This program can be further gamified by declaring that the first X number of employees to forward a valid phishing email will win a $50 gift card. One way to increase the effectiveness of this contest would be to involve security champions in a competition to see whose team wins the most gift cards. A security champion is an employee with an interest in security who boosts the security message at the team, peer, or department level. These techniques ensure the audience's attention in the training and while opening and reading emails at their job.

Periodic Content Reviews

Security threats are a bit like fashion trends and last about a season before a new one comes along. That means that a SETA program's content, while partially unaffected as security principles are usually relatively static, will need to be updated in portions to keep the program current. You may have noticed a recurring theme in this book, that of *continuous improvement*. Nowhere is that more important than in a SETA program. It is crucial to stay abreast of the latest cyber risks to your organization and make sure the appropriate awareness and training are added to your SETA program.

Program Effectiveness Evaluation

The first step toward continuous improvement is measurement. Without measurement, you cannot determine the effectiveness of your SETA program. An effective measurement program begins with a thorough understanding of the goals of the program being measured. You cannot determine whether the program is effective if you don't understand what success looks like. Once the goals are understood, measurement can take place to determine whether the objectives of the program have been met. Those same measurements can be used to determine whether the program is successful.

Summary

This chapter concludes the first of the eight domains. Risk management, threat modeling, SCRM, and security awareness, education, and training – commonly referred to as SETA – were discussed in this chapter. For risk management, how to apply risk management concepts was covered. You also examined threat modeling concepts and methodologies, and SCRM concepts. Finally, you explored SETA programs. The next chapter will begin looking at the asset security domain – specifically, asset and privacy protection.

Further Reading

- Strupczewski, Grzegorz. *Defining cyber risk*. Safety Science 135 (2021): 105143. `https://packt.link/ocLlP`

- Shevchenko, Nataliya, et al. *Threat modeling: a summary of available methods*. Carnegie Mellon University Software Engineering Institute Pittsburgh United States, 2018. `https://packt.link/zcf4n`

- Ho, William, et al. *Supply chain risk management: a literature review*. International Journal of Production Research 53.16 (2015): 5031-5069. `https://packt.link/ZJ7uq`

Exam Readiness Drill – Chapter Review Questions

Apart from a solid understanding of key concepts, being able to think quickly under time pressure is a skill that will help you ace your certification exam. That is why working on these skills early on in your learning journey is key.

Chapter review questions are designed to improve your test-taking skills progressively with each chapter you learn and review your understanding of key concepts in the chapter at the same time. You'll find these at the end of each chapter.

> **How to Access These Materials**
>
> To learn how to access these resources, head over to the chapter titled *Chapter 24, Accessing the Online Resources*.

To open the Chapter Review Questions for this chapter, perform the following steps:

1. Click the link – `https://packt.link/chapter04`.

    Alternatively, you can scan the following **QR code** (*Figure 4.3*):

Figure 4.3: QR code that opens Chapter Review Questions for logged-in users

2. Once you log in, you'll see a page similar to the one shown in *Figure 4.4*:

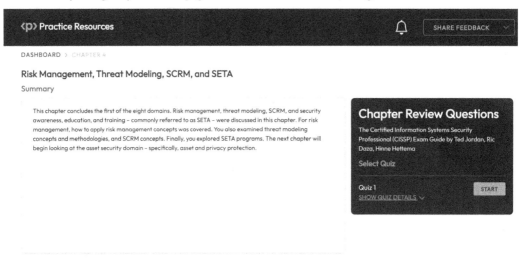

Figure 4.4: Chapter Review Questions for Chapter 4

3. Once ready, start the following practice drills, re-attempting the quiz multiple times.

Exam Readiness Drill

For the first three attempts, don't worry about the time limit.

ATTEMPT 1

The first time, aim for at least **40%**. Look at the answers you got wrong and read the relevant sections in the chapter again to fix your learning gaps.

ATTEMPT 2

The second time, aim for at least **60%**. Look at the answers you got wrong and read the relevant sections in the chapter again to fix any remaining learning gaps.

ATTEMPT 3

The third time, aim for at least **75%**. Once you score 75% or more, you start working on your timing.

> **Tip**
> You may take more than **three** attempts to reach 75%. That's okay. Just review the relevant sections in the chapter till you get there.

Working On Timing

Target: Your aim is to keep the score the same while trying to answer these questions as quickly as possible. Here's an example of how your next attempts should look like:

Attempt	Score	Time Taken
Attempt 5	77%	21 mins 30 seconds
Attempt 6	78%	18 mins 34 seconds
Attempt 7	76%	14 mins 44 seconds

Table 4.1: Sample timing practice drills on the online platform

> **Note**
> The time limits shown in the above table are just examples. Set your own time limits with each attempt based on the time limit of the quiz on the website.

With each new attempt, your score should stay above **75%** while your "time taken" to complete should "decrease". Repeat as many attempts as you want till you feel confident dealing with the time pressure.

5
Asset and Privacy Protection

The asset security domain is one of the most important for cybersecurity as it helps define what you are trying to protect and how much effort you should put into protecting it. This domain is all about assets and their security, how to classify them, how to handle them, and most of all, how to protect them. An asset is defined as anything of value to an organization. This encompasses people, partners, equipment, facilities, reputation, and information. Cybersecurity is typically most concerned with "information" or, more specifically, digital assets. This concept will appear repeatedly throughout this book because it is crucial to what a cybersecurity professional does.

By the end of this chapter, you will be able to answer questions on the following:

- Identifying and classifying information and assets
- Establishing information- and asset-handling requirements
- Provisioning resources securely

As already touched on in the previous chapters, there is no such thing as "100% secure." This is largely because there's no such thing as unlimited resources, so it is important that you make best use of resources The first task in protecting assets is to prioritize which ones are most important. The first step in those efforts is creating a comprehensive digital asset inventory. Then, you assign a value to each asset in that inventory. With an inventory like that, you can then make better business decisions when directing efforts to secure assets. To that end, this chapter will discuss the identification and classification of digital assets, how to handle assets based on their classification, and how to allocate the necessary resources to a given asset.

Identifying and Classifying Information and Assets

There is a saying that's long been popular in cybersecurity that highlights the importance of building a comprehensive digital asset inventory: "You cannot protect what you do not know that you have." The problem with **digital assets** is that they are usually intangible; you cannot see or physically touch them. It can be challenging to explain to stakeholders that they must expend resources to protect intangible assets. In order to properly inventory digital assets, you must first understand the organization, the digital assets it relies upon, how they are used, and where they are stored. This is a concept covered in *Chapter 3, Security Policies and Business Continuity* in the previous domain under **Business Impact Analysis (BIA)**.

This brings us to a very important aspect, which is the distinction between data and information. To grasp this distinction, consider the old saying, "Data is just ones and zeros, but information is actionable." Essentially, context is what transforms data into information. One of the most effective ways to ensure a comprehensive digital asset inventory is by first considering the business processes that power the organization. Once those are understood, you can move on to the IT processes that those business processes rely on. Then, the final step is to understand the information and the information systems that those IT processes rely on. Once that is understood, it is a fairly straightforward process to inventory and prioritize the digital assets yielded from that effort.

Again, the point of a comprehensive **digital asset inventory** is to help prioritize your efforts; in order to do that, you have to be able to rank the importance of each entry in your asset inventory. Once identified, how do you go about ranking or prioritizing a digital asset? There are several ways to go about this, but probably the most common is asking yourself how the organization as a whole would be affected if the confidentiality, integrity, or availability of that asset were to be adversely impacted. The greater the impact, the higher the priority, and the greater the expense to secure the digital asset in question or mitigate any adverse outcome. Remember to always consider the interdependencies between various digital assets during this process; they are not always standalone assets but have interdependent relationships that enhance the value of other assets. A negative impact on one asset could have a negative impact on other assets as well.

Data Classification

An additional layer to aid the ranking of digital assets is an analysis of the kind of information contained within the asset. This is called **data classification**. For example, a database containing marketing information may not rank as high as a database containing customer records or intellectual property. This, of course, depends upon the organization. Particularly, how the organization creates value determines what assets are most valuable to it. So, in order to understand your organization's digital assets, it is important to be familiar with the kind of information that business processes and IT processes are using and that is subsequently contained within digital assets.

This extra layer allows additional classification. This classification may relate to the sensitivity or significance of the data to the organization's value creation process or the organization's requirement to keep that information confidential. The main purpose of classifying digital assets is to provide another level of granularity when prioritizing or ranking how best to spend resources, protecting a given asset or asset class.

Asset Classification

In order to determine **asset classification**, you must first understand the classification level of the data contained within the asset. Ultimately, the asset classification level will determine the countermeasures that are deployed to protect a given asset. It is important to have a documented classification process in order to prevent two different people from coming up with a different classification level for the same asset. The following list contains a few critical steps that should be included in any classification process:

1. Document classification levels.

2. Define criteria for each level of asset classification.

3. For each level of asset classification, document a minimum of countermeasures to be deployed to protect assets within that classification level.

4. Identify the data owner(s) and custodian(s) for assets.

5. Set intervals and procedures for the review of the classification of all assets.

6. Define criteria for the reclassification, classification, or destruction of an asset.

While not an exhaustive list, this represents a bare minimum that should be considered in any organization. This asset classification process leads handily into the discussion of asset-handling requirements.

Establishing Information- and Asset-Handling Requirements

In order to establish **information- and asset-handling requirements**, the first step is to look at all the ways that you can store information and the form that assets may take. Usually, information and digital assets are intangible, but some information assets can be tangible. Some examples of physical information and assets include paper, such as printed information. Another important physical information asset is the actual storage media – for example, backup storage media, whether magnetic or optical, **hard disk drives** (HDDs), servers, laptops, or anything that might contain valuable information.

Another category of assets that is also physical is mobile devices. Mobile assets can contain information of interest and comprise phones and tablets. These can be either employee-owned under a **bring-your-own-device** (BYOD) policy or company-owned.

An additional information storage location you must consider is cloud storage. Cloud storage can not only take forms such as cloud service offerings such as hosting but also cloud service storage options such as Dropbox, Box, OneDrive, and others.

This section has covered storage locations common to most of today's typical organizations. To make sure you have covered all possible digital asset locations, you must understand your organization's business processes, related IT processes, and the information it stores and produces. Now, you can begin to consider the creation and handling requirements for a given asset: how to create it, how to maintain it, and how and when it should be destroyed.

Provisioning Resources Securely

This section will discuss the secure provisioning of digital assets. To properly discuss this topic, you must first examine a few foundational subject areas such as information and asset ownership, the idea of asset inventory, and the larger concept of asset management over the course of an asset's life cycle.

Information and Asset Ownership

Asset ownership is not the same as the classical ownership that you might think of. In most cases, every asset within an organization is owned by the organization, but in this case, ownership is related to who is responsible for the care and feeding of a given asset. It would probably be more helpful to think of an asset's owner as its manager—that is to say, the person responsible for making day-to-day operational decisions about the given asset's security throughout its life cycle.

Asset Inventory

As touched upon previously, a reliable **asset inventory** is critical to the challenge of protecting assets. You cannot protect what you do not know you have. However, some of the most valuable assets cybersecurity is charged with protecting are intangible. Inventorying tangible assets is fairly straightforward because you can see and touch them. They are items that tend to have serial numbers, such as laptops, servers, IP cameras, routers, switches, wireless access points, and printers.

Intangible assets, on the other hand, are a little more difficult to inventory and even easier to ignore, miss, or forget. Examples of intangible assets include **intellectual property**, **databases**, **customer lists**, and **employee information**—in short, data stored digitally. However, not all data is an asset; some can even be a liability. Further, not all data has the same value to the organization or requires the same expense to protect. Therefore, utilizing a systematic process to inventory intangible digital assets is imperative for two reasons: the first to ensure that you don't miss any critical assets, and the second to ensure that you have the necessary context to accurately evaluate any given asset.

Asset Management

Asset management can be a particularly complex topic because it is a function performed in just about every industry or specialization. A few examples of asset management include intellectual property, software, engineering, finance, enterprise, and many other types of asset management. For each industry, the term asset has a different meaning. Cybersecurity is largely concerned with intangible digital assets that can affect the confidentiality, integrity, and availability of organizational value creation.

This chapter has discussed assets, their identification and classification, their handling requirements, and how to provision them securely in detail. The key concept to understand in asset management is the idea that digital assets have a life cycle, and effective management entails understanding the various stages of that life cycle. There are many different ideas about how many stages this life cycle has, from as few as 4 to as many as 14. At their core, all models have variations of the following four stages: planning, creation, use, and retirement, with interchangeable names for each of these stages (that is, the stages may go by different names, but what happens in each stage is the same).

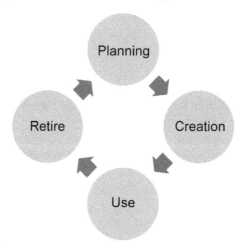

Figure 5.1: Digital asset life cycle

The first stage is **planning** and is sometimes referred to as the **business case** or the **inception phase**. This is where the initial need for the digital asset is realized and identified. This is also the stage where the data owner is identified (a concept that will be covered in detail in *Chapter 6, Information and Asset Handling*). During this stage, the risks and rewards of the creation of this data asset are considered along with the resources that will be required to generate and maintain it over its life cycle.

The next phase is **creation**, which is sometimes referred to as the purchase or acquisition phase. The plan from the previous stage is considered and put into motion. Whether this entails a purchase or a value added to existing digital assets to create the desired product, the digital assets' integration into the organization's existing business processes and potential effects are considered. This stage can sometimes seem a bit like planning part two; at this point, the organization's senior management reviews the work done and either approves or declines the creation/acquisition of this digital asset.

The next phase, **use**, is also known as the **operation** or **maintenance phase** and is necessarily the longest. This is the phase where IT and security staff review the requirements of this digital asset and consider how it should operate in the organization to be successful. They must also consider how to integrate this digital asset without causing negative effects on existing processes and assets, all the while considering the security of this and all other digital assets.

The final phase is **retirement**, which is also known as the disposal or destruction phase. It is what happens when an organization deems that an asset no longer provides value to it commensurate with the cost of maintaining it. At this point, the organization must consider how to remove this asset without negatively impacting existing assets or processes. Once the asset has safely been removed from production systems, its value versus its risk must be considered along with how it will be properly disposed of. Sometimes, a digital asset can be sold as it may have value to another organization. More often than not, the risk or liability to the organization outweighs that value and its careful destruction must be considered. Of course, local laws and regulatory compliance must always be considered at this stage in the digital asset life cycle. At this stage, a replacement for this asset might be necessary, which of course starts the entire life cycle again.

Summary

In this chapter, you reviewed asset security, specifically the identification and classification of assets and their provisioning and handling. Further, you reviewed the reasons for data and asset classification and the importance of determining asset ownership, as well as the difference between tangible and intangible assets. The next chapter will dig deeper into some of the specifics of asset management, such as data roles, the asset management life cycle, and the determination of appropriate controls for digital assets.

Further Reading

- Toygar, Alp, C. E. Rohm Jr, and Jake Zhu. *A new asset type: digital assets.* Journal of International Technology and Information Management 22, no. 4 (2013): 7. `https://packt.link/tEt7A`

- Laney, Douglas B. *Infonomics: how to monetize, manage, and measure information as an asset for competitive advantage.* Routledge, 2017. `https://packt.link/XzXCM`

- Chaisse, Julien, and Cristen Bauer. *Cybersecurity and the protection of digital assets: assessing the role of international investment law and arbitration.* Vand. J. Ent. & Tech. L. 21 (2018): 549. `https://packt.link/oilaM`

Exam Readiness Drill – Chapter Review Questions

Apart from a solid understanding of key concepts, being able to think quickly under time pressure is a skill that will help you ace your certification exam. That is why working on these skills early on in your learning journey is key.

Chapter review questions are designed to improve your test-taking skills progressively with each chapter you learn and review your understanding of key concepts in the chapter at the same time. You'll find these at the end of each chapter.

How to Access These Materials

To learn how to access these resources, head over to the chapter titled *Chapter 24, Accessing the Online Resources*.

To open the Chapter Review Questions for this chapter, perform the following steps:

1. Click the link – `https://packt.link/chapter05`.

   Alternatively, you can scan the following **QR code** (*Figure 5.2*):

Figure 5.2: QR code that opens Chapter Review Questions for logged-in users

2. Once you log in, you'll see a page similar to the one shown in *Figure 5.3*:

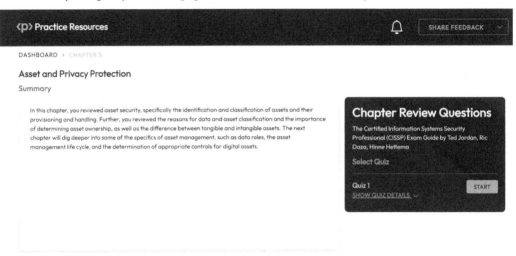

Figure 5.3: Chapter Review Questions for Chapter 5

3. Once ready, start the following practice drills, re-attempting the quiz multiple times.

Exam Readiness Drill

For the first three attempts, don't worry about the time limit.

ATTEMPT 1

The first time, aim for at least **40%**. Look at the answers you got wrong and read the relevant sections in the chapter again to fix your learning gaps.

ATTEMPT 2

The second time, aim for at least **60%**. Look at the answers you got wrong and read the relevant sections in the chapter again to fix any remaining learning gaps.

ATTEMPT 3

The third time, aim for at least **75%**. Once you score 75% or more, you start working on your timing.

> **Tip**
> You may take more than **three** attempts to reach 75%. That's okay. Just review the relevant sections in the chapter till you get there.

Working On Timing

Target: Your aim is to keep the score the same while trying to answer these questions as quickly as possible. Here's an example of how your next attempts should look like:

Attempt	Score	Time Taken
Attempt 5	77%	21 mins 30 seconds
Attempt 6	78%	18 mins 34 seconds
Attempt 7	76%	14 mins 44 seconds

Table 5.1: Sample timing practice drills on the online platform

> **Note**
> The time limits shown in the above table are just examples. Set your own time limits with each attempt based on the time limit of the quiz on the website.

With each new attempt, your score should stay above **75%** while your "time taken" to complete should "decrease". Repeat as many attempts as you want till you feel confident dealing with the time pressure.

6

Information and Asset Handling

This chapter will continue the discussion on asset security—specifically, how to manage digital assets over the course of their life cycle. The chapter will delve into the usage and destruction phases of the data life cycle, explaining the information you need to collect and the questions you need to answer to properly oversee digital assets. Additionally, you will examine the destruction phase and the criteria for when and how to carry out the destruction of a digital asset. Finally, you will learn about security controls and compliance requirements, which indicate how digital assets should be managed during their life cycle.

By the end of this chapter, you will be able to answer questions on:

- Managing the data life cycle

- Ensuring appropriate asset retention (e.g., **end of life (EOL)** and **end of support (EOS)**)

- Determining data security controls and compliance requirements

This chapter will start with managing the data cycle.

Managing the Data Life Cycle

This section will cover two main areas: data roles and what to do with your data once you have it. As you look more carefully into data roles, such as **owners**, **controllers**, **custodians**, **processors**, and **users**, also known as **subjects**, you will gain an understanding of the primary functions and their significance. Next, you will look at the different stages within the **data life cycle**—namely, **collection**, **storage**, **maintenance**, **retention**, **remanence**, and finally, **destruction**.

Data Roles

To effectively grasp the concept of the data life cycle, you must first understand data roles because you need to be aware of who is responsible for what aspects of the stages, which will be discussed in this section. Without the accountability provided by data roles, it would be difficult to ensure the protection and proper management of your data. These are the key data roles in a data life cycle:

- **Data subject**: This is a simple data role that is very intuitive. As the title suggests, the data subject is the person the data is about. While this role is not directly involved in the data life cycle, it is vital, especially with respect to the concept of privacy and regulatory compliance.

- **Data owner**: One of the most important data roles is that of the data owner. The data owner is usually the person who brought the data into existence in the first place and is also responsible for setting the policy regarding that data. *Policy*, in this context, refers to the classification of the data, as well as the protocol for handling disclosure requests. In a nutshell, that means defining who gets access to the data directly or indirectly.

- **Data controller**: Data controllers are an extension of the data owner; they instantiate and enforce the policy set forth for the data by the data owner. If a particular class of data happens to not have a data owner defined, that role is taken by the data controller. So, the data controller either implements or sets a policy for the data—in particular, data that is sensitive in nature.

- **Data processor**: Data processors are typically the folks entrusted with the daily care and feeding of the data. Depending on how the organization is structured, they could be your IT folks. They ensure that the policies set by the data owner or data controller are adhered to. This involves making sure that only the people who are supposed to access the data can access it, the data is protected in the manner it's meant to be as per the policy, and regular audits are performed to ensure that only appropriately authorized people are accessing the data per the policy.

- **Data custodian**: Data custodians are a specific variation of the data processor; they are more technical and tactical in nature than data processors. Custodians not only implement security controls as per the policy for a given data type but also ensure that the security controls are functioning as designed. This verification process is achieved through an audit of the controls and the data type. These audits are usually performed by a data custodian and occasionally verified by an external third-party auditor when necessary for regulatory compliance or certification. The production of valid artifacts from the audit process is a required deliverable of the custodian's function.

You can't have a data life cycle without the collection of data, but data collection can be complicated. The following section will cover the importance of data collection and some of the issues associated with it.

Data Collection

Before the advent of **data privacy laws**, organizations used to collect and store every bit of data they could get their hands on, on the off chance that it could be worth something in the future. A few years ago, it was common to hear phrases such as data is the new oil. However, since the **California Consumer Privacy Act (CCPA)** was enacted, data collection has changed such that the strategy is now frowned upon and often illegal. This is most obviously the case when the data is **personally identifiable information (PII)**, which refers to information that can be used to distinguish or trace an individual's identity—such as name, social security number, **date of birth (DOB)**—but there are other forms of data that are also protected.

The most important consideration about data collection and data privacy is the residence/citizenship of the data subject as that determines the applicable regulatory compliance and jurisdiction. The two primary factors that data owners and data controllers should consider when creating a policy about a type of data are the specifics of the data subject that is being collected and the residency/citizenship of that subject. Some examples of specifics include the subject's full name, social security number, credit card number, address, DOB, and healthcare information. Some residency information to be aware of is any information related to citizens of the **European Union (EU)**, residents of the state of California, or the state of Massachusetts, as these locations tend to have the most stringent privacy protection and penalties.

Data Location

After the data is collected, you have to consider where you are going to store it. Certain types of information have certain restrictions on where they can be stored geographically. **Data residency** (also known as **data localization**) laws state that data about a data subject must remain within the country to which the data subject belongs. These laws are based on the concept of data sovereignty, which applies to data subjects and certain types of information collected about them. Those types of information typically include PII, financial data, healthcare data, and national security data.

You might not think this would be as big as it is, but with international companies collecting data about subjects of all residencies around the world and storing it using cloud architecture, it can be difficult to pinpoint exactly where a piece of data is stored exclusively. This is due, in particular, to the high availability of cloud architectures, which commonly have them replicating and distributing data among international data centers, unbeknown to the data owner. Unless there are strong and accurate data classification rules and policies in place, data stored in a cloud architecture could land the organization in serious trouble regarding its data localization practices.

Data Maintenance

One of the core tenets of cybersecurity is **integrity**. **Data maintenance** entails preserving the integrity of the data collected over the course of its life cycle. Consider a simple example: your organization wins a new client, and the point of contact for this new client is somebody named Bob Jones. All of Bob Jones's details (such as his title, address, phone numbers, spouse's name, and so on) are entered into the organization's **customer relationship management** (**CRM**). Over the course of the entire business relationship, Bob changes titles, changes his phone numbers, and even divorces and remarries. This data subject's record requires constant maintenance to keep the information within it current. If the information is not updated, there could be possible damage to the relationship if a new salesperson were to take over the representation of this company and call Bob and ask about his former spouse.

Data Retention

So far, you have learned about the collection of data, its storage location, and the maintenance of its integrity. Now, you can focus on how long to keep the data. This may seem like a simple issue, but you will find it may not be in an organization's best interest to keep data longer than it needs to. Certain kinds of data have a regulatory compliance requirement to preserve that data for a certain amount of time. Also, there are situations that mandate the preservation of data, such as ongoing litigation. Outside of those two situations, it's best to have a policy spelling out exactly how long to keep data and how to destroy it and follow an audit procedure to ensure that the data retention policy is followed exactly.

It may be tempting to look at all of your compliance requirements about data retention to see which has the longest requirement and make that requirement your policy. However, this can be detrimental to your organization for two reasons, the least of which is storage cost. The more imperative reason is that the more data you keep unnecessarily, the more arduous it is to comply with things such as e-discovery orders. E-discovery refers to court orders that require you to preserve and produce often large amounts of data in very short amounts of time. So, the more stored data you have, the more expensive it is to comply.

A more cost-effective solution would be a granular classification scheme with a corresponding retention policy attached to each classification level. This could begin with classification schemes for data with mandated retention periods, followed by classification schemes for data that does not have a mandated retention period. For the latter, business requirements would dictate the ideal minimum retention timeframe. In order for organizations' retention policies to maintain their effectiveness, they would have to be regularly audited.

Data Remanence

Confidentiality is another one of the core tenets of cybersecurity. While this is important for the entire data life cycle, it is never more pronounced than in a discussion regarding **data remanence**. When you delete a file, it is typically not actually gone, but the table that points to that file (that is, the entry for that file) has been cleared. Until all the places where that data was stored have been overwritten with new data, it is not truly gone. This makes it relatively simple for unauthorized parties to access or even restore the deleted information. This is known as data remanence.

Depending upon the sensitivity of the data stored on the media, more aggressive means of erasure may need to be employed. There are four basic strategies to reduce the risk of data remanence: **physical destruction**, **degaussing**, **encryption**, and **overwriting**. Degaussing refers to the removal of magnetic fields in a magnetic storage media, such as traditional hard drives. These magnetic fields on magnetic storage media are used to store data as ones and zeros; when the magnetic charges are removed, the data is lost permanently. Physical destruction, as its name implies, just means running any storage media through a device (such as an industrial shredder) that will destroy it completely and render it unusable.

The last two are considerably more elegant and easier to deploy. Starting with overwriting, for many years, the US **Department of Defense (DoD)** required any storage media containing sensitive data to be overwritten seven times before it was considered safe. Most modern operating systems also have this function available for civilians when erasing data; the operating system overwrites deleted files or entire **hard disk drives (HDDs)** seven times with ones and zeros. Finally, encryption prevents the risk of data remanence through a process known as **crypto-shredding**. Simply put, any sensitive data is encrypted with a complex key that is then discarded or destroyed, rendering that data unretrievable.

Data Destruction

Eventually, all data must reach the end of its life cycle, which means its **destruction**. The most important part of this stage of the life cycle is to make sure that when you dispose of data, it is actually destroyed and unreadable. Essentially, it really doesn't matter how a data breach occurs, whether due to hacking or from data retrieved from an old laptop with an HDD that was not properly deleted. The damage to the organization would be the same, except the second scenario mentioned is probably the least expensive to prevent. The most challenging aspect of this stage of the life cycle is to maintain a secure and reliable inventory of every location where your data is stored; that means backups, archives, hot standbys, warm standbys, and so on. Paired with this secure and reliable inventory should be policies and procedures for the proper destruction of any of these data stores, and audits to make sure that the data destruction is happening as it should be.

The next section will take you through another important aspect of asset security—namely, EOL and EOS.

Asset Retention

It may be tempting to keep assets indefinitely, but there are security ramifications to be considered. For example, most organizations lease their laptops/desktops for between 18 and 36 months. After that lease expires, they can either return them or keep them at no additional cost. Again, the temptation might be to extract additional value from such assets, even though they have been completely amortized off of the books. Value can be extracted just by putting them in storage as emergency spares or perhaps by selling them as used equipment. But both options come with risks to the organization.

Consider the first option—keeping them as emergency spares in storage. Suppose one day somebody's machine breaks and they need a quick replacement. Depending on how old the machine is, it may not be able to run modern software. In this case, you have kept one or several hundred machines around for nothing. It is always recommended to use new machines because they come with replacement warranties.

Now, consider the other option—attempting to sell the surplus equipment as used equipment. While it seems this could be a valid option, you always have to consider the risks of data remanence. Whenever there is a change of custody of an asset, data remanence precautions should be taken, whether that custody change comes from a gift or a sale. These two scenarios are examples of asset retention policies' EOL.

EOL is a concept created by **original equipment manufacturers (OEMs)** to describe the time that they expect their product to stop performing at the quality level at which the product was originally sold. EOL is usually similar to the time period of the warranty coverage on a given product. EOL is an important date to be aware of as maintaining a product beyond its EOL increases its risk to the organization of failure from many potential sources, including security. OEMs are in business to make bigger and better products, and they cannot do this when they are constantly supporting old technology.

An additional concept helps extend the finality of EOL—that is, EOS. You can think of it as a grace period beyond EOL. EOS refers to the time period from EOL to the point at which the manufacturers will no longer support a product (that is, when they stop making patches and replacement parts for a given product line). The objective of EOS is to give customers who are not ready to transition to a new warranty-covered product additional time. They need that additional grace period to make plans and facilitate an effective transition.

Data Security Controls and Compliance

Since you have already read about the importance of an accurate digital asset inventory and the proper categorization of that inventory, you can now look at the protection of said inventory. *Chapter 5, Asset and Privacy Protection*, discussed the **asset life cycle**, what a digital asset is, and the importance of the proper classification of a digital asset. In particular, this section will help determine the quantity of resources to apply when securing a given asset. This section deals with how to determine the proper security controls for a given asset at a given classification level and compliance requirement.

Data States

One of the first considerations regarding the protection of any digital asset is the digital asset's change state —that is, when the asset is being used by an application, when it is in transit over a network, and when it is at rest in storage. Each one of the states has a different control and strategy for the protection of the assets.

Let's begin by looking at the data state known as **at rest**. This is the state in which the data spends most of its life. It includes any time it is stored on any digital media such as **solid-state drives** (**SSDs**), optical devices such as **CDs** or **DVDs**, or magnetic media such as traditional HDDs or backup tapes. **Data at rest** poses its own set of unique security challenges. Primarily, data at rest can move and be stored in places that you're unlikely to think of, such as USB keys or even printed pages.

This is due largely to the fact that data at rest is transportable, and it is difficult to prevent physical access to it, which makes its protection even more difficult to ensure. Consider the common case of a laptop in the back seat of a rental car. All it takes is a broken window, and somebody has access to the laptop and all the data on its hard drive, even if they don't have the passwords. The most common countermeasure for data at rest is encryption, which can render data unreadable by unintended parties; fortunately, it is in common use, highly reliable, and inexpensive.

The next data state is **in transit**. This refers to the time when data is moving from storage to other computing resources via either a **local area network** (**LAN**), a **wide area network** (**WAN**), or the internet. Like data at rest, data in transit is vulnerable to breaches, particularly when it travels over networks you have no control over, such as the internet. Once again, encryption comes to the rescue, offering a ubiquitous and cost-effective control to protect your data in transit. This typically takes the form of a **virtual private network** (**VPN**), which can be built upon one of several different technologies—most commonly, **IPSec** or **HTTPS**, among several others. The data in transit encryption provided by a VPN helps prevent unintended parties from reading data captured in transit.

Now that you have covered two of the three data states, you should also consider **data in use**. This is the state that is probably the most difficult to protect, is most often overlooked or ignored, and is the one state you cannot simply protect through encryption. Data in use is the state where information lives in volatile memory such as **random access memory** (**RAM**), memory caches, or your CPU registers. The common belief is that in order to take advantage of a data-in-use attack, one would require physical proximity to the CPU and RAM of a system, so the risk is relatively low. However, there are attacks known as side-channel attacks that can take advantage of unencrypted data in use. Be sure to check out the *Introduction to Side-Channel Attacks* link in the *Further Reading* section at the end of this chapter. In the following section, you will learn about how selecting a proper standard can help determine the most effective security controls.

Standards Selection

As was discussed in *Chapter 3, Security Policies and Business Continuity*, **cybersecurity standards** are described as a documented accumulation of best practices created by industry experts to protect organizations from cyber threats. Perhaps the most important aspect of standards is best practices; they save us from having to reinvent the wheel, and you get the benefit of numerous other people's past experiences, both successful and not. Using a standard as the foundation of your security practice is a great start.

It is also important to pick the appropriate standard for your given situation. Some standards are more appropriate for international versus domestic companies, and others are more suitable for companies that operate in different, highly regulated industries such as finance or healthcare. You may even say that choosing the right standard is half the battle. If you choose properly, a lot of the headaches of maintaining regulatory compliance and checking all the appropriate boxes are resolved for you.

Scoping and Tailoring

Apart from choosing the proper standard, it is also important to take the pieces of a good standard that apply to your situation and customize them for your organization. This practice is known as **scoping and tailoring**. The easiest way to narrow down and select the right standard for your organization is to be aware of what others in your industry are using as their standard. There is a good chance that this is what regulators in your industry are accustomed to dealing with and what they expect everyone to use.

Once you know which standard is best to use, you can begin trimming and shaping that standard to fit the exact requirements of your organization. Unless your standard is also a framework that must be followed meticulously to ensure certification, you're generally free to pick and choose from the portions that make sense to you, thus tailoring the standard to the needs of your organization.

Data Protection Methods

To round out this domain on asset security, this section discusses some methods you can use to protect your digital assets. Specifically, you'll be learning about digital rights management, data loss prevention, and a cloud access security broker. To begin learning about the important foundational concept of **digital asset management** (**DAM**), you need to keep in mind the concept of accurate digital asset inventory, which was covered in the previous chapters. However, you have to go a step further than the initial stage, which is data collection, because a digital asset inventory is not a static concept.

As discussed earlier, data has several states. It moves and can be accessed by different people at different times. One of the foundational requirements of DAM is knowing what you have in your digital asset inventory, where it is at all times, and who is accessing it. The importance of proper categorization of assets in your digital asset inventory was also discussed earlier. Categorization is invaluable to access and storage management. You also depend heavily upon the categorization to apply the proper countermeasures and controls to a given asset so that you can limit access to the asset and provide an effective audit trail to that access. Categorization is also important for other data life cycle decisions, including the storage and destruction of a digital asset.

Digital Rights Management

It's one thing to protect your digital assets when they are on your network, within your organization, and under your control, but how do you protect them when they're outside of your control? This is particularly challenging when your digital asset is also your product such that letting that asset out of your control is part of your business model. This is where **digital rights management** (**DRM**) comes into play. DRM refers to the management of legal access to digital content.

Whether you know it or not, you have all probably been exposed to DRM in some form or another. One example is **Adobe Creative Cloud**, which is a software suite for everything from editing movies to photos, graphics, and other media. This software used to come as a box of five to seven DVDs, and now it is a subscription you buy and download from the internet. Your monthly subscription allows you access to install and utilize an application in their larger suite. Each time you open one of their applications, the software checks your subscription to make sure you're licensed and paid up to the moment. The obvious downside to this form of DRM is its dependency on internet connectivity.

Another form of DRM you may be familiar with is the use of a classic activation key for your software, which is an approach that does not have the limitation of requiring internet access to function. Yet another is that which is applied to e-books or music files that prevents more than the intended recipient from using the documents or music. Similar kinds of DRM technology can be used for copy protection or access protection on documents, such as PDFs or Word documents, to make sure that their access is limited to an intended party.

Data Loss Prevention

Another important data protection method is **Data Loss Prevention** (**DLP**). In order to really understand its significance, you need to first grasp what it is and the problem it is intended to solve. DLP is the effort to detect and prevent the leaking of sensitive data from your realm of control. By now, you can probably guess what we mean by "sensitive" data; this could be PII, financial data, intellectual property, or anything of value in the form of a **digital asset**. Depending upon the importance of digital assets to a given organization, DLP can be both expensive and difficult to implement.

Data loss is exceptionally common in most organizations and, contrary to popular belief, it is not malice but rather neglect that is the cause. Take a look at a couple of common causes for inadvertent data loss. Consider an email conversation among people discussing restricted data that is contained in or attached to an email. All recipients are authorized to view this data. However, at some point in the conversation, some unauthorized person external to the organization or the group becomes added to the email conversation, so the sensitive data gets shared inadvertently.

In order to limit such inadvertent data loss scenarios, you have to have an understanding of the nature of the problem. Employees are generally focused on their jobs and not on the potential security ramifications of their actions. Sometimes, unsanctioned data loss is done on purpose, as is the case with a disgruntled worker who emails the organization's customer list to themself before quitting. While there are technological countermeasures for DLP via email and other vectors, they are not as effective as proper categorization, effective policies, and security awareness training.

Cloud Access Security Broker

Now that you've reviewed DLP, you may have noticed that these controls only make sense in a private environment where you have complete control. What about cloud environments, which are becoming increasingly prevalent? This is where a **cloud access security broker** (**CASB**) comes into play. Put simply, a CASB is a system or a set of tools that gives you security, visibility, and control in cloud environments. A CASB lets you extend your visibility and the enforcement of your security policy into the cloud environment.

Consider an example. Suppose you have an administrative assistant who is asked to take notes in a publicly held board meeting of a company. The assistant takes notes using Microsoft Word via Microsoft Office 365. Therefore, these notes (which likely contain information about the company's performance before they publicly released information about their quarterly earnings) would be stored in the cloud. According to the **Securities and Exchange Commission** (**SEC**), this information is considered confidential and is thus required to be controlled and protected until it is released to the public. In this case, a CASB could be configured to extend and enforce this organization's policy to the **cloud provider** (Office 365).

In general, CASBs work in one of two ways; they either use an **application programming interface** (**API**) or proxy mode. Proxy mode works as you would expect; all traffic is routed through a device that monitors requests to make sure that all requests to cloud providers adhere to policy requirements. The proxy mode is neither as common nor as efficient as API mode as it cannot guarantee that all traffic goes through the proxy, as in the case of remote workers. When worker traffic is coming from your internal network, it is a simple matter to redirect it through a proxy, whereas when remote workers are connecting to a cloud service, it is impossible to tell where they are coming from at any moment in time, thereby making it difficult to force their traffic through a proxy.

Using a CASB in API mode alleviates the concern about where your users are when they are accessing cloud services. In API mode, the cloud provider exposes its API to your security system so it can be an extension of your security system, reporting on the activity of your users and any violations of your policy. This information is typically ingested by some kind of **Extended Detection and Response (XDR)** platform along with many other data feeds from any other security platforms, creating situational awareness at any moment.

Summary

In this chapter, you completed your review of asset security—specifically, how to manage digital assets over the course of their life cycle. You reviewed some of the particulars in the usage and destruction phases of the data life cycle. You learned what information you need to collect and the questions you need to answer to properly oversee digital assets. Additionally, you explored the destruction phase and the criteria for when and how to carry out the destruction of a digital asset. Further, you examined the security controls and compliance requirements that influence how you should manage digital assets during their life cycle. Finally, you reviewed how to control/prevent DLP in traditional environments as well as cloud environments. In the next chapter, you will begin learning about *Domain 3, Security Architecture and Engineering. Chapter 7* will focus on secure design principles and controls, being the first of three chapters covering *Domain 3*.

Further Reading

- Liu, Simon, and Rick Kuhn. *Data loss prevention.* IT professional 12, no. 2 (2010): 10-13. https://packt.link/4PkWp

- Standaert, François-Xavier. *Introduction to side-channel attacks.* Secure integrated circuits and systems (2010): 27-42. https://packt.link/CkB59

- Liu, Qiong, Reihaneh Safavi-Naini, and Nicholas Paul Sheppard. *Digital rights management for content distribution.* In Conferences in Research and Practice in Information Technology Series, vol. 34, pp. 49-58. 2003. https://packt.link/vMAgo

Exam Readiness Drill – Chapter Review Questions

Apart from a solid understanding of key concepts, being able to think quickly under time pressure is a skill that will help you ace your certification exam. That is why working on these skills early on in your learning journey is key.

Chapter review questions are designed to improve your test-taking skills progressively with each chapter you learn and review your understanding of key concepts in the chapter at the same time. You'll find these at the end of each chapter.

> **How to Access These Materials**
>
> To learn how to access these resources, head over to the chapter titled *Chapter 24, Accessing the Online Resources*.

To open the Chapter Review Questions for this chapter, perform the following steps:

1. Click the link – `https://packt.link/chapter06`.

    Alternatively, you can scan the following **QR code** (*Figure 6.1*):

Figure 6.1: QR code that opens Chapter Review Questions for logged-in users

2. Once you log in, you'll see a page similar to the one shown in *Figure 6.2*:

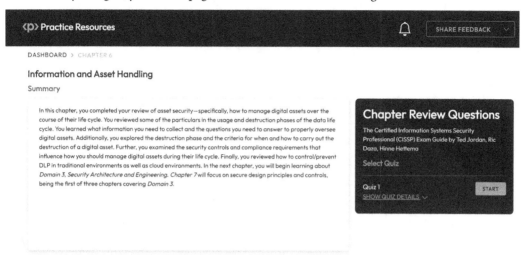

Figure 6.2: Chapter Review Questions for Chapter 6

3. Once ready, start the following practice drills, re-attempting the quiz multiple times.

Exam Readiness Drill

For the first three attempts, don't worry about the time limit.

ATTEMPT 1

The first time, aim for at least **40%**. Look at the answers you got wrong and read the relevant sections in the chapter again to fix your learning gaps.

ATTEMPT 2

The second time, aim for at least **60%**. Look at the answers you got wrong and read the relevant sections in the chapter again to fix any remaining learning gaps.

ATTEMPT 3

The third time, aim for at least **75%**. Once you score 75% or more, you start working on your timing.

> **Tip**
> You may take more than **three** attempts to reach 75%. That's okay. Just review the relevant sections in the chapter till you get there.

Working On Timing

Target: Your aim is to keep the score the same while trying to answer these questions as quickly as possible. Here's an example of how your next attempts should look like:

Attempt	Score	Time Taken
Attempt 5	77%	21 mins 30 seconds
Attempt 6	78%	18 mins 34 seconds
Attempt 7	76%	14 mins 44 seconds

Table 6.1: Sample timing practice drills on the online platform

> **Note**
> The time limits shown in the above table are just examples. Set your own time limits with each attempt based on the time limit of the quiz on the website.

With each new attempt, your score should stay above **75%** while your "time taken" to complete should "decrease". Repeat as many attempts as you want till you feel confident dealing with the time pressure.

Secure Design Principles and Controls

Being a **Certified Information Systems Security Professional (CISSP)** means carrying the responsibility of architecting systems that secure environments using researched and tested standards and methods. Models that work well for one organization may not work well for another, so knowledge of security models and their selection are also important. For example, security professionals who manage a military base may decide to install barbed-wire fencing and check the identification cards of everyone who enters the base. On the other hand, security professionals who design a commercial enterprise might design a four-foot fence for the parking lot and have visitors just sign a registry to enter the workplace.

By the end of this chapter, you will be able to answer questions on the following:

- Researching, implementing, and managing engineering processes securely
- The fundamental concepts of security models
- How to select controls based on system security requirements

We will start by looking at design principles.

Researching, Implementing, and Managing Engineering Processes Securely

In this section, you will review 11 design principles that are critical when employing secure architectures. These models are listed as follows:

- Threat modeling
- Least privilege
- Defense in depth
- Secure defaults

- Fail securely

- Separation of duties

- Keep it simple

- Zero trust

- Privacy by design

- Trust but verify

- Shared responsibility

The CISSP candidate must know the basics of these secure design principles and understand that multiple principles are used when securing an organization's architecture. The following sections delve into each model in detail.

Threat Modeling

Threat modeling is the process of identifying and enumerating *potential threats*. Security firms and departments use threat modeling to address potential attacks against their information systems. **Pen testing** (short for **penetration testing** – not to be misconstrued as testing with an ink pen) utilizes an ethical hacker to attack the company to uncover vulnerabilities before the "bad guys" find them. Ideally, all discovered vulnerabilities are mitigated before an adversary can find them.

Wargaming divides organizational security groups into a red team (*the attacker*), and a blue team (*the defender*). This exercise uncovers vulnerabilities, such as pen testing, and allows the use of planned scenarios, or *playbooks*, that red teams follow and blue teams work to mitigate. Further details on pen testing, wargaming, and other assessments are covered in *Chapter 14, Designing and Conducting Security Assessments*.

Note

Threat modeling may not always involve running attack simulations such as penetration testing and wargaming. A security professional can install mitigations for commonly known attacks without running a simulation – for example, installing input validation on a web server to mitigate injection attacks.

Penetration testing and wargaming is a five-step process from the point of the attacker. These are the same five steps a black-hat hacker would take during a live attack:

1. **Reconnaissance (recon)**: Collecting data to learn about the target. The attacker keeps a log of what they learn about their target. Active recon involves directly interacting with something the target owns. Passive recon is information through an intermediary.

2. **Scanning**: This generally involves using technical tools to learn about the target, such as running Nmap to study which network ports are open on the target or conducting a Nessus vulnerability scan.

3. **Gaining access**: The attacker takes control of a network device or server or can pivot from that device to launch attacks on other targets.

4. **Maintaining access**: The attacker wants to stay persistently in the environment by creating backup accounts, or backdoor access – that is, they utilize obfuscated or clandestine means.

5. **Covering tracks**: The attacker removes any signs that they were on the victim's network to keep themselves from being detected.

Threat modeling helps the firm gain insights about the attacker (*who*, *what*, *when*, *where*, *why*, and *how*), their vulnerability profile, and where to prioritize mitigations.

Microsoft STRIDE

Software developers face several potential threats, so Microsoft developed a classification system for their most common threats during their software development process. This aligns with Microsoft's **Trustworthy Computing** directive to ensure that their developers consider security during the design phase. The common threat classification system is summarized as follows:

- **Spoofing**: The attacker poses as someone or something else. For example, they may pretend to be tech support.

- **Tampering**: The attacker modifies data (such as data from files, programs, apps, and even MAC addresses) – for example, a pharming attack where victims end up at forged websites because IP addresses were altered.

- **Repudiation**: The attacker can deny involvement – for example, an email from a forged account that cannot be traced back to the attacker

- **Information disclosure**: The attacker releases private organizational information – for example, a client's credit card and CVV number.

- **Denial of service**: The attacker makes data or a system unavailable. For example, in a ransomware attack, computer files are encrypted until the victim pays for decryption, after which their data is made available again.

- **Elevation of privilege**: The attacker exploits a flaw to gain access and achieve a higher level of control. For example, the attacker acquires the root or administrator password to a web server because the default password was never changed.

These threats will be covered in more detail in *Chapter 8, Architecture Vulnerabilities and Cryptography*. Other methodologies to create models are designed by non-profit organizations or incorporate principles of the **software development lifecycle (SDLC)**, such as OCTAVE and VAST, respectively.

OCTAVE

The **Operationally Critical Threat, Asset, and Vulnerability Evaluation (OCTAVE)** model was created specifically for cybersecurity threats. Developed by the Carnegie Mellon University Software Engineering Institute and **Computer Emergency Response Team (CERT)**, this model focuses on non-technical risks within the organization. IT departments work together within an organization and qualitatively address security needs by reviewing the risks of their most critical assets.

VAST

The **Visual, Agile, and Simple Threat (VAST)** modeling system is an updated system that also models infrastructure and DevOps, works within Agile environments, and provides consistent output for developers. Unlike STRIDE, which focuses on the attacker, or OCTAVE, which focuses on IT threats, VAST supports both application and operational threat models as part of the SDLC.

Least Privilege

Least privilege is the process of allocating an employee or contractor the minimum rights and permissions they need to do their job. For example, a product designer will have rights to operate their computer-aided design software but will not have rights to access corporate financial records because they do not need that data to perform their tasks. Another example would be an engineer who has the keys to their office but not any other office. In many cases, a janitor has keys to all offices to perform their job function of keeping everything clean and neat but they may not have a computer account because that is not necessary for their job.

Defense in Depth

A **defense-in-depth** approach, or the use of multiple layers of protection, allows security professionals to keep environments secure by maintaining confidentiality, integrity, and availability. Good security is the result of using the appropriate controls in the correct ways. For example, the security lighting in a parking lot or car park should light the area fully; otherwise, an attacker can hide in unlit areas. Good governance allows the organization to create baselines of security effectiveness and better evaluate security performance systematically.

Compare protecting your organization to royalty protecting their kingdom. The castle sits on a high hill, which makes it difficult for attackers to access it and keeps it safe from attacks. At the far edge of the kingdom, sentries keep guard to provide early warning of attackers. If attackers defeat the sentries, they still have to get beyond the piranha-infested moat. If the attackers get past the moat, they must get past the warriors on the outer wall. If they defeat the warriors, they have to enter the fort, either by bashing down the huge fort doors or climbing the walls.

Each layer proves to be more difficult and costly to the attacker, as shown in *Figure 7.1*, reducing the success of a breach. The royals mitigate the risk of these threats by designing their fortress to prevent and thwart these attacks. However, there is no such thing as zero risk. Attackers learn from their mistakes and design tools and weapons to defeat each layer of defense as skilled archers might do against the sentries; they might build a bridge for the moat, bring ladders to scale the walls, and use a battering ram to destroy the fort's doorway to get inside the kingdom:

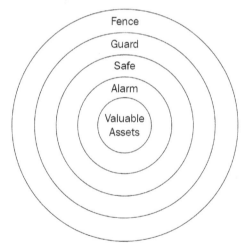

Figure 7.1: Protecting the kingdom, layers of protection

Organizations protect their assets in a similar manner. Security professionals use **administrative controls**, **physical controls**, and **technical controls** as layers of protection to reduce the risk of cybersecurity attacks. The administrative controls include the following:

- Personnel screening
- Job rotations
- Security policies
- Education and training programs
- Acceptable use policies

Another layer of protection includes physical controls, such as the following:

- Barbed-wire fencing
- Media storage facility
- Security guards
- Fireproof safe
- Evidence storage

Finally, technical controls add another layer of defense. These include the following:

- Multi-factor authentication
- Firewall and intrusion prevention system configuration
- Setting up access control rules
- Enabling audit logging features

In comparing organizations of today to a kingdom, you'll find that one of the biggest issues handled by organizations is dealing with hackers exploiting corporate networks. To mitigate these risks, corporate leaders set up firewalls to block traffic from unwanted users. If attackers are successful in defeating these defenses, the next layer of defense recognizes that the attacker's data is malware and is stopped by an **intrusion prevention system** (**IPS**). If the attacker defeats this safeguard, then the malware is stopped by the server's **host-based IPS**.

Protecting private customer records, such as tax identification numbers, is the point of multiple layers of defense. The attacker has to be successful and skilled at defeating multiple countermeasures since each layer provides more security for the organization. Company breaches are very common where private client records are stolen, but if organizations follow the fundamentals of good governance, strong policies, and system hardening, they can make themselves less of a target and protect their assets better.

Secure Defaults

Ideally, when a technician or security professional deploys a new device, it should have the best security options set by default. For example, when deploying a new **virtual private network** (**VPN**), the default account should be deleted or the default password should be changed. Also, the account should be set up with dual authentication where the user enters a password and a six-digit code texted to their phone. Such a measure would have better secured Colonial Pipeline, which was hacked in 2021 due to a poorly secured VPN. (The details about this can be found in *Hackers Breached Colonial Pipeline Using Compromised Password* in the *Further Reading* section.)

Fail Securely

When systems fail, they must be designed to fail in a secure manner. For example, if a security guard gets sick and has to leave their post, a replacement should be available to unlock the door as people enter and exit the building. This is considered *fail secure* because, if there were no replacement, door locks would not allow anyone to enter or exit the building. However, if the server room catches fire, the exit doors must disable authentication, or *fail open*, so that the staff can exit quickly. Staff safety supersedes data security in all cases as far as the CISSP is concerned because of the ISC2 Code of Ethics. However, there may be situations in the military where data security supersedes human life and systems must fail securely. You can read about this in more detail in the *Further Reading* section about how the Soviets declined US assistance in saving the sailors in one of their submarines.

Separation of Duties

Many of the security models used by information system security professionals today are taken from other industries where security is critical. The banking, accounting, and finance industries deployed **separation of duties** (**SoD**) to prevent fraud. One practiced accounting method is that the person who writes the checks for payables is not the same person who signs the checks; otherwise, an insider could pay themselves through a shell corporation.

Information security professionals also need to deploy SoD processes for high-security operations. One such area is data backups—the individual who performs the backup should be different from the person who verifies the backups. Does this make procedures more efficient? Unfortunately, as with most security measures, efficiency slows. However, it is a small price to pay for security against fraud that could bankrupt an organization.

Keep It Simple

The economy of mechanisms in system design helps increase security because fewer possibilities for errors exist. As a result, system design, testing, and production become simpler because there are fewer functions that can fail. As a CISSP, think **KISS** (for **keep it simple, SysOp**) during system planning.

Trust but Verify

In the late 20th century, information security professionals used a model where the network was like M&M candies—hard and crunchy on the outside but with a soft center. This model is known as **trust but verify**, where users are authenticated to gain access to a network and can work anywhere within the network, limited only by their rights and privileges. This was because most devices within a network were secure.

However, with mobile technology, such as smartphones, tablets, and the **internet-of-things (IoT)** users and devices are moving between different networks, and the boundaries of a secure network have become much more fluid and less defined. These devices often connect to the network from various locations, sometimes through insecure or public networks, and they might not always be under the organization's direct control. As a result, relying on "trust but verify" is no longer sufficient because it assumes that once inside the network, the user or device is trustworthy. A better approach is **zero trust**, which we will look at now.

Zero Trust

Once a user authenticates to a network, they are normally trusted to access the entire organization's environment based on their privilege. However, normally, the user's identity is no longer re-challenged. Zero Trust corrects this using the philosophy *Do not trust anything by default*.

For example, suppose a user walks away from their computer to attend a meeting after logging in to the corporate network. An intruder could sit and work at the user's computer and the network would assume that the authenticated user is requesting data and not the intruder. With **zero trust architecture (ZTA)**, the user (or intruder, in this case) would have to re-authenticate every few minutes to assure the network that the trusted user is requesting access.

Zero Trust not only verifies users but also devices. Devices such as switches and routers, processes, assets, services, and other components are verified using an additional firewall, data loss prevention, and network segmentation within the organization's secure network.

Privacy by Design

Dr. Ann Cavoukian, the Information and Privacy Commissioner of Ontario, Canada, coined the term **privacy by design**. This involves implementing protections for **personally identifiable information (PII)** and **personal health information (PHI)** in the planning, design, development, and implementation phases of information systems.

The **General Data Protection Regulation (GDPR)**, the **Health Insurance Portability and Accountability Act (HIPAA)**, the US **Federal Trade Commission (FTC)**, and other data protection privacy regulations also adopt the privacy by design philosophy to better protect users' identities.

Shared Responsibility

According to the US Department of Homeland Security, cybersecurity is a shared responsibility between users and IT security professionals. IT security professionals might add a firewall and intrusion detection systems to secure a network. Additionally, users can follow the following security steps to further secure the environment:

- Don't open suspicious-looking emails or attachments
- Use complex passwords with a minimum of 12 characters including special characters
- Keep systems patched and updated
- Limit personal information on public social networking sites

In cloud-based architectures such as the SUSE Cloud Service, Red Hat Cloud Access, Ubuntu Cloud, and others, both the cloud service provider and their customer share security responsibility. For example, customers using a **Software-as-a-Service (SaaS)** model may have to harden authentication, and the cloud service provider must keep the operating system updated with security patches (more details in *Chapter 8, Architecture Vulnerabilities and Cryptography*). The system could be breached by hackers if either party mishandles their responsibility. Responsibilities are defined within the **service-level agreement (SLA)** provided by the cloud vendor, as shown in *Figure 7.2*:

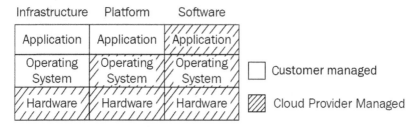

Figure 7.2: Cloud service provider responsibility model

The CISSP candidate will be tested on several security models used commonly within government and corporate organizations. Some models prioritize confidentiality, others integrity, and even availability. These are discussed in the next section.

Understanding the Fundamental Concepts of Security Models

Security models are designed to prioritize access to data based on confidentiality, availability, or integrity. In this section, we discuss popular security models and whether they focus on the confidentiality, integrity, or availability of data.

Bell-LaPadula

The Bell-LaPadula model prioritizes confidentiality over integrity and availability in systems such as **mandatory access control** (MAC). This is probably best explained by an example using *Figure 7.3*. The diagram shows three levels of clearance: top secret, secret, and confidential. If the user (data subject) has secret access, they can read secret and confidential documents (objects) but not top secret. This is called the **Simple Security** property, also known as *read down, no read up*.

The data subject can save objects at their clearance level or higher but cannot save documents at lower levels. This is called the **Star Security** property, also known as *write up, no write down*.

The **Strong Star Security** property means that the data subject can only write objects at their assigned security clearance, so *no write up or write down*. These three properties help maintain the confidentiality of objects. However, Bell-LaPadula does not include the need-to-know feature. Need-to-know further limits access to data; for example, an army general does not need to know top secret aspects of naval ships. The following figure shows three levels of clearance—top secret, secret, and confidential:

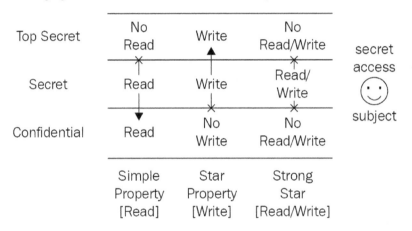

Figure 7.3: The Bell-LaPadula confidentiality model

A memory tool to help you remember this concept is **WURD**. This means **write-up, read-down**.

The following section will discuss a system that prioritizes integrity over confidentiality and availability.

Biba

The Biba security model prioritizes integrity over confidentiality and availability. This is probably best explained using *Figure 7.4*. This diagram shows three levels of integrity using three different types of blog websites.

One example of a low level of integrity includes social networks. Suppose users forward chain letters, over Twitter/X or Facebook, stating that a billionaire is giving $1,000 to everyone that forwards the message. Fact-checking organizations such as snopes.com will have a higher level of integrity in this case because they research information for accuracy before posting it to public blog sites. Similarly, the said billionaire will have the highest level of integrity because they could post on their blog site that they are not giving away money for such a purpose.

In the preceding example, if the data subject works at the fact-checking company, they will rely on posts and update their records based on what the billionaire donor says and not on social networks. This is called the **Simple Integrity** property, or *read up, no read down*. The **Star Integrity** property allows Snopes to update posts on social networks, but Snopes cannot update posts made by the billionaire donor, thereby maintaining object integrity. The Star Integrity property applies *write down, no write up*:

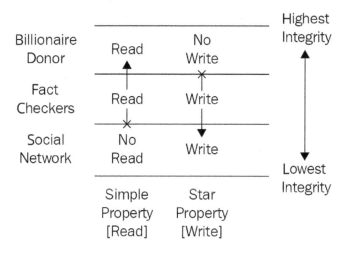

Figure 7.4: The Biba integrity model

Clark-Wilson

The Clark-Wilson integrity model not only includes features of Biba that prevent unauthorized outsiders from tampering with data files but also includes features to mitigate tampering from insider attacks or authorized users. Clark-Wilson also maintains internal and external consistency of transactions.

The Clark-Wilson model has three integrity goals, as follows:

- Prevent unauthorized users from making modifications on the system
- Authorized users should be prevented from making improper modifications
- There should be internal and external consistency by way of well-formed transactions

Figure 7.5 demonstrates how, if a subject wants to withdraw Bitcoin from their bank account, they will not have direct access to their funds. They would have to enter a username and a password to get access to their funds. Next, SoD is enforced by the system as the subject can only withdraw as much Bitcoin as they own, an example of *external consistency*. *Internal consistency* is performed by a separate process, which ensures that the object is in terms of the correct currency, for example, that the values are in Bitcoin as opposed to dollars:

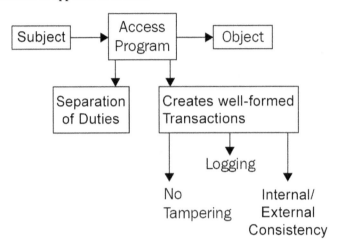

Figure 7.5: The Clark-Wilson integrity model

Brewer and Nash

The Brewer and Nash model is designed to prevent conflicts of interest by using a system called **ethical walls**. These walls are designed to separate corporate insider information from brokerage services, accounting receivables and payables from auditors, or secret plans from competitors.

For example, consider an advertising company that creates content for two different auto companies. This advertising company would use the Brewer and Nash model to ensure that competitive data for the Alpha Motors team does not leak to the Beta Motors team, and vice versa. This prevents intentional or unintentional conflict of interest from either automotive team.

Furthermore, in the brokerage industry, CEOs work with brokers daily, providing them with information on the financial performance of their organizations. Ethical walls managed within the Brewer and Nash confidentiality model block the communication of that information to shareholders until appropriate, as shown in *Figure 7.6*:

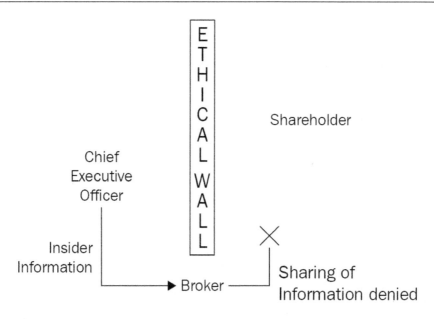

Figure 7.6: The Brewer and Nash confidentiality model

Graham-Denning

The Graham-Denning model focuses on how subjects and objects are securely generated and destroyed and how proper access rights are selected. It also provides eight sets of rules to preserve these interactions. These are eight basic actions that the Graham-Denning model calls **protection rules**. They are the following:

- How to securely create an object
- How to securely create a subject
- How to securely delete an object
- How to securely delete the subject
- How to securely provide the read access rights
- How to securely provide the grant access rights
- How to securely provide the delete access rights
- How to securely provide the transfer access rights

Figure 7.7 is based on an **access control matrix** where subjects and objects interact with each other. Rules within the matrix are called **preconditions**, so if subject X wants to delete a specific object Y, they must be the owner of the object and have the right to delete it. Graham-Denning calls the access control matrix the **monitor**, which was later termed the *reference monitor* by the James Anderson model:

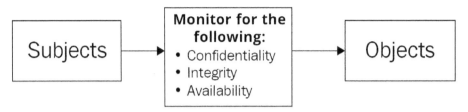

Figure 7.7: The Graham-Denning model

James Anderson

James Anderson coined the term **reference monitor** in 1972 when he presented his model at the Charles Babbage Institute. His key point is that objects must be secured within an operating system using some type of access control policy. A reference monitor has four important **NEAT** properties. According to these properties, the reference monitor must be the following:

- **Non-bypassable** so that an attacker cannot circumvent the mechanism and disregard the security policy

- **Evaluable**, which means it can be analyzed, verified, and tested

- **Always invoked** so that it is active at all times

- **Tamperproof**, otherwise, an attacker can exploit it and disrupt the security policy

The James Anderson security model, as shown in *Figure 7.8*, shows how the reference monitor is utilized for subjects to access objects:

Figure 7.8: The James Anderson model

Harrison, Ruzzo, Ullman

The **Harrison, Ruzzo, Ullman** (**HRU**) model focuses on integrity and access rights into the system. It's mostly similar to Graham-Denning except that it also shows that systems can be secured using an algorithm. The HRU model is similar to database transactions because, just as in a database, if a transaction fails, the entire sequence fails. HRU formalized the access control model, as shown in *Figure 7.9*:

Figure 7.9: The HRU model

Now you will look at some controls and frameworks for system security requirements.

Selecting Controls Based on System Security Requirements

Applying frameworks and layered security, and fine-tuning security needs are key roles of the security professional. This section discusses the different types of controls and frameworks, and where to apply them. The candidate will have to be prepared to answer questions about the following:

- Determining security requirements versus controls
- Identifying security controls
- Noting layered controls
- Choosing the security controls
- Noting major security control frameworks
- Tailoring or scoping security controls
- Evaluating security controls

The conversation will now shift toward reviewing security requirements versus controls.

Determining Security Requirements versus Controls

When considering security controls, the security professional needs to assess the countermeasures and safeguards that will mitigate risks to the confidentiality, integrity, and availability of systems under their charge. These systems not only include information systems but also networks, operational technology, and IoT technology. The security controls put into place will impact the performance of the system. Some examples of these performance impacts are the following:

- The behavior of people will change as they observe fences and signs. To understand this, consider the example of people only being able to access a parking lot through the gate entrance.

- Processes will perform poorly due to increased logging and accounting of data.

- Specific technologies may be totally eschewed. For example, the use of an IoT door lock is avoided because there is a high risk of an attacker being able to bypass the lock over the internet.

After security professionals determine requirements through risk analysis, they must next determine the appropriate security controls to protect the organization.

Identifying Security Controls

There are three major security classes described in ISO and NIST that security controls can be selected from. These are as follows:

- **Administrative** or management controls, such as policies and agreements
- Physical **operational** controls, such as fencing and door locks
- **Technical** or logical controls, such as those with an on and off switch

There are also six different security control categories. The first three, listed as follows, generally trigger *before* the exploit occurs:

- Directive controls specify acceptable rules of behavior, such as signs
- Deterrent controls discourage the violation of a security directive, such as fake cameras
- Preventive controls block an undesirable event, such as door locks

The following security control categories trigger *after* the exploit:

- Detective controls log an event or report it to an administrator.

- Corrective controls are similar to preventive controls, but they reduce the impact of an undesirable event. An example of this would be a water sprinkler system activating if a room catches fire.

- Recovery controls restore assets, such as recovering data from backup tapes or returning access controls to normal operations.

Some controls work for multiple types. A security camera is both a deterrent and a detective device.

Noting Layered Controls

Similar to what was stated in the *Defense in Depth* section previously, the security professional can consider implementing layered security controls in their systems as well. For example, a guard in the lobby inherits the security benefits of signage and fencing that deter most intruders. These inheritable controls don't just exist for physical or operational applications but also for administrative, management, and technical applications. For example, antivirus software inherits the security benefit of a firewall that blocks most of the malware before it even enters the network.

Choosing the Security Controls

When selecting controls, it is important to use the most appropriate controls to manage a particular system. For example, an endpoint device would benefit from antivirus software, passwords, and screen lockout features. Similarly, the security operation center would benefit from multi-factor authentication, a turnstile, and a sign-in sheet. Without these security controls, attackers could exploit the systems.

Noting the Major Security Control Frameworks

Control frameworks standardize the implementation of security controls and provide ways to evaluate security controls to verify effectiveness. These standardized frameworks have been researched and studied thoroughly by successful firms and universities and are superior to the ones that organizations attempt to create on their own. Organizations worldwide follow the standards, compare results, and offer tested updates.

Frameworks, in general, have the common features of **Plan, Do, Study, and Act** (**PDSA**) influenced by Dr. W. Edwards Deming's product quality model. The cycle never ends and promotes continual security improvement. *Planning* involves what is to be accomplished. *Doing* actualizes the plan. *Checking* monitors the process and results. *Acting* reviews the results and takes steps to improve security.

For example, the dozen or so **International Organization for Standardization** (**ISO**) standards are generally used by for-profit organizations, and specifically, the 27001 standard provides requirements for information security management systems. The *planning* phase analyzes the security issues for the organization and the *doing* phase implements security policies and procedures. Once the controls are put in place, they are measured to ensure that security needs are being met. Finally, in the *acting* phase, improvements are deployed to eliminate additional threats.

Similar to ISO, the **National Institute of Standards and Technology** (**NIST**) has security requirements designed specifically for government agencies. Publication SP 800-53 provides a catalog of privacy and security controls for information systems to protect their organizations and assets and ensure the effectiveness of these controls in maintaining security and privacy.

The **Control Objectives for Information and Related Technology (COBIT)** framework focuses on security operation centers and ensuring the information system strategy aligns with organizational goals.

The **Cloud Security Alliance (CSA)** released the **Security, Trust, Assurance, and Risk (STAR)** registry as a requirement for privacy and security controls for cloud computing. The registry allows cloud providers to show clients their security posture. The CSA registry includes a list of trusted cloud providers that meet STAR requirements.

Frameworks need to be selected based on the nature of the organization, and its security goals. Some organizations might have different regulatory requirements; for example, an organization that accepts credit cards must run vulnerability scans annually or after every major update according to PCI DSS requirement 11. The way an organization functions and the kind of business needs determine which control framework should be utilized.

Tailoring or Scoping Security Controls

Once an organization decides on a security control framework, they may find that the framework contains features they do not need. Removing those components from the framework is known as **scoping**. For example, consider that the organization works in an area where the customers never enter their physical business location; so, there is no need for the organization to secure the lobby or search customers' bags.

Tailoring involves modifying a baseline to make the framework more suitable for an organization's needs. For example, a framework might state that password updates need to be done every 90 days. However, because the organization works with proprietary records, they decide to require password updates every 30 days instead. Another organization may decide that passwords only need to be updated annually because they use *two-factor authentication* and work with unclassified records.

Evaluating Security Controls

Once security controls are placed throughout an organization, they must be tested for effectiveness. NIST designs three methods to evaluate security controls. These methods are as follows:

1. Test the security device.

2. Interview the staff and get their opinions on the effectiveness of the security controls and how the device is operating.

3. Review logging data to ensure that the device has been deployed properly.

Security controls do not need to be evaluated by these steps only; they could also be combined. For example, a security professional could test a firewall and then interview the network administrator on its effectiveness. In the end, they can review the instruction manual to ensure it is deployed correctly and functions properly.

Summary

This chapter covered security design principles and controls, and the importance of not only installing security controls but also testing them for effectiveness. Security professionals need to apply security design principles that include threat modeling, least privilege, defense in depth, secure defaults, failing securely, SoD, keeping things simple, Zero Trust, privacy by design, trust but verify, and shared responsibility.

Multiple systems are used to secure and access data, including Bell-LaPadula, which focuses on confidentiality, and Biba, which focuses on integrity. Clark-Wilson deploys most features of Biba and prevents tampering. A system of ethical walls, which helps to prevent conflicts of interest, constitutes the Brewer and Nash security model.

Security professionals must remember that one of the most important security principles is a layered defense model, and then scope and tailor controls as needed depending on the framework used for their organization.

The next chapter discusses how these security processes are deployed onto servers, industrial control systems, cloud architectures, and even IoT. There is also a detailed discussion on encryption and cryptanalytic attacks.

Further Reading

- *Hackers Breached Colonial Pipeline Using Compromised Password*, `https://packt.link/RRZlV`, Bloomberg News, June 4, 2021.

- *School District still using Default Login for Armin Account Surprised to learn its Site has been Hacked*, `https://packt.link/eZNxj`, Techdirt News, January 8, 2014.

- *Russia: US Offers Help to Rescue Submarine*, `https://packt.link/QCarP`, Radioisotope, Aug 8, 2000.

- *Developing a Framework to Improve Critical Infrastructure Cybersecurity*, `https://packt.link/otEb3`, NIST Whitewater, April 4, 2013.

Exam Readiness Drill – Chapter Review Questions

Apart from a solid understanding of key concepts, being able to think quickly under time pressure is a skill that will help you ace your certification exam. That is why working on these skills early on in your learning journey is key.

Chapter review questions are designed to improve your test-taking skills progressively with each chapter you learn and review your understanding of key concepts in the chapter at the same time. You'll find these at the end of each chapter.

> **How to Access These Materials**
>
> To learn how to access these resources, head over to the chapter titled *Chapter 24, Accessing the Online Resources*.

To open the Chapter Review Questions for this chapter, perform the following steps:

1. Click the link – https://packt.link/chapter07.

    Alternatively, you can scan the following **QR code** (*Figure 7.10*):

Figure 7.10: QR code that opens Chapter Review Questions for logged-in users

2. Once you log in, you'll see a page similar to the one shown in *Figure 7.11*:

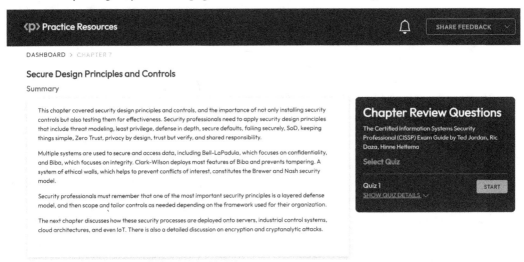

Figure 7.11: Chapter Review Questions for Chapter 7

3. Once ready, start the following practice drills, re-attempting the quiz multiple times.

Exam Readiness Drill

For the first three attempts, don't worry about the time limit.

ATTEMPT 1

The first time, aim for at least **40%**. Look at the answers you got wrong and read the relevant sections in the chapter again to fix your learning gaps.

ATTEMPT 2

The second time, aim for at least **60%**. Look at the answers you got wrong and read the relevant sections in the chapter again to fix any remaining learning gaps.

ATTEMPT 3

The third time, aim for at least **75%**. Once you score 75% or more, you start working on your timing.

> Tip
>
> You may take more than **three** attempts to reach 75%. That's okay. Just review the relevant sections in the chapter till you get there.

Working On Timing

Target: Your aim is to keep the score the same while trying to answer these questions as quickly as possible. Here's an example of how your next attempts should look like:

Attempt	Score	Time Taken
Attempt 5	77%	21 mins 30 seconds
Attempt 6	78%	18 mins 34 seconds
Attempt 7	76%	14 mins 44 seconds

Table 7.1: Sample timing practice drills on the online platform

> Note
>
> The time limits shown in the above table are just examples. Set your own time limits with each attempt based on the time limit of the quiz on the website.

With each new attempt, your score should stay above **75%** while your "time taken" to complete should "decrease". Repeat as many attempts as you want till you feel confident dealing with the time pressure.

Architecture Vulnerabilities and Cryptography

This chapter covers the commonalities between different computing architectures, including servers, cloud computing, and the **Internet of Things (IoT)**. Because these computing devices share common architectures, some common threats and vulnerabilities must be considered. In this chapter, we will begin by looking at IT architecture as a whole, and discuss engineering standards and processes and different architectures.

In this chapter, you will explore the different types of encryption processed by computers, including their needs and purposes, and when and where they can be used. You will also learn about the different types of encryption attacks and mitigations.

By the end of this chapter, you will be able to answer questions on:

- The security capabilities of information systems
- Assessing and mitigating the vulnerabilities of security architectures and designs
- Selecting and determining cryptographic solutions
- Methods of cryptanalytic attacks

In the first section, you will review information system security capabilities. Let's start with this foundation.

Security Capabilities of Information Systems

Today's information systems include individual servers, cloud-based systems, IoT, industrial control systems, and virtualized computing. This diverse range of architectures means there is also a diverse range of policies, practices, and technologies, and security professionals are tasked with protecting these architectures from ransomware attacks, loss of privacy data, and other exploits.

This section describes practices that are used by security professionals to properly deploy security engineering policies, standards, processes, and technologies for organizations.

Security Engineering Standards

Although every organization and IT architecture is different, there are many commonalities in the risks they face. Though targeted attacks are common, the methods that malicious actors use are replicated across the IT landscape. To create consistency and best practices in cybersecurity, organizations such as the US **National Institute of Standards and Security (NIST)** and the **International Organization for Standardization (ISO)** have created voluntary frameworks. Such standards are regularly updated and help security professionals and organizations understand current risks and how best to implement them. ISO and NIST information system security engineering standards are well respected and widely adopted. These include the *NIST SP 800-160 – Systems Security Engineering* and the *ISO 15026 - Systems and Software Engineering* standards, which both apply to recommendations for securing an organization as a whole.

The NIST SP 800-160 framework proposes 14 methods to improve cybersecurity resiliency. These methods occur most commonly in the CISSP exam and are shown in *Table 8.1*, which is adapted from the NIST Special Publication *Developing Cyber-Resilient Systems*. It is available for free at `https://packt.link/cStHN`.

Technique	Description
Adaptive Response	Implement flexible actions to manage risks.
Analytic Monitoring	Continuously monitor and analyze various properties and behaviors.
Contextual Awareness	Maintain updated representations of mission or business posture considering threats.
Coordinated Protection	Ensure protection mechanisms work together effectively.
Deception	Mislead adversaries to protect critical assets or expose compromised ones.
Diversity	Use varied technologies to minimize common vulnerabilities.
Dynamic Positioning	Relocate functionality or resources dynamically.
Non-Persistence	Create and retain resources only when needed.
Privilege Restriction	Limit privileges based on user attributes and environmental factors.
Realignment	Adjust systems to align with mission needs and evolving threats.
Redundancy	Provide multiple instances of critical resources for protection.
Segmentation	Separate system elements by criticality and trustworthiness.
Substantiated Integrity	Verify the integrity of critical system elements.
Unpredictability	Make random or unpredictable changes to enhance security.

Table 8.1: Methods to improve cybersecurity resiliency

The ISO 15026 framework consists of four parts including concepts and vocabulary, assurance cases, system integrity levels, and assurance in the life cycle. Each part details requirement levels for the standard; for example, 15026-3 specifies integrity levels and requirements for systems, software projects, and dependencies.

Security Engineering Processes

Security engineering processes are composed of technical, management, facilitation, and accreditation mechanisms. When deploying technical processes, security engineers must consider the organization's mission, scope, and risk of loss. They must also understand who the stakeholders are and their security priorities. The security professionals must also design systems to meet the requirements of the stakeholders, and design system architectures with reduced vulnerabilities.

For example, a security professional working in a healthcare environment needs to ensure the security of patient data across an organization to protect confidentiality, comply with local data legislation, ensure data can be efficiently accessed, and protect against data tampering, which could cause significant health impacts. Stakeholders include management, regulatory bodies, and patients.

Within systems engineering, many technologies get integrated into others; for instance, IoT and desktop computers may share the same network. In the case of the healthcare provider, PCs and mobile computing devices such as iPads may integrate with IoT devices such as patient monitoring equipment.

A validation and verification process must be conducted to ensure that security requirements have been met. This could take place as regular security testing on computers, patient monitors, and network systems, as well as regular pen testing. Security management processes include proper planning to determine which teams will perform which tasks and ensure that the organization's security processes meet the organization's goals.

Security facilitation processes include life cycle considerations of not only the technology but also the policies and processes surrounding the technology. Also, security considerations must support the organization's objectives. Facilitation of technology management includes determining the return on investment and human resources for skilled, trained staff to work with the technologies.

Finally, accreditation processes are put in place for suppliers to audit their security, and quality objectives are defined for technologies that are part of the process. The CISSP professional must keep change management records updated to maintain the security history as the life cycle of the technologies extends.

Universal Security Models

While designing security engineering processes, systems engineers look primarily at two fundamental security models: the hardware model and the software model. The hardware model focuses on the physical setup of IT systems. The software model focuses on OSs, file and data management, and applications. Systems designed with these features and architectures in mind reduce the likelihood of cyberattacks.

The hardware model, as shown in *Figure 8.1*, starts at the bottom with the firmware layer. This layer includes security features such as password protection and device management. For example, firmware often has tamper alerts as part of its security mechanism. An alert can be set up to inform a security engineer if a storage device is removed and replaced. This alert could potentially catch a malicious actor switching out a clean hard drive with an infected one, leaving the system open to exploits.

The next layer contains a component called the **trusted platform module** (**TPM**). This is a hardware security feature that manages **public key infrastructure** (**PKI**) certificates, public keys, and private keys of users and components, such as the hard drive and network card. (Security features around RAM and input/output are covered later in this chapter and in *Chapter 6*.)

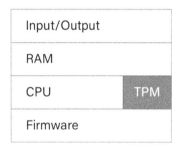

Figure 8.1: The universal computer hardware model

Combined with the hardware model, the software model adds a layer of security. To understand how this works, we should understand the universal software model. In *Figure 8.2*, we can see how, on top of the CPU (hardware) and the kernel (the fundamental part of the OS), there are libraries, which include general code that is used across different applications. The outer layer is the application layer, where most users do the majority of their interactions.

The kernel is the foundation of the OS – it is the part of the OS code that it always in the computer memory and provides and manages the software and hardware resources that are available to the users. It will load and run programs, schedule when the programs run, set their priorities, and manage memory. The kernel also works with the filesystem manager to maintain the locations of files on the hard drives and their integrity. Because it has full access to the CPU and other hardware, it is the most important to protect.

The system achieves this protection by operating in two basic modes: kernel mode and user mode. In kernel mode, the OS executes critical and low-level tasks with full access to all hardware and system resources. In user mode, full access is blocked and all processing is done through applications that only have access to the CPU through the OS. The switch in modes is done constantly in real time by the CPU based on processing needs.

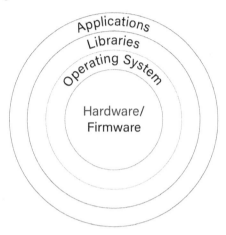

Figure 8.2: The universal OS Software model

Users run their applications and services in user mode, which restricts access to system-level resources. For example, when a user accesses a text file, the request is made in user mode, but the kernel, which manages system resources and links hardware and software, will securely manage access to the file so as not to harm the hardware or other applications.

It is important to consider the universal hardware system and the universal OS at the same time, as the kernel (software) and the CPU (hardware) work together to make the system secure.

A memory manager, which is a part of the kernel running off the CPU, will manage and allocate the physical or virtual memory of the system; this is known as memory management. Similarly, the security monitor enforces the access control policies on how the subject should interact with an object. The **input/output** (**I/O**) manager, another kernel component, handles the communication between the OS and different devices, such as network cards and hard drives.

CPUs implement security by having two different modes: user mode and kernel mode. In kernel mode, the CPU is operating at the highest privilege level, also known as a supervisor state. In this mode, programs have unrestricted access to all system resources and underlying hardware. Any CPU instruction can be executed, and every memory address can be accessed. This mode is reserved for low-level operations performed by drivers, which require direct interaction with hardware.

User mode, which is also known as a problem state, limits access to data and hardware based on the privilege that is given to the user. This is where all user programs execute. In this mode, programs do not have direct access to RAM and hardware. This restriction is necessary to prevent programs from overwriting each other's memory, which would compromise system stability and security. If a program needs to access hardware or system memory, it makes a call to the underlying API provided by the OS. In Windows, all processes, except for the system process, run in user mode.

The **user interface** (**UI**) also allows users to interface with the computer in an understandable manner, such as the standard Windows environment. Most people will experience computing systems in user mode, though depending on needs, experience, and seniority, they may get differing levels of access, such as the ability to install new applications or access other system components.

These hardware-level security mechanisms work in conjunction with the OS's kernel, which enforces access control policies, manages memory securely, and provides secure interfaces for applications to interact with the system. It is important to ensure the memory is secure as this is one of the most vulnerable parts of an infrastructure.

Mitigating with Access Control Techniques

Having user mode and kernel mode is a fundamental security feature of all IT architecture. However, there are numerous other threat mitigation techniques used by systems engineers. Many of them are access control techniques, designed to create barriers to vulnerable parts of a system. For example, one of the primary concerns for information systems is the buffer overflow attack on memory. These attacks attempt to run code within memory, where it can be executed by a hacker to take control of the system. One way to mitigate these attacks is to deploy an OS technology named **data execution prevention** (**DEP**), which shuts down programs attempting to run into unauthorized locations in memory.

Another access control technique designed to protect information systems is a feature called process isolation. This is a feature that is built into most smartphones (and is now being used on other types of information systems, such as desktops and laptops) that allows processes to run and access data from specific locations in memory. By isolating processes, the system ensures that each process operates in its own designated memory space, preventing unauthorized access or interference with other processes. This enhances security by limiting the potential impact of malicious software or errors within a single process. For example, a web browser can access only the data and resources it needs to function, such as web pages and cookies stored for the browser. It cannot access data from a messaging app or banking app.

OSs also restrict access to system users based on a range of criteria, such as job role, seniority, or whether the user is an administrator or not. A security kernel built into the OS acts as the reference monitor and provides security to the hardware and software. The security kernel will allow a user to access files they own and will deny the user access to records they do not own. When a user has Windows Administrator rights or Linux/Unix root rights, they have full administrative privileges over the entire system and can access any object. A best practice is to give users only the least privilege that they need to do their jobs. It is not recommended to use administrator or root account privileges when performing normal user functions; this privilege should only be used when the user needs to implement an administrative function.

Mitigating Threats with Other Techniques

While protections that work to stop attackers from accessing the system can be effective, modern security also considers the protection of data even when barriers don't work. Encryption techniques mean that even when systems are accessed, data can be obscured so as to be rendered unusable by malicious actors. There are various encryption techniques, which are discussed later in the chapter, and they can be applied to data in motion while being processed or transferred, or when data is at rest, such as when sitting on a hard drive. The goal is to protect the confidentiality and integrity of the data. For example, if an attacker steals a user's hard drive, even if they install that hard drive into another system, it should be useless because the data will be encrypted.

In support of these encryption techniques, the TPM, introduced in 2019, is a hardware component built onto the motherboard that provides security services for the system, such as generating and storing encryption keys. The TPM includes a cryptographic processor that performs cryptographic functions, enhancing the system's ability to secure data. Systems available before 2010 can get TPM features through a hardware add-on device called the **hardware security module** (HSM). By integrating such hardware solutions, systems can ensure robust encryption management and further safeguard data against unauthorized access.

There are various types of security software that protect data in OSs. These include anti-malware, host-based intrusion prevention systems, host-based firewalls, file integrity monitoring systems, and configuration monitors.

Anti-malware protects systems from viruses and other types of malware that can negatively affect system performance. Host-based intrusion prevention systems act as network-based intrusion prevention systems but operate exclusively on endpoint systems. Host-based firewalls block inbound and outbound network threats based on rules defined within the firewall. File integrity monitors detect when files have been modified by comparing known hash values to a baseline. Configuration monitors ensure that policies are being followed – for example, that proper software versions are installed, and that unlicensed and banned software is not installed.

Mitigating Threats in System Virtualization

You saw earlier how OSs work on top of hardware systems. However, common in today's IT architecture are virtualized systems, which have their own threat mitigations. Virtualized systems are simulated or *guest* computers that share hardware from a primary or *host* system, including hard drives, memory, network cards, and other hardware. These systems allow developers to work in different environments, testing their applications on many OSs (for example, a guest Windows, Red Hat Linux, and Unix system on top of a single SUSE Linux host system), all from the same computer. *Figure 8.3* shows an example setup of **virtual machines** (**VMs**), with the hardware (i.e., the computer system), the host OS, and the hypervisor, which is the software that creates and manages the VMs. There are three VMs, all running different apps on potentially different OSs.

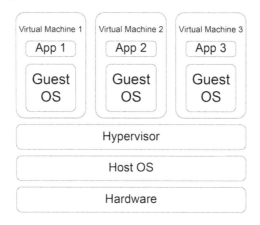

Figure 8.3: Virtual machine model

Virtual systems support the creation of test sandbox (or *sheep dip*) computers. This is a clean, uninfected system used to test untrusted applications before installing them onto production systems. The clean image gets reserved as a special backup called a snapshot. If a corrupt application is detected, you can simply revert to a clean snapshot (that is, a system image) and add the application to the *deny* list. Snapshots are significantly quicker to restore than reverting to backup tapes or backup drives.

Since these are virtual computers, they need to be secured in a similar way to traditional computers, including ensuring regular patches and updates, modifying default logins and passwords, and disabling unused services. Virtual systems must be added to the configuration management database as well because an insecure VM could have negative consequences for the entire organization if exploited. Finally, security engineers must add VM escape mitigations so that attackers cannot attack the host machine or the network from the VM.

You have seen how different security architectures mitigate threats using access control, cryptography, sandboxes, and monitoring. However, not all IT systems are equal, and the mitigation techniques needed will vary depending on size, design, and purpose. As we will see in the next section, a large part of a security professional's job is to assess the threats and corresponding mitigation techniques of individual architectures.

Assessing and Mitigating the Vulnerabilities of Security Architectures and Designs

Information system devices all use the universal computer hardware model shown in *Figure 8.1*, but each is otherwise designed for its own specific purpose. Whether the technology is designed for a single user or it supports millions of users, the model is the same. In this section, we break down the specific security needs for each technology, from client-based systems through to the cloud, and how threats and vulnerabilities are managed in their unique way.

Client-Based Systems

Client-based systems support several applications on a single device, for example, a smartphone, smartwatch, or workstation. Such devices do not require networks and can run without networking support.

Such devices depend on the individual to manage the security. Most of these individuals are not sophisticated when it comes to computers, and often fail to add security features such as adding a password or using biometrics. If these devices are lost or stolen, they are simpler for an attacker to compromise.

Server-Based Systems

Server-based systems provide services to client-based systems that need access to resources from the server. Some of these services include file transfer and remote logon or access to a web browser, and their high networking speeds provide superior quality of service.

Security on such systems is enabled through the application. For example, to use a file transfer protocol, a user must enter a password for a user account or share their public key to gain remote access through **Secure Shell (SSH)**.

When using a web browser, there might be multiple tiers of connections. This is known as an N-tier architecture. For example, when a user accesses the web browser, they will connect through the web server, which might support some type of Java application, which then sends data to a database. This would be a four-tier architecture (an N-tier architecture with four layers): the web browser, web server, Java application, and database.

The advantage of N-tier architectures is they allow developers to support a broad array of devices from one location. However, with one UI acting as a gateway to multiple locations, it needs to be completely secure. Password strength and security are the first consideration, but beyond this, there is always a risk of a malicious actor using the login to enter harmful commands that attack application layers. This is known as a command injection attack and will be discussed in more depth in *Chapter 23, Secure Coding Guidelines, Third-Party Software, and Databases*. It is mitigated with input validation (i.e., the UI will only allow inputs of strict formats), and is discussed in more detail later in this chapter.

Database Systems

Database systems can contain information about an organization's customers and users. Aggregation and inference attacks make it possible for an intruder to gain information by only reviewing segments of data. For instance, they may access details of customers and purchases from one database, and purchases and shipping addresses from another. Individually, both sets of data could be considered low security, but when combined, they reveal names and addresses. This process is known as aggregation. Inference involves using data to deduce information. For example, a group of patients in a trial might only include patient IDs and ZIP codes. However, from the ZIP code locations, an intruder could infer where the trial is taking place by identifying the closest hospital.

Mitigation comes down to standard access and encryption, but while designing systems, engineers should keep aggregation and inference in mind and remember that even seemingly low-security information can be dangerous if combined with other information. Data protection techniques are covered in the *Encrypting Data at Rest and Data in Transit* section later in this chapter.

Cryptographic Systems

Cryptography is a powerful security feature; however, it relies on the fact that an attacker does not have access to information that will allow them to decrypt protected data. Cryptographic systems maintain public keys, private keys, and certificates for individuals within an organization. Keeping these keys protected and secure is of great importance.

Restricting access to proper administration is the most important method to use to protect the keys, and enabling encryption systems around these so that the systems mitigate remote attacks. As discussed earlier, TPMs store keys and passwords because they store cryptographic keys in a hardware module, which is inherently more secure than software-based storage. This prevents keys from being extracted through software attacks.

Industrial Control Systems

Industrial control systems (also known as operational technology) exist within factories, water treatment facilities, nuclear facilities, and more. Many of these systems were built to operate in isolated environments – that is, not connected to local networks on the internet. Because of this, many standard access controls are not present. They might also not have the capability to update security software depending on new threats. For this reason, security has to be reconsidered.

As these critical systems have become connected to the cloud, cybersecurity issues are of great concern. An attacker could affect the water treatment system, or even cause a nuclear power centrifuge to explode (see the article on the Stuxnet attack listed in the *Further Reading* section).

Operational technology includes devices **programmable logic controllers (PLCs)**, **distributed control systems (DCSs)**, and **supervisory control and data acquisition (SCADA)** systems. PLCs and DCSs control processes on factory floors – for example, how much product should go into a box of cereal, or how much paint to spray on a car door.

SCADA systems are used to determine how much chemical treatment should go into water to purify it, and how much product should go into a nuclear facility to safely create nuclear power for their users.

Though general security techniques can be effective, mitigating attacks sometimes depends on the individual system. Typical considerations are the isolation of specific critical parts of the system from larger networks, physical and logical access barriers, and strict controls of code that can be transferred to systems that do not have other robust security controls.

Cloud-Based Systems

Cloud-based systems allow users to access all of their computing services over a network. This is similar to server-based systems, but the management of the hardware is handled remotely. There are three forms of cloud computing services. These are known as **software as a service (SaaS)**, **platform as a service (PaaS)**, and **infrastructure as a service (IaaS)**.

Systems administrators who are looking to manage several OSs (for example, different versions of Linux and Windows) may consider IaaS, where the cloud provider manages the hardware only and the systems administrator manages the applications and the OS.

Software developers who are only looking to create websites for new applications may consider PaaS, where the cloud vendor handles the hardware as well as the management of the OS, whether it be Linux, macOS, Unix, or Windows.

End users may want to consider SaaS to access just their email or word processing capabilities. The cloud vendor is going to manage the hardware, the OS, and the application.

The differences in these systems are shown in *Figure 8.4*. Securing these systems depends on a shared model where the cloud provider and customer must work together to secure their environment. For example, with a SaaS system, a cloud vendor will provide security for the hardware, the OS, and applications as needed, and the user will manage the security of their login name and password, utilize two-factor authentication, and secure their encryption keys.

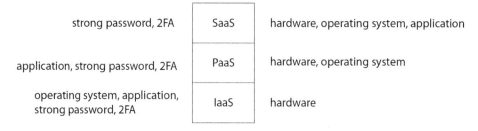

Figure 8.4: Comparison of Cloud Service Models

Distributed Systems

Distributed systems are systems in which computers are combined to resolve a single issue or work together to solve a single problem that requires high computing resources – if you were to place several inter-communicating computers across the network to jointly crack passwords, for example.

When you consider the security of distributed systems, you're essentially considering the individual machines, or nodes, and how that impacts the system. Malware that enters one node could quickly spread across the system. Individual nodes might have different hardware or OSs so security patching can be complex. Data is transferred regularly, so there is an increased risk of eavesdropping or man-in-the-middle attacks. Challenges involve maintaining physical security to ensure that an attacker cannot add a compromised node into the system, ensuring correct access management, and isolation from other networks. Because systems can be spread across different jurisdictions, local data laws could differ across the system so should always be taken into account.

IoT

IoT systems are devices that generally have a single purpose, such as a thermometer or camera, but can connect to the internet through an IP address. Some other examples of IoT devices include televisions, refrigerators, and alarm systems.

Despite their connection to the internet, suppliers generally do not build them with security in mind; they open a host of security challenges. Some IoT systems have hardcoded passwords, no firmware updates, and do not instruct users to change the default login name and password. Some systems do not have encryption features, and there are very few patches and security updates. Because of this, you should always consider security implications before components are built into a system. Do not rely on default security patches and settings, and always change default passwords. Consider isolation away from main networks.

Microservices

The microservice architecture breaks down applications into small components that provide a specific service or function. On a website, for example, the purchase function or tax calculation might be a microservice. If that function needs to be modified, it can be modified without changing the entire website. These services result in websites with better performance, better fault tolerance, and fewer disruptions. Microservices can be used by multiple applications, and are often connected using API, so connections should be clearly defined and well-secured.

Some of the threats that can target microservices come from the container configuration. Are those containers that home the microservices isolated from others, and are the libraries supporting those containers secure? Improvements to such systems can be managed with strong authentication, encryption, and avoiding the use of hardcoded certificates.

Serverless

In contrast to microservices, which can be running all the time, serverless systems run only as needed and stop functioning when not in demand. For example, if a user uploads an image and it needs to be resized automatically, the serverless system will perform the resize function and then stop running as soon as the job is done. Support for serverless platforms can be hard to find because they are new to the marketplace. Portability is also an issue. For example, transferring serverless systems from Azure to Google Cloud may require certain data to be rewritten.

Containerization

Containers facilitate consistent OSs and allow user applications to be built on the fly. The container's consistent build makes it a lot simpler to keep systems patched and updated.

Securing the OSs requires focusing on authentication, authorization, and isolation between containers. Other than those any OS might face, other threats could be insider attacks where the user attempts to install older, vulnerable applications and libraries onto a container.

Embedded Systems

Embedded systems are computing devices that require low power and may have mechanical components as well, such as those needed to operate traffic lights or control components in an assembly line. In general, these systems do not have an IP address, so remote attacks via the internet may be difficult. But more and more embedded systems are becoming similar to IoT systems.

The software in an embedded system is generally closed source, and not available for source code reviews, so there is the potential for a back door to be written into an embedded system if the supplier is untrusted, or has poor security practices.

High-Performance Computing Systems

High-performance computing (HPC) systems are generally composed of devices called supercomputers and these are designed to run trillions of instructions per second. Applications include aerodynamics, fluid dynamics, weather forecasting, and molecular modeling. Also, supercomputers are used to help break encryption and crack passwords.

Securing such systems involves strong authentication and physical security, ensuring that only approved personnel can work with the systems. Enabling multi-factor authentication can mitigate insider attacks and attacks via the internet. Also, using a jumpbox-style bastion host with Secure Shell to encrypt the connections can be a huge plus for performing secure updates to HPC systems.

Edge Computing Systems

Edge computing helps improve network performance for the end user. This is done by getting data closer to the user. Looking at *Figure 8.5*, the primary systems reside in the cloud as edge computing gets data closer to the user, so that the user has near instant access to their files, music, or videos.

Because these edge systems are distributed, security is similar to that which is in the cloud. There must be strong authentication and encryption. But, because these systems are on the edge of the network, it also results in better privacy because transmission of secure traffic is minimized since it is not coming from the cloud.

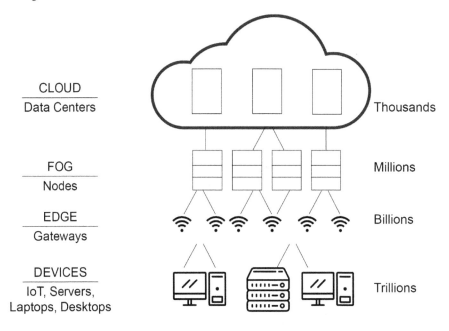

Figure 8.5: Comparison of cloud, fog, and edge computing

Virtualized Systems

Virtualized systems (also known as VMs) emulate a computer system on some type of host hardware, as shown in *Figure 8.6*. A hypervisor is running to support the VMs and allows for the virtual systems to maintain isolation between the virtual environments.

Securing these systems involves many of the same steps as securing a regular computer. This includes adding virus protection, malware protection, and enabling strong authentication. Some of the other important matters include systems to mitigate VM escape, where an attacker can attack a single VM, and then attack other VMs that are sharing the same hypervisor.

VM sprawl can also be an issue as it can be pretty straightforward to create additional VMs and lose track of the systems configured within the network. Users may forget about these systems, but persistent hackers find them, exposing the network to potential attacks. Also, there is potential that the hypervisor is a single point of failure in that multiple VMs are shared on one host OS; if that system fails, then all the VMs fail.

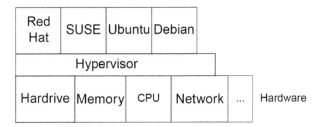

Figure 8.6: Virtual computer model

As you have seen, there are numerous system architectures that you may come across in your professional life, and though there are best practices that apply to nearly all systems, it is important to always consider the unique vulnerabilities of each system when deciding on the best security solutions. In the following sections, we will look at some of the security solutions, and what to consider when implementing them, in greater detail.

Selecting and Determining Cryptographic Solutions

When it comes to selecting cryptographic solutions, do not reinvent the wheel by attempting to create a custom encryption solution. Popular solutions tend to be so because they have been tried and tested; if you start from scratch, you will not be able to test your solution adequately. When selecting encryption types, use standard ones approved by NIST or other organizations that certify and accredit encryption models. They are tested and have a lot of experience behind them for data at rest and data in transit.

Depending on how data is classified determines which quality encryption to use. Top secret data may require AES encryption. Publicly available data may need no encryption at all. Data at rest systems, such as hard drives and backup tapes, must be encrypted if they contain critical data.

Make certain to protect private keys. Keep them backed up in case of data corruption. If private keys are compromised by hackers, generate new keys and place the compromised keys into escrow as they will still be required to decrypt past records.

As before, different scenarios will require differing solutions. Next, we look at some common scenarios that require cryptography – portable devices, email, and web applications.

Portable Device Security

Removable media and portable devices such as laptops, tablets, and smartphones are vulnerable to loss and theft. Use encryption to protect private customer data, employee data, and other records contained on these devices. Other mitigations include risk avoidance by not allowing certain records onto portable devices.

Encryption solutions are available on Linux, macOS, and Microsoft Windows through packages such as open-sourced TrueCrypt, Microsoft's BitLocker, and the Encrypting File System. macOS offers FileVault encryption for their hard drives. These tools integrate well with the HSM or TPM, which are hardware devices used to maintain encryption keys.

Email Security

Asymmetric encryption and certificate solutions are best for securing email messages. These *data in transit* solutions allow the following:

- Encryption and decryption to protect the messages
- Signing and verification to trust that messages are authentic
- Hashing to protect message integrity and ensure no modifications have taken place

Two of the best tools for encrypting emails are Pretty Good Privacy and certificates with S/MIME. These are discussed in detail in the following subsections.

Email and Pretty Good Privacy

Pretty Good Privacy (**PGP**) is an encryption application that allows users to send encrypted messages using asymmetric keys. Should a sender desire to send a private message to Chang, that sender would encrypt the message using Chang's public key.

The user could obtain Chang's public key from a myriad of public key servers such as `https://pgp.mit.edu`. There, the user would enter Chang's email address and obtain Chang's public key for message encryption. When Chang receives the message, they will decrypt it with the only key they have (their private key, because the message is only for Chang).

If Chang is receiving the message from Montrie and wants to ensure the message is from Montrie, Montrie needs to sign the message with their private key. Chang can visit `https://pgp.mit.edu`, obtain Montrie's public key, and confirm the message is from Montrie. Email clients such as Thunderbird, Outlook, Opera Mail, and Mailbird complete these processes automatically.

If you are using open-sourced OpenPGP for privacy, underneath the hood, it uses SHA1 for hashing, CAST for 128-bit cryptography, and Diffie-Hellman for key exchange. The paid version uses MD5 for hashing, IDEA for cryptography, and RSA for key exchange.

Email Certificates with S/MIME

Secure/Multipurpose Internet Mail Extensions (S/MIME) uses the same system as e-commerce websites, such as X.509 certificates to manage email security. This differs from PGP because now Montrie and Chang each have their own certificate that links to some certificate authority. This provides better security because the messages are managed by the TLS protocol, discussed in more detail in the next section, instead of by the less secure SSL protocol, which can be vulnerable to downgrade attacks.

Despite S/MIME's important security benefits, it is not as widely used because S/MIME certificate management becomes arduous as more users are added.

Email clients such as Thunderbird, Outlook, Opera Mail, Mailbird, and others work with S/MIME but usually require an extension.

Web Application Security

Like email, individuals can transfer sensitive data across the web. Developers of successful web applications can invite the world to their online store, social network, news site, and so on. But this also allows hackers access to their data.

Attackers commonly target credit card details, **personally identifiable information (PII)**, and **personal health information (PHI)**. A successful breach of these types of data allows attackers to commit fraud via identity theft. To mitigate such attacks, web applications must be secured by multiple methods including encryption and input validation.

Today, the **Transport Layer Security (TLS)** v1.3 protocol is considered the best because it uses the strongest levels of encryption and hashing. TLS replaces **Secure Socket Layer (SSL)** as the security protocol. SSL is vulnerable to an attack called the **Padding Oracle On Downgraded Legacy Encryption (POODLE)** attack, which allows attackers to downgrade SSL 3.0 to an older version, thereby exposing the private key. The POODLE attack is so serious that the **Payment Card Industry Data Security Standard (PCI DSS)** now requires at least TLS for online credit card transactions.

Web application developers must also secure their databases. Wherever a customer can input data into the web application, such as their name, address, phone, or password, there is an opportunity for an attacker to breach the website. Attacks on data input fields have the general term of **injection attacks** and are the number one threat stated by the **Open Web Application Security Project (OWASP)**, an organization that studies and publishes best practices for web applications.

Successful injection attacks allow hackers to take control of web applications and make the website do what they want, including website defacement and hidden field manipulation to significantly reduce the prices of store products. An example of an injection attack might be an entry in the web application for a five-digit USA ZIP code. What the web app expects is a valid five-digit US ZIP code, but a hacker might input five special characters (for example, \ / ' ; *) to test the web app for vulnerabilities. If the application does not reject the input, throws an error message with sensitive information, or behaves unexpectedly, the attacker might have exposed a vulnerability. Attempting invalid input into a web application is called **fuzzing**.

A special class of fuzzing attack into SQL databases is called **SQL injection**. This is the passing of SQL commands into the web application to gain access to, manipulate, or harm a database. Instead of the attacker using a login name to access their web account, they enter the following:

- Login: `trying' or 1=1--`
- Password:

The password field is intentionally left blank.

What the database expects is this:

```
SELECT id FROM users
WHERE username = 'trying' and password = 'abc123'
```

But the SQL injection attack pollutes the authentication process so that what the app actually receives is this:

```
SELECT id FROM users
WHERE username = 'trying' or 1=1-- and password = 'abc123'
```

The double dash component of the string means any data that follows the double dash is ignored by the computer. The - - part is for human readability only. This is called *commenting the code* or simply *comments*. In this case, the only executed part is this:

```
WHERE username = 'trying' or 1=1
```

Because 1=1 is always true, and or means that the username doesn't have to be true, the statement returns true. By using this double dash, the attacker is abusing the comments and tricking the web application into ignoring the password and the attacker gains entry to the web application. At this point, a skilled attacker has breached the website and can download credit card details, PII, and PHI if available.

To mitigate the risk of such injection attacks, the web application developer needs to code systems so that they never automatically trust outside data input. A system needs to be put in place to verify the data before it passes to the database. This process is called **input validation**.

With input validation, a username similar to the example shown here is entered:

- Login: tcboonyapataro
- Password: ********

The web application will first verify that a few rules are passed:

- Does the login name meet the minimum and maximum length requirements?
- Does the login name use only allowable characters, such as A-z, 0-9, -, and _?
- Are there NO disallowed characters, such as ', ", /, \, =, and so on?

If the answer to all of these questions is yes, the username and password are passed to the database; otherwise, the packets are logged and redirected to security for further investigation. Using input validation, the application could reject any input that includes the double dash used in the previous SQL injection.

Fuzzers are tools that test code for fuzzing vulnerabilities. They input garbage data into all user data entry points of the application and test how the application fails. Fuzzy testing is an important step of the software development process because this vulnerability could be devastating, causing loss of national card IDs, social security numbers, credit card information, and so on. These and other types of attacks are discussed in *Chapter 22, Securing Software Development*.

Encrypting Data in Transit

As more and more highly classified data is passed from system to system on the public internet, so too grows the necessity of encryption to protect that information from being leaked to attackers via man-in-the-middle or *on-path* attacks. This includes data being sent by **voice over IP (VoIP)** or wireless internet.

It's also imperative that message integrity is maintained and message alterations are detected. Data encrypted on a network is encrypted either by end-to-end or link encryption. Both methods are discussed next.

Link Encryption

Link encryption is carried out by service providers over satellite, telephone, or T-3 communication lines. Link encryption not only encrypts the data payload but also the routing headers, so the user also has **traffic flow confidentiality** (TFC) because IP addresses are hidden. TFC is a security measure designed to protect the patterns and characteristics of network traffic from being analyzed and exploited. It aims to obscure the volume, timing, and flow of communication to prevent an attacker from inferring sensitive information, even if they cannot decrypt the content of the traffic itself. *Figure 8.7* shows how decryption and re-encryption occur at each router to hide the address headers. Each router represents a potential vulnerability in the system, as there are instances where data is decrypted and then re-encrypted, leaving it temporarily unencrypted and readable. Other disadvantages of link encryption include the complexities of key management and performance because link encryption slows traffic due to multiple encryption and decryption processes.

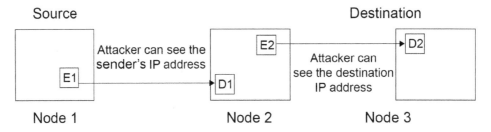

Figure 8.7: Link encryption

End-to-End Encryption

Virtual private networks (**VPNs**) and other end user programs such as **SSH**, **secure copy protocol** (**SCP**), **secure file Transfer protocol** (**SFTP**), and **secure Telnet** (**STelnet**) maintain data confidentiality by encrypting data before passing through the network and then decrypting once the data reaches its destination, as shown in *Figure 8.8*. VPN performance is faster than link encryption because data is encrypted and decrypted only once, and there are fewer keys than in link encryption, making management simpler. However, users might lose traffic flow confidentiality.

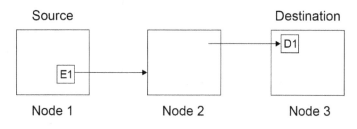

Attacker can see the sender
and receiver's IP address

Figure 8.8: End-to-end encryption

Other Security Methods for Data in Transit

For web traffic, protect sensitive data such as usernames, passwords, credit card details, and so on with TLS (TLS v1.2 or, better yet, TLS v1.3 is ideal). Use SSL v3 as a last resort because it has been compromised with a downgrade attack.

Secure email with PGP encryption or the S/MIME protocol, which includes features such as digital signing and encryption.

VPNs use **Internet Protocol security** (**IPSec**) to ensure secure communications over public internet networks and support data integrity and confidentiality.

Encrypting Data in Use

Eventually, data needs to appear in plain text to be usable, which makes it difficult to secure. Screen guards can help protect the user from shoulder surfing. Ongoing research is focused on developing CPU security enclaves to protect data in memory and using homomorphic encryption to keep data encrypted in RAM until it is displayed on the screen.

> Exam Tip
>
> Encryption of data in use is not covered in the CISSP exam. Additional resources are listed in the *Further Reading* section at the end of the chapter.

Cryptographic Life Cycle

Being able to identify the right places and methods to apply encryption is a key part of any security assessment. But understanding the ways things are encrypted and the limitations is also important. As computers become more powerful, it becomes easier for systems to work out private keys. As with all tools in cybersecurity, cryptographic tools must stay updated and maintained. Keys may get stolen and systems may get compromised, so they should be continually monitored and refreshed with clear policies and guidelines in place to dictate how this is done.

Security engineers must continually review cryptography requirements by answering the following questions:

- Which algorithms are best for the organization?

- How often keys should expire?

- Are there better key management methods?

Security managers need to stay on top of the latest research on cybersecurity through data from NIST, OWASP, and others. Their research focuses on attacks on encryption algorithms, mitigations, and improved algorithms.

Different Cryptographic Methods

Valuable data must be protected from attackers. There are multiple ways to do this, but what if the data gets into the offender's hands? If the data in the files is "scrambled" or encrypted, it will be unreadable to the hacker and the data remains confidential. Or, what if our log files notify us that our system has been attacked? How can we ensure that files have not been altered? Now, an integrity check needs to be run and the quickest way to determine whether files have been altered is to compare the serial number of the file or hash. Multiple algorithms can be used to review file integrity, and the better ones prevent "collisions" or situations where the same hash is labeled on two different files. Cryptography services help to support the confidentiality and integrity legs of the three-legged stool of support for security. Other utilities provided by cryptography include authenticity, non-repudiation, and access control. There are two major forms of encryption: symmetric and asymmetric. They each have their advantages. For example, symmetric encryption speeds are much faster than asymmetric encryption speeds, but key management is much easier for asymmetric systems if you have more than 100 users. Finally, without asymmetric encryption, users would not be able to securely do online shopping, and companies such as Amazon would not exist. The next section will cover the different types of encryption, hashing, and certificates used to maintain confidentiality, availability, and trust.

Symmetric Encryption

In symmetric encryption, the same key is used for encrypting and decrypting data. As you will see, there are different types with different complexities. The complexity of an encryption algorithm is important as it increases the **work factor**, which is the effort (or computational power) needed to crack the code.

The following example is the one most people are familiar with as it is quite easy to do with just pen and paper. Some friends in school have passed secret messages using a simple substitution cipher (monoalphabetic cipher) by replacing a letter in the alphabet with another, as shown here:

Plain	A	B	C	D	E	F	G	H	I	J	K	L	M	N	O	P	Q	R	S	T	U	V	W	X	Y	Z
Key	Z	Q	L	M	R	S	N	W	B	C	G	E	D	A	F	H	O	P	I	J	T	Y	K	U	X	V

Consider a secret letter that reads as follows:

Ciphertext: NZDBAN HZPJX IZJTPMZX

This would decode to the following:

Plaintext: GAMING PARTY SATURDAY

Simple substitution ciphers can be an effective method for sending secret messages, but they do have two vulnerabilities, which are that they can be solved easily if the key is not properly protected and that there is a low work factor.

As anyone with the key will be able to decode the message, keys must be tightly secured to ensure they do not pass into the hands of someone unintended. Likewise, if an attacker manages to intercept some of the messages themselves, they may be able to crack the code without much work using "frequency analysis." The attacker cracks the code knowing that all English words have vowels, most single-letter words are "a," and most three-letter words are "the."

ROT3 cipher: To create a cipher that is difficult to solve (or has a high work factor), Julius Caesar came up with an algorithm (ROT3) that simply involved shifting the letters. The number of letters shifted is known as the shift value. Caesar's key had a shift value of 3, and if used these days, would look like the following:

Plain	A	B	C	D	E	F	G	H	I	J	K	L	M	N	O	P	Q	R	S	T	U	V	W	X	Y	Z
Key	X	Y	Z	A	B	C	D	E	F	G	H	I	J	K	L	M	N	O	P	Q	R	S	T	U	V	W

Today, this would be easy to crack, but only the wealthy had access to education in Caesar's time, so many people did not know how to read, let alone decode secret messages. The work factor can be increased by giving each letter a different shift value.

Transposition cipher: Transposition ciphers simply involve changing the order of the message. Here is an example:

> `Plaintext: TESTING SATURDAY`

This becomes the following:

> `Ciphertext: ESTING SATURDAYT`

Pig Latin is an example of a transposition cipher, in which one transposes the first character to the end of the word and adds "ay" as a final syllable, as follows:

> `Ciphertext: ESTINGTAY ATURDAYSAY`

Careful listeners can decode the messages, but distracted listeners find them difficult to decode.

One-time pad cipher: The one-time pad is a polyalphabetic cipher that is considered to be unbreakable. It starts with a poem, phrase, or book that both parties agree to use. Here is an example:

> `Key: ROSES ARE RED, VIOLETS ARE BLUE`

Each letter of the alphabet has a number from 0 (zero) through 26 (twenty-six):

0	1	2	3	4	5	6	7	8	9	10	11	12	13	14	15	16	17	18	19	20	21	22	23	24	25	26
	A	B	C	D	E	F	G	H	I	J	K	L	M	N	O	P	Q	R	S	T	U	V	W	X	Y	Z

The key phrase "ROSES ARE RED" is then converted into numbers:

Key		R	O	S	E	S		A	R	E		R	E	D	
Value		18	15	19	5	19	0	1	18	5	0	18	5	4	

Now, let's send a simple secret message, **AAAAA**. The results are the following values:

Plain		A	A	A	A	A		
Value		1	1	1	1	1		

In real life, a user would not send a message such as **AAAAA**, but this example shows how symmetric encryption is much stronger when using a polyalphabetic cipher.

Using addition combined with modulo math results in the ciphertext, as follows:

Plain		A	A	A	A	A		
Value		1	1	1	1	1		

Key		R	O	S	E	S		A	R	E		R	E	D	
Value		18	15	19	5	19	0	1	18	5	0	18	5	4	

Value Sum	19	16	20	6	20		
Code		S	P	T	F	T	

Unlike our grade school or ROT3 examples, this is much harder to crack. In this simple example, the letter **A** may represent any of four possible letters, which will only grow in complexity as other messages are created and passed.

To decode the message, the recipient performs simple subtraction.

At this point, both parties agree to never use that part of the poem, phrase, or book again, and their next message will start from the next available character. In this example, the next message is **BBBBB**:

Plain							B	B	B	B	B		
Value							2	2	2	2	2		

Key		R	O	S	E	S		A	R	E		R	E	D	
Value		18	15	19	5	19	0	1	18	5	0	18	5	4	

Value Sum							2	3	20	7	2		
Code							B	C	T	G	B		

That is why this cryptosystem is called the one-time pad – because one uses the key space only once. This provides spectacular randomness for ciphertext and, even if the conversion book or "pad" is intercepted, the attacker does not know which page or line the messengers are using. Cracking this system has a huge work factor and is nearly impossible to crack.

Steganography: Steganography is a system where one can send secret messages in plain sight. In *Figure 8.9*, you will note two pictures that look alike, but one of them hides a secret message.

Original image

Original image plus hidden data

Figure 8.9: Steganography in action

> **Note**
>
> The image in *Figure 8.9* is taken from the following website: `https://www.researchgate.`
> `net/figure/Example-of-Image-Steganography-23-Network-`
> `Steganography-Taking-cover-objects-as-network_fig3_302977532`.

This allows attackers to use public and visible systems such as Instagram, VK, Weibo, or Twitter/X to post secret messages. Other ways to conceal messages include null ciphers, web pages, and covert channels. Organizations also use steganography for digital watermarking to help track offenders who are using their photos without permission.

Null cipher: Null ciphers are secret messages hidden within text. For example, this is a common method for sending secret messages in and out of prison. Consider the following example:

> Sheryl eats today under painted models every early Tuesday.

This decodes to the following when looking at only the first letter of each word

> SETUP MEET

Of course, all these examples are extremely simple ciphers to show how they work. Modern IT systems use increasingly complex algorithms with much longer keys, as you will see in the next section.

Encrypting Data at Rest and Data in Transit

Symmetric ciphers on electronic systems can encrypt data at rest on a hard drive, and in motion such as on a Wi-Fi network or VPN. Special algorithms were developed to encrypt data at rest. The best considered today is AES 256. This is a 128-bit block cipher combined with a 256-bit key, making it one of the strongest encryptions available, and the strongest algorithm for data at rest, as far as the CISSP exam is concerned. A block cipher encrypts data one block at a time instead of at a bit or byte at a time like a stream cipher.

The RC4 encryption algorithm is the only stream cipher tested on the CISSP exam. Encryption is performed one bit or byte at a time. RC4 encrypts data in motion systems such as Wi-Fi using WEP or WPA security. Because of the weak implementation of RC4 on WEP security systems, it is easy to crack with free online tools such as Aircrack-ng, so a stronger implementation of RC4 on Wi-Fi was created, called WPA. Instead of taking a few minutes to crack like on WEP systems, it can take months to crack WPA systems. WPA systems combine RC4 with TKIP, which gives the system its strength. Also, older WEP-based hardware systems could upgrade to WPA on the same Wi-Fi router with a firmware update.

Asymmetric Encryption

Asymmetric algorithms use two keys per individual to provide encryption as well as other capabilities such as signing and non-repudiation. Each user has a public key that is available to the world and a private key only known to them. For example, if a co-worker wants to send a secret message to Alice, they will use Alice's **public key** to encrypt the message. Alice will use her **private key** to decrypt the message. The important difference with symmetric encryption is that you cannot use the encryption key to decrypt the message.

> **Note**
>
> The trick to remembering this for the CISSP exam is recognizing that only the recipient's keys (Alice's) are used when sending a secret message.

Now that Alice has received the message, she wants to be certain that the sender has not been spoofed or impersonated. A sender can **sign** their message using their private key. So, if Cheng signs a message with his private key, Alice can verify the message is from Cheng using his public key. When Cheng signs the message with his private key, it creates an encrypted hash that is decoded and matched during the verification process. A match means that Cheng cannot repudiate that the message came from him. This is known as **non-repudiation**. The trick to remembering this for the CISSP exam is recognizing that only the sender's keys (Cheng's) are used for signing and verification.

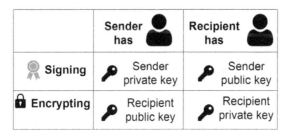

Figure 8.10: Asymmetric encryption and signing process

Figure 8.10 shows which keys are used for which service. The signing, encryption, decryption, and verification of messages can be done in a single process. This is done automatically through the email client being used, whether that client is Thunderbird, Outlook, or any other.

Asymmetric encryption algorithms include **Rivest-Shamir-Adleman (RSA)**, **Diffie-Hellman (DH)**, ElGamal, and **Elliptic Curve Cryptography (ECC)**. These algorithms use either large prime numbers or logarithms of large exponentiated numbers to create strong private keys with very high work factors, taking up to 300 years to crack using conventional methods. Some examples follow.

The **RSA algorithm** was created by Rivest, Shamir, and Adleman in 1977 and is now used not only for encryption and signing but also for online shopping. Without RSA, there would be no Amazon, Tmall, or Alibaba.

The **DH key exchange algorithm** is an asymmetric algorithm used strictly for symmetric key exchange. The encryption and decryption time for asymmetric algorithms can be tens of hours longer than symmetric encryption, so if only a small portion can be handled asymmetrically, the process can benefit from the much faster symmetric encryption and decryption speeds. This algorithm allows two parties to generate a shared secret key, which can be used for encrypted communication. However, if an eavesdropper can intercept the communication, they will not be able to get the key, as it has been encrypted.

For example, as can be seen in *Figure 8.11*, Alice and Cheng need to pass a shared secret symmetric key to have a secret conversation because they desire the speed benefits of symmetric cryptography. The conversation begins with them agreeing on some key value. A hacker who successfully performs a **Man-in-the-Middle (MITM)** attack will also be able to obtain a copy of this key value.

To protect this key value, each user's system encrypts the key value through its own private key to attain a new key value result. But since Alice's and Cheng's keys are different, the resultant key value is now different for both of them, and this new key is unusable.

If the spy is listening via MITM, they have two resultant keys that have nothing to do with the final secret session key. That resultant key is passed through their private keys again and now Alice and Cheng have a final secret usable session key that only they know. For maximum security, this is done for only one session.

DH is often used with PKI for online shopping sites to secure transactions. Also, DH is used with SSH and SCP for the purpose of encrypting remote client-server user sessions. It is an effective replacement for programs such as Telnet or FTP, which do not encrypt user sessions or file transfers.

Diffie - Hellman

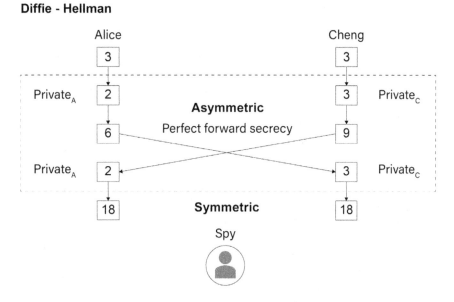

Figure 8.11: Diffie-Hellman symmetric key exchange process

The asymmetric **ECC algorithm** uses logarithms to create very strong encryption and digital signatures with shorter key lengths. This makes it a desirable algorithm for power-saving devices such as smartphones, smart cards, other wireless devices, and more.

The asymmetric **ElGamal algorithm** is an extension of the DH encryption algorithm that provides digital signatures and confidentiality.

Symmetric versus Asymmetric Algorithms

Asymmetric algorithms allow confidential messages to be passed to individuals, but the decryption speeds are a lot slower than symmetric encryption.

If key management is important, then asymmetric algorithms win here if working with larger teams. Key management for asymmetric keys is simply *2 x (the number of users)*. So, if there are 100 users, 200 keys need to be managed. The number of symmetric keys managed for 100 users is almost 5,000 because each user needs a copy of the other 99 users' keys to communicate with them privately. This is clarified with the following equation:

(the number of users) x (the number of users – 1) / 2

Security administrators need to balance what is most important for their organization when weighing key management and performance.

Hashing Algorithms

Although it uses a cryptographic method, hashing is different from encryption. Rather than being a method to hide data that is being sent, it will verify that the data received has not been tampered with. Hashing provides a type of serial number, hash value, or a **message digest** to a specific input. The sender can put the data through an algorithm, and when that data is sent to someone else, then can run the data through the same algorithm and compare the two outputs. From the result, the user can confirm the integrity of the content (i.e., that it has not been altered).

From *Figure 8.12*, after Sergei downloads `Super_Video_Game.iso` online, he will run the ISO image through the hashing algorithm. The *Super Video Game* website informs him that the hash value must be `2da492mdKLKjq`. If the message digest does not match `2da492mdKLKjq`, this could mean several possibilities. He may have only received a partial download of the game, or if he obtained the game at an alternate website, he may have a version with malware.

Either way, if it does not match `2da492mdKLKjq`, he must not use this ISO.

Case A. (model with Super Video Game ISO> [hash algorithm] > 2da492mdKLKjq

Case B. (model of Super Video Game with Malware > [hash algorithm] > jJKwa43095ksm

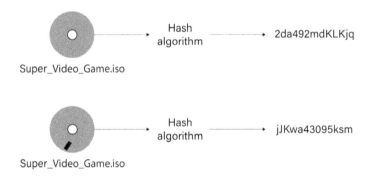

Figure 8.12: Hashing example

Hashing algorithms have five properties. First, the output of the algorithm is unpredictable, meaning even a small change in the input will create an entirely different result. Second, there are no duplicates, so different inputs giving the same output should be highly improbable enough to be impossible. Third is irreversibility – you can't work out the input from the hash. The entire input should be used to calculate the hash so that small changes or missing sections can still be detected. Finally, the same input should always give the same output, so that different parties can check for integrity.

The CISSP candidate is required to understand the basics of the MD5 and SHA hashing algorithms. The following subsections discuss these two algorithms.

MD5 Hashing Algorithm

Message Digest 5 (MD5) is one of the most widely used hashing algorithms worldwide. It generates a 128-bit digest. Suppose you are downloading Firefox version 7 from `http://releases.mozilla.org/pub/firefox/releases/7.0/`. The website will show you a list of **MD5SUM** hashes for their files.

When downloading the program and running it through the MD5 hashing algorithm, the value must match the MD5SUM listed on the Mozilla website. If not, that means the file has been altered in some manner. From their website, the hash for the `./win32/en-US/Firefox Setup 7.0.exe` file is `027f02accaae40e9d799bfa8480ecede`.

After the user downloads the file, they run it through an MD5SUM hash tool such as the Linux `md5sum` command or the Microsoft PowerShell utility called `Get-Filehash` using `./win32/en-US/Firefox Setup 7.0.exe` as the input file. If the output from the command is `027f02accaae40e9d799bfa8480ecede`, then the file is okay to use because it is consistent with the hash listed on the website.

If the hash is any other value, do not use the file. It may only be a partial download or a malware-infected file due to downloading from a poorly managed mirror. MD5 hashing is still used today but because of its short digest, it is prone to collisions (i.e., different inputs creating the same hash). For better security, it is recommended to use the longer SHA hash algorithm.

SHA Hash Algorithm

The **Secure Hash Algorithm (SHA)** is also widely used and growing in popularity because it results in significantly fewer collisions than other algorithms. Because it has a message output, or digest, of 160 bits, it takes longer to compute than MD5, but as computers become faster, this becomes a less important factor.

SHA has other output lengths, such as SHA-224, SHA-256, SHA-384, and SHA-512, where the number denotes the digest size in bits, so SHA-224 computes a 224-bit hash. Again, the longer the hash, the fewer the collisions, but the longer the compute time.

Considering the Mozilla example, the process will be the same as MD5 when matching the hash after downloading the file, but the security technician will use an SHA-1 hashing tool to match the hash value, such as Linux's `sha1sum` command.

SHA-3 was released in 2015 but computes differently from SHA-1 and SHA-2. The goal is that SHA-3 can directly replace SHA-2 in existing applications because it's stronger and faster.

> **Note**
>
> Knowledge of the hash algorithm programs such as `Get-Filehash` or `sha1sum` will not be in the exam.

Stronger Encryption with Salts and Initialization Vectors

Hashing is used to protect users' passwords. Although this means that systems do not save lists of user passwords, it does create other security issues. When a user enters a password, it's not the password that is saved but a hash representation of that password. On Linux systems, including macOS, which is Unix-based, these hashes are saved in a file called `/etc/shadow`, and on Windows systems, the file is called `C:\Windows\System32\Config\SAM`.

For example, the hash for the password `pa3 CI$$P Now!` looks like this:

    1yXOIL5K7$/BGGIeQapbRdmnBDXbdNM

Now, assume that users Chang and Pradeep use the same password of `pa3 CI$$P Now!`. Their hashes would appear as the preceding one or as follows:

    1yXOIL5K7$/BGGIeQapbRdmnBDXbdNM

A malicious actor or hacker keeps a table of hashes and passwords and begins to notice a pattern. The hacker sees this:

    1yXOIL5K7$/BGGIeQapbRdmnBDXbdNM

And, according to the table, they know this means the password is `pa3 CI$$P Now!`.

This specific problem can be mitigated using two different techniques: salting and initialization vectors.

Salts

What salting will do is provide variance if users use the same password by adding some random values. The salt value is randomly generated by the OS. For example, Lisa, Chang, and Pradeep happen to use the same password, and then some "salt" gets added by the OS.

	Password	Salt	String to hash	Resultant Hash
Lisa	pa3 CI$$P Now!	yMj	pa3 CI$$P Now!yMj	/BGGIeQapbRdmnBDXbdNM
Chang	pa3 CI$$P Now!	Ppj	pa3 CI$$P Now!Ppj	zZrej38epbMdmn256YHgte
Pradeep	pa3 CI$$P Now!	oO8	pa3 CI$$P Now!oO8	2mHgeuapbrMmnBDfjmYhq

> **Note**
> Password salting is not available in Windows 10 or Windows 11.

Password salting mitigates dictionary and rainbow table attacks because attackers would have to compute many more hash tables to correspond to the many salting possibilities.

Initialization Vectors

Initialization Vectors (IVs) are similar to salting but for symmetric and asymmetric encryption. IVs are added to keys to provide randomness to identical inputs. For example, the input `Gold is here!` will provide a different result because of random IVs:

Plain Text	Key	Initialization Vector	Resultant Cipher Text
Gold is here!	Mxobswzsq,c,*	WqI9.&=k	,Ki29saMKw2=8
Gold is here!	Mxobswzsq,c,*	o(3$cNex	ncoe423Mms3#A

Table 8.2: Initialization vectors

The longer the salt IV, the stronger the hash or encryption because there is more variance with the hashing or encryption results.

An earlier Wi-Fi system called WEP used initialization vectors to help secure the Wi-Fi password. At 24 bits, the IV on WEP systems is too short and allows a hacker to determine the Wi-Fi password quickly. The author was able to crack a Wi-Fi password within minutes. Newer versions of Wi-Fi systems, called WPA and WPA2, use better methodologies and longer IVs, making them much more difficult for hackers to crack. The author gave up after a month using modern tools.

Further discussion of password and encryption cracking will be discussed in the upcoming *Understanding methods of cryptanalytic attacks* section. For now, let us continue looking at security tools and methods with public key infrastructure.

Public Key Infrastructure

Companies such as eBay, Tmall, and Shopee would not exist if it were not for **Public Key Infrastructure (PKI)**, digital certificates, and digital signatures. These tools allow individuals to shop online securely and allow clients to trust their vendors because of background checks completed by **Certificate Authorities (CAs)**. CAs ensure trust by conducting background checks on the owner or top managers of the company, thereby "certifying" that the vendor can be trusted.

The following section will cover how encryption, hashing, and signing are implemented to support PKI as well as how the system protects online shoppers.

Encryption, Hashing, and Digital Signatures for PKI

In order to protect an online shopping session, the transaction must be protected from hackers, completed successfully, and from a verified vendor.

These are accomplished through asymmetric encryption, symmetric encryption, hashing, and digital signatures managed by a system called security protocols. The security protocols used to manage online transactions are called SSL and TLS.

The hosting website will set which protocols to allow at their online store. Then, these define the asymmetric encryption (for example, RSA or DH) that will be used to manage the early parts of the transactions, such as ensuring the online store is a valid and verified entity and hiding a symmetric key used for the protection and completion of transactions such as the login, password, and final purchase. (Remember that symmetric encryption is significantly faster than asymmetric encryption, so for best performance, the latter parts of the transaction are completed symmetrically.)

Digital Certificates and CAs

Soon, you will pay exam fees to sit and successfully pass the CISSP exam. ISC2 verifies and validates that you are qualified through your passport or driver's license. After you complete the exam, and your references, experience, and sponsorship are accepted by ISC2, you will receive your CISSP certificate. The certificate shows the name of the certifying organization ISC2, that you are the certificate holder, and your date of certification and expiration, and it is signed by ISC2.

You will use your certificate to further your career. Employers verify that you hold the certificate by checking with the authority, ISC2, and ISC2 provides validation that you are the certificate holder.

Digital certificates work in a similar manner; they are ways to verify things are what they say they are over the internet. For example, you decide to open an online store to sell smartphones. First, you need to purchase a domain name (such as `https://google.domain`) from a firm. This costs about $15 annually. You find that `https://www.supersmartphone.com` is available and set up your new online store.

Other providers of domain name registration include the following:

- 1and1
- Bluehost
- Domain.com
- GoDaddy
- Name.com
- Namecheap
- Network Solutions
- Register.com
- Yahoo

A CA provides a proofing process to verify and validate the vendor. Because we trust the CA, we can trust the online store. Some reputable CAs are as follows:

- Comodo

- DigiCert

- Entrust

- GeoTrust

- Global Sign

- Symantec

- Thawte

The proofing process varies from CA to CA, but most require a copy of a driver's license or passport, business address, personal address, phone number, and a notarized signature. This information is collected by the **Registration Authority (RA)**, which is the face of the CA. Some RAs will visit the business location for further validation. Once the vendor is approved, they will receive a signed digital certificate from the CA. Now, when the public visits their shopping website, they can trust the online website because their digital certificate is approved by the CA.

More importantly, if an attacker hears of your fantastic success of selling smartphones and creates a "spoofed" website that looks and appears just like `https://www.supersmartphone.com` but their domain is listed as `https://www.supersmartphone.com`, users will be warned that this is not the valid `supersmartphone.com` website. If they continue to shop there, they shop at their own risk.

The digital certificate is simply a verification of your shopping site's public key back through the CA. After your certificate is signed with the private key of the CA, they update the world's web browsers with the public key of your online store. Shoppers electronically match this to your digital certificate and can shop in confidence because it is signed by the CA.

The details of your digital certificate follow the X.509 v3 international standard:

- X.509 standard to which digital certificate conforms

- Serial number from certificate originator

- Signature algorithm ID – how the CA signed the certificate

- Issuer name – CA that issued the certificate

- Starting date and time of certificate validity

- Ending date and time of certificate validity

- Subject name – the owner of the public key of the certificate
- Subject public key – owner's public key
- Extensions

Digital Certificate Verification Process

Figure 8.13 shows the steps of the verification process that occurs when a shopper visits your online web store, as explained here:

1. The customer visits `https://www.supersmartphone.com`.
2. TLS v1.2 or SSL v3 is selected based on the customer's web browser.
3. RSA or DH is selected for asymmetric session key exchange.
4. 3DES or DES is selected for the symmetric encryption algorithm for data in transit.
5. SHA-1 or MD5 is selected as the hashing algorithm.
6. `https://www.supersmartphone.com` responds, confirming the preceding conditions, and provides their certificate.
7. Optional: If the client has a certificate, it will be sent to your online store at this point.
8. The client checks to see whether `https://www.supersmartphone.com` is on the **Certificate Revocation List** (**CRL**). If so, the customer gets a warning. If not, the customer server moves to *step 4* in the figure.
9. Asymmetric encrypted session key exchange occurs here or DH negotiation.
10. Symmetric encryption is enabled for data in transit.

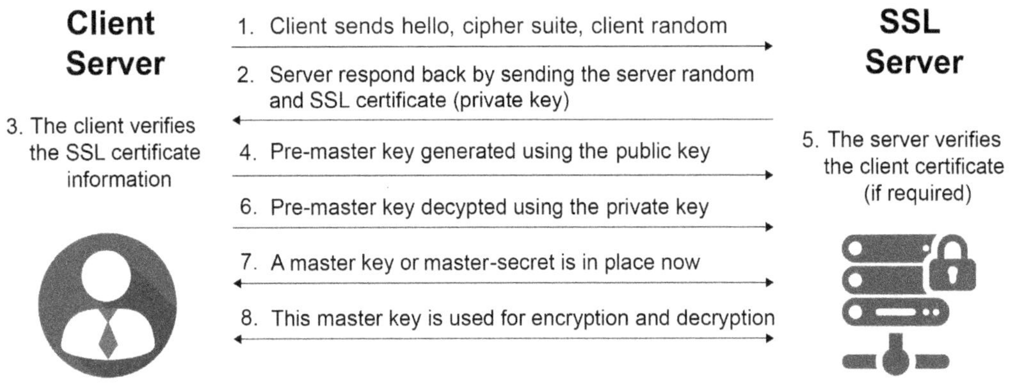

Client Server

1. Client sends hello, cipher suite, client random
2. Server respond back by sending the server random and SSL certificate (private key)

3. The client verifies the SSL certificate information

4. Pre-master key generated using the public key
6. Pre-master key decypted using the private key
7. A master key or master-secret is in place now
8. This master key is used for encryption and decryption

SSL Server

5. The server verifies the client certificate (if required)

Figure 8.13: TLS handshake protocol

Understanding Methods of Cryptanalytic Attacks

As you have seen, there are various ways that cryptography can be used to keep your data and systems secure. Encryption and verification continually develop as technology and methodology progress. But as defensive systems are becoming more powerful, attackers are finding new ways to break encryption systems. There are two basic forms of attacks on encryption to break a cipher. These are active and passive attacks. With an active attack, the hacker has access to the victim's network or system. Such attacks make it more likely the attacker will be discovered. Passive attacks break encryption largely through MITM attacks away from systems the victim owns. Attackers can attack different components of encryption, for example, the cipher, key, algorithm, or message. Common encryption attacks follow, starting with brute force attacks:

- **Brute force attack**: When an attacker launches a brute force attack, they are going to try and use every possible character to break the encryption. Brute force attacks are commonly used to attack passwords where the attacker tries every possibility to get into the system.

  Brute force attacks are mostly successful when the user uses a dictionary word as a password, also known as a dictionary attack. For example, it is easier to crack the password `elephant` than `El3ph@nt!`. The best way to prevent this type of attack is with complexity. If a user uses a long and complex password, it could take centuries before the password is discovered using a brute force attack.

- **Ciphertext-only attack**: With ciphertext-only attacks, the attacker acquires several encoded messages from the victim and uses those messages to figure out the key. It requires several messages for this technique to work and can take a long time or be almost impossible with complex ciphers.

- **Known plaintext attack**: With known plaintext attacks, the attacker has both the plaintext and the ciphertext of the message and uses both those pieces to determine the key. Messages may all start the same (for example, with "the morning weather report") and end the same (for example, "be careful out there"). With enough messages, the attacker can determine the key that is used to create the message.

- **Frequency analysis attack**: A frequency analysis attack is a type of known ciphertext attack based on the frequency of letters, how often they appear, and how often they are grouped. For example, in English, if there is a common grouping of three letters in the encoded message, this could be "`THE`" or "`AND`" and the cipher could be worked out starting with that. Single-letter words are likely "`A`" or "`I`." Pairs of letters are usually the words "AN" or "TO." With enough data, the attacker can figure out the key.

- **Chosen ciphertext attack**: Chosen cipher text is a method for attacking asymmetric encryption. The attacker has access to the decrypted message and some of the encrypted message. They use knowledge of the encryption system to figure out the private key.

- **Implementation attack**: An implementation attack targets the weaknesses of an encryption device. They often exploit poor configurations, setup errors, or inherent vulnerabilities. For example, WEP-based encryption systems within early Wi-Fi routers had weak IVs that led to poor security implementation. Attackers could tap into wireless transmissions to the WAP and discover the Wi-Fi password. The implementation was improved and upgraded from WEP to WPA, WPA2, and WPA3, which became harder to attack.

- **Side-channel attack**: Side-channel attacks listen for electromagnetic frequencies, timing information, and power consumption to help decode messages. For example, electromagnetic frequencies emitted from a copper line can be read from a decoder and converted back into digital data, resulting in data loss.

- **Fault injection attack**: In a fault injection attack, attackers alter the input fed into an encryption algorithm and monitor its behavior to detect faults. By causing faults during the execution of cryptographic algorithms, attackers may reveal information about secret keys or plaintext data through faulty outputs.

- **Timing attack**: The timing attack is similar to the side-channel attack and focuses on the time that it takes to compute certain components of messages to help discover the key. For example, encrypting a 0 is faster than encrypting a 1.

- **MITM attack**: The MITM (or on-path) attack is usually a passive attack on data in motion through a network (but could also be an active attack). The attacker listens to the encrypted data, collects it, and attempts to crack the message using brute force, frequency analysis, or other methods. In a passive attack, the interceptor only eavesdrops on the communication. In an active attack, the interceptor attempts to alter or block the message.

- **Pass-the-hash attack**: A pass-the-hash attack compromises active directory systems by copying the hash the user uses to authenticate rather than their password because it is the hash that validates the user, not the password. The attacker duplicates this hash to authenticate as the user. Implementing Kerberos has helped to mitigate pass-the-hash attacks. More details are in *Chapter 13*.

- **Kerberos exploitation attack**: The golden ticket is a forged ticket-granting ticket stolen from the **Kerberos domain controller** (**KDC**). The attacker is able to fake being the domain administrator and thereby gain access to the entire system. This attack can be mitigated by routinely updating the Kerberos passwords on the KDC.

- **Ransomware attack**: Ransomware attacks occur when an attacker (usually through a phishing attack) convinces the victim to run an application that encrypts all the files on their system. If the user wants to regain access to those files and does not have good backups, they will need to pay the ransom fee to obtain the private key for their files to be decrypted. Attackers may demand a recovery fee of $250 or more in cryptocurrency. One of the largest ransomware attacks occurred with the Colonial Pipeline in the US, where the ransom was eight million dollars.

Summary

Understanding IT systems architecture, whether on a laptop, server, phone, or IoT device, is important for a security administrator because it enables them to identify potential vulnerabilities, implement appropriate security controls, and ensure that the entire ecosystem is protected against threats, regardless of the platform or device being used. Encryption is a core part of how information passes through architectures and, properly applied, it is an important tool for any secure system.

In this chapter, you covered system architectures as well as the security issues around virtual systems and containers. You saw how encryption is used across architectures to protect information and how devices are protected with more encryption. You saw different encryption solutions and which ones are better to use, when to use them, and their best applications. This chapter also covered vulnerabilities with encryption, and that it has a life cycle and is important to update encryption methods and keys on a timely basis. It also discussed the differences between symmetric and asymmetric encryption and showed how it could be attacked and used against victims in ransomware attacks.

With these skills, you can identify weaknesses, properly configure systems, and mitigate against attacks. Appropriate cryptographic solutions safeguard data, ensuring regulatory compliance, and managing risks associated with data breaches and unauthorized access. These combined skills are foundational to protecting information systems and maintaining the confidentiality, integrity, and availability of data. The next chapter will discuss how your physical environment impacts the security of your organization.

Further Reading

- *Stuxnet*, `https://packt.link/2Wv81`, NJCCIC Threat Profile, August 10, 2017

- *Symmetric quantum fully homomorphic encryption with perfect security*, `https://packt.link/Eai7k`, NASA Astrophysics Data System, December 1, 2013

- *PKI Fundamentals*, `https://packt.link/zvwOh`, US Department of the Treasury, April 4, 2021

Exam Readiness Drill – Chapter Review Questions

Apart from a solid understanding of key concepts, being able to think quickly under time pressure is a skill that will help you ace your certification exam. That is why working on these skills early on in your learning journey is key.

Chapter review questions are designed to improve your test-taking skills progressively with each chapter you learn and review your understanding of key concepts in the chapter at the same time. You'll find these at the end of each chapter.

How to Access These Materials

To learn how to access these resources, head over to the chapter titled *Chapter 24, Accessing the Online Resources*.

To open the Chapter Review Questions for this chapter, perform the following steps:

1. Click the link – `https://packt.link/chapter08`.

    Alternatively, you can scan the following **QR code** (*Figure 8.14*):

Figure 8.14: QR code that opens Chapter Review Questions for logged-in users

2. Once you log in, you'll see a page similar to the one shown in *Figure 8.15*:

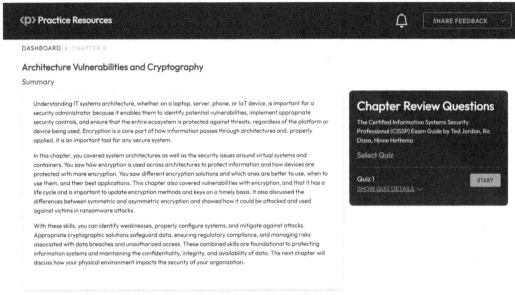

Figure 8.15: Chapter Review Questions for Chapter 8

3. Once ready, start the following practice drills, re-attempting the quiz multiple times.

Exam Readiness Drill

For the first three attempts, don't worry about the time limit.

ATTEMPT 1

The first time, aim for at least **40%**. Look at the answers you got wrong and read the relevant sections in the chapter again to fix your learning gaps.

ATTEMPT 2

The second time, aim for at least **60%**. Look at the answers you got wrong and read the relevant sections in the chapter again to fix any remaining learning gaps.

ATTEMPT 3

The third time, aim for at least **75%**. Once you score 75% or more, you start working on your timing.

> **Tip**
> You may take more than **three** attempts to reach 75%. That's okay. Just review the relevant sections in the chapter till you get there.

Working On Timing

Target: Your aim is to keep the score the same while trying to answer these questions as quickly as possible. Here's an example of how your next attempts should look like:

Attempt	Score	Time Taken
Attempt 5	77%	21 mins 30 seconds
Attempt 6	78%	18 mins 34 seconds
Attempt 7	76%	14 mins 44 seconds

Table 8.3: Sample timing practice drills on the online platform

> **Note**
> The time limits shown in the above table are just examples. Set your own time limits with each attempt based on the time limit of the quiz on the website.

With each new attempt, your score should stay above **75%** while your "time taken" to complete should "decrease". Repeat as many attempts as you want till you feel confident dealing with the time pressure.

9
Facilities and Physical Security

All IT infrastructure exists in the physical domain. Where components such as servers are defined by code, that code still exists on a machine. Cloud-based services, such as infrastructure or software as a service, still have code on hardware. Likewise, the security of your data also comes with physical concerns. Malicious actors can attempt to break into data centers to insert listening devices or damage and destroy equipment. Critical cabling and electronics can be destroyed by accidental fires, and data centers are susceptible to the effects of natural disasters such as floods and earthquakes.

In this chapter, you will learn how to secure information systems by securing the premises. This is known as physical security. Risks of attacks on the valuable data inside your organization's premises come from various sources, including wireless spoofing, social engineering, and data exfiltration through USB devices and smartphones.

Locations where networking systems are installed require mitigations so that only authorized personnel have access. They also require defenses against external attacks, including worm-based ransomware and natural disasters. These threats can be minimized through careful consideration of site design, facilities, operations, and the implementation of security measures such as fencing, guards, and person traps.

By the end of this chapter, you will be able to answer questions on the following topics:

- Applying security principles to site and facility design
- Designing site and facility security controls
- Creating restricted- and work-area security systems
- Implementing utilities and **heating, ventilation, and air conditioning (HVAC)** measures

You will start by reviewing the security principles of site and facility design.

Security Principles in Site and Facility Design

If an attacker gains physical access to an organization, they can steal data, corrupt data, deploy rogue devices, or obtain confidential records. Holders of the CISSP certification and security professionals in general must understand the value of introducing mitigations to protect the organization's site and facility against physical attacks.

Physical controls limit access, restrict unauthorized individuals, and monitor individuals in specific locations within the organization and how they interact with items of value. Controls such as security guards or barriers can be placed to limit access to buildings, equipment, the **security operations center** (**SOC**), human resources, legal areas, network centers, financial zones, and more.

Which areas need to be prioritized for physical access controls and the controls to be used are decided through qualitative or quantitative risk assessment. There might also be some legal and regulatory requirements when it comes to safeguarding the facility and its assets.

Physical controls depend on the same security fundamentals as those for securing networks or data. There should be processes for authentication, such as listing who is allowed access to certain buildings or rooms and proper badging techniques, so that individuals can continually be validated. Also, there should be systems in place to validate authorization, such as ID badges, within controlled environments. Finally, sign-in/sign-out sheets, video recordings, or other systems should be used at entry and exit points to verify individuals. Such systems also aid the investigation of breaches.

Each secure building or room should be considered a separate security zone with some type of security control. For example, office suites and research and development labs can each be secured separately by requiring electronic key cards to enter each one.

External Boundary Security

Just as properly securing IT infrastructure takes knowledge and planning, organizations must think carefully about planning their physical sites to ensure that staff and visitors are protected, and the discovery and deterrent of threats are facilitated. The following preventative controls are all important components of the external boundary for the organization:

- **Signs**: Using signs on exterior fencing and elsewhere on the property helps to deter intruders. Signs could include "No trespassing," "Beware of dog," "Highly secure environment," "Warning: high voltage," and "Do not enter." Such signs are deterrents for the casual intruder as they enforce the notion that security is tight.

- **Site layout**: Site layout security improves if an organization is on a hill and surrounded by a waterway, for example, because these create natural barriers. Sometimes the best practice to secure an area is to not advertise that it's a secure area, thereby making it less attractive to potential intruders. One way to do this is to make the building look very simple but maintained. Use **crime prevention through environmental design (CPTED)** concepts to protect the environment. These include keeping the lawn maintained, repairing broken windows, keeping bushes and plants trimmed, having the building far from the road, having the building on a high hill, and using thorny plants near the windows and bushes near the exteriors of the building to block the vision of intruders. Create honeypot buildings that look attractive to attackers to distract them from the secure areas.

- For best security, keep cars as far from secure facilities as possible, mandating that all cars are to be parked as far away as possible. If parking must be near a building, it should be parallel to the building and not facing the building, as shown in *Figure 9.1*. When parking faces the building, it increases the likelihood of a person accidentally rolling into the building and breaking through a window or wall.

Figure 9.1: Recommended parking space near buildings

- **Defense from vehicles and pedestrians**: Barricades are used to prevent access to facilities, buildings, work rooms, and offices. Barricades that protect the facility's grounds include different types of fencing, mantraps, turnstiles, and bollards.

- **Securing facilities with fencing**: K-rated fencing is specifically designed to stop vehicles from penetrating a protected area. The "K" rating is a classification that indicates the fence's ability to withstand an impact from a vehicle of a certain weight traveling at a specified speed, such as the trucks shown in *Figure 9.2*:

Figure 9.2: K-rated fencing

Eight-foot fencing with razor-sharp barbed wire defends a facility's grounds against all except the most determined attacker; such fencing is used at correctional facilities, as in *Figure 9.3*. Four-foot fencing protects the grounds from casual intruders. Fencing needs to be transparent so that security guards can see through it and monitor impending attacks, but it also needs to be strong enough that it is difficult to cut or saw through. For better security, fencing can be electrified and a perimeter alarm system can be installed to detect cutting and climbing:

Figure 9.3: Penitentiary fencing

- **Bollards**: Bollards are designed to stop vehicles but allow people to walk through. Bollards are shown in *Figure 9.4*. Bollards can either be placed statically in the ground with cement or elevated when necessary:

Figure 9.4: Static bollards

- **Person traps**: A person trap, also known as a man trap, is designed to protect an organization from offenders by verifying their motives before granting them entry to the organization. It is normally two doors on either side of a vestibule. As shown in *Figure 9.5*, the individual enters through the first door. Once that door closes, the other remains locked until security personnel establish the individual's reason for being at the location. Some mantraps have metal detectors in case an offender is carrying a weapon.

  If the individual is approved, the other door will open and allow them access. If they are denied, both doors remain locked until authorities arrive, or they are free to go from the door they entered.

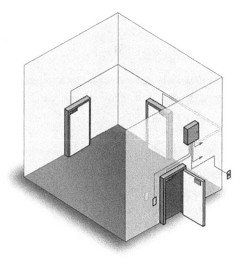

Figure 9.5: Person trap

- **Turnstiles**: Turnstiles allow entry to only one person at a time, as shown in *Figure 9.6*. They are effective in mitigating piggybacking and tailgating attacks. Piggybacking is when an employee allows a person entry into a facility at the same time as themselves, such as allowing through a delivery person on their security pass. Tailgating is when a person gains entry without the employee's consent, for example, quickly slipping in behind an employee when they have electronically unlocked a door.

Figure 9.6: Turnstile

- **Grounds and parking security**: Security lighting is critical to deterring crime and detecting attacks on the grounds or in the parking lot. Proper security lighting will have no dark spots, and cameras will have a larger field of view to witness activities in the parking areas. Good lighting and cameras will better allow guards to detect intruders and their vehicles.

Personnel Access Security Controls

Once inside an organization's building, it is still important to restrict and monitor the movement and access of people to prevent internal threats and accidents and mitigate against malicious actors who may have gained entry. Personnel access security controls are measures implemented to manage and mitigate risks associated with individuals accessing information systems and sensitive data within an organization

High-security areas such as the SOC, the networking facilities room, and backup tapes should be within the central part of the building, making them hard to access by someone who has just entered. They should ideally be several levels below the earth to protect these assets from disasters. When that is not possible, highly secure zones should be made as far away from building entrances as possible.

Public-access areas, or **demilitarized zones (DMZs)**, should not be near or past security zones. These DMZs should be very visible and have video cameras to deter incidents. In areas where there is a lot of movement of people, make sure there is video recording and limited access to network ports and computer equipment. Good use of surveillance in these areas is important. Minimizing traffic flow between secure zones mitigates attacks because it makes attackers more visible. It's important to make sure that near security zones, entering and exiting traffic never crosses.

Shoulder-surfing attacks are when someone watches another person input passwords, codes, or other personal information. Be cautious of where computers and their displays are placed. Security and data leaks could occur if the public can easily access others' names, addresses, phone numbers, and so on. In highly secure areas, use one-way glass to mitigate these types of attacks.

Physical and Electronic Locks

In order to secure a zone, locks must be deployed to deny offenders access. For better security, doors should be self-locking so they do not depend on an individual to engage the lock.

Physical locks disallow movement of the door handle unless a key is provided. However, these can be broken into using lock picking. Mitigations include anti-lockpicking systems, deadbolts, and multi-factor authentication via electronic locks.

Electronic locks operate with a key to employ two-factor authentication, where the individual uses a key and enters a PIN. Other forms of electronic locks use a proximity reader, biometric reader, or card scanner. A few examples of locking mechanisms are shown in *Figure 9.7*.

> **Note**
>
> *Bump key* is mentioned in the exam from time to time. Attackers can defeat some locks by inserting a bump key into the lock and then tapping it with a hammer. Bump keys exploit the mechanics of pin-tumbler locks and are a concern for security professionals because they can be used to compromise locks quickly and with minimal skill.

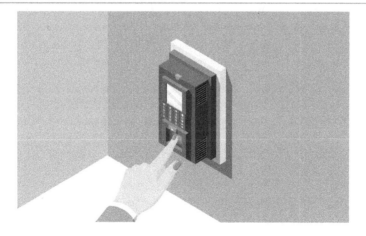

Figure 9.7: Door locks

Attacks on card readers include cloning and skimming, which are discussed in the next section.

Mitigating card cloning and skimming

Card skimming is when an attacker makes a copy of an existing card so that they can identify as the victim to steal their money, gain access to the victim's accounts, and so on. Unencrypted cards that are lost or stolen can be physically duplicated with a card-cloning machine. When a card is lost or stolen, the user should report this as soon as possible so that the old credential can be disabled and a new card can be issued. Security administrators need to watch for attempted entries at odd times of the day after a card has been reported lost or stolen; this is a sign that it is likely the attacker is seeking entry to the facility.

Skimming is when a counterfeit card reader is used to obtain a card's security details, which are then duplicated. Skimmer card readers are made to appear like regular card readers so they are hard to recognize.

Often, these vulnerabilities can be found in public areas that are used frequently, such as ATMs or self-service petrol or gas stations. These attacks generally target smart cards that transfer tokens without encryption, making the magnetic strip vulnerable. The magnetic strip stores data in an unencrypted format, which can be easily read and copied by skimming devices or other malicious tools.

Site and Facility Alarm Systems

When looking into security mitigations for a facility, the security professional must first consider the facility's vulnerabilities. Vulnerabilities could include windows, doors, and even rooftops since adversaries can break in. Mitigations include locks, alarm systems, bars over windows, and barbed-wire fences on rooftops. Also, be aware that attackers can enter secured rooms through false ceilings or raised floors, so detection sensors need to be placed in these areas.

Different types of alarm systems include circuit-based, motion detection, noise detection, proximity, and duress. Circuit-based systems trigger when the circuit connection is broken, as shown in *Figure 9.8*.

Figure 9.8: Circuit-based alarm system

These are commonly placed at doors and windows. The open-circuit alarm type is vulnerable to being cut within the circuit, so closed-circuit alarm types provide more security.

Motion detection alarms are triggered when there is a movement within their security zone. Motion detector sensors can use reflection, microwave technology (which can sense movement through walls), or infrared technology (which detects changes in heat).

Alarms that are triggered by sound are known as noise-detection alarm systems; sounds can be picked up by multiple means, including microwaves. False positives can be reduced by using AI-based analysis to identify specific sounds.

RFID systems can be used to track inventory systems; for example, when an alarm is triggered, this could indicate equipment, clothing, toys, and so on are being stolen.

Duress alarms are meant to be triggered by an employee under stress. Users trigger duress alarms when under attack by kidnappers or other offenders. These alarms can be in the form of concealed sensors, wireless pendants, or alarms within handsets or smartphones. For example, many smartphones offer an SOS feature that texts a person's contacts an SOS message when the user presses the power button five times. Some alarm systems offer a duress PIN code that allows the user to quietly contact the authorities.

In general, alarms can be set up to sound alerts, causing the offender in most cases to flee. But alarms could also be linked to monitoring systems or can be made to contact law enforcement or security companies. Silent alarms, as shown in *Figure 9.9*, contact security personnel but do not alert the offender, increasing the likelihood of them being apprehended.

Figure 9.9: Silent alarm system

Surveillance Systems

After the primary layers of defense, such as barriers and locks, are compromised, a second layer of defense involves surveillance systems. Surveillance can focus on zones or on different security areas. Security guards can be placed near these locations and, depending on the asset values, guards can be armed or unarmed. Guards can monitor important checkpoints, validate the identification of individuals, allow or deny access to visitors, and monitor sign-in sheets for building entrances.

Guards can also provide visual verification and judgment when analyzing potential security events. Just the presence of guards is sufficient to deter intruders or attackers, similar to fake cameras. Make sure that guards have the correct security clearance level to protect the zones they are guarding; for example, a guard with classified clearance cannot work in an area with top-secret clearance. Training and background checks of security guards are also important.

Closed-Circuit Television

Another way to survey different security zones is using **closed-circuit television** (**CCTV**). CCTV comes at a much lower cost than assigning guards in each security zone because it allows one guard to monitor multiple locations from a single point. Another advantage of CCTV is that actions can be recorded. However, response times can be longer since the security guard usually monitors the cameras from several meters or even kilometers away.

CCTV cameras are connected to a multiplexer using a coaxial cable, and live-feed or static images from the cameras are displayed on multiple screens. The guard operating the cameras can record images to tape or hard drive. Modern camera systems are linked to internet protocol networks using a data cable or fiber optics.

Machine Learning and AI

While surveillance systems such as CCTV have been commonplace for decades, modern data innovations can also help to make systems more robust. Machine learning and AI allow the implementation of smart physical security and robotic systems. These provide motion recognition, object detection, robot sentries, or drones. Motion systems allow a camera to be configured with gait identification technology. This technology allows cameras to identify movements that don't match that of authorized individuals and alert guards.

Object detection systems can detect changes in the environment such as a missing computer or if a switch or a rogue wireless device has been installed on a wall or table.

Robot sentries are surveillance systems that are mounted onto drones. These detect people over wide areas either on the ground or in the air. They can also use computer vision to recognize specific threats.

Reception Area Personnel and Person Badging

Organizations must introduce a challenge policy to mitigate social engineering attacks by potential threats seeking to gain access to the building. Personnel needs to be trained on how to properly challenge individuals for badging by learning to ask why a person is not wearing their badge and ensuring that at entry points, identification is shown.

At high-security entry points, the security staff should be armed with weapons, in line with all local and national laws and guidance, to protect people and the organization.

Cabling Wiring and Distribution Facilities

Throughout all modern office campuses exist interconnected systems of wiring and cabling that supply communications networks, Wi-Fi hubs, servers, and interconnected electronic security systems such as CCTV. Much thought should be put into how to make these systems secure, as intruders can cut cables, rewire, or place listening devices.

Distribution facilities provide communications cabling and wiring to multiple areas of an organization. The network data and voice channels come from the main distribution facility and then branch off into intermediate distribution facilities. The main distribution facility feeds the important facilities of the building's infrastructure, including the data center and intermediate distribution facilities. Main distribution facilities provide data services to smaller facilities, such as wiring closets, a type of intermediate distribution facility.

Wiring closets house the cabling from the main distribution facility; they are the intermediate distribution facility. It's important to secure this room with door locks because if hackers gain access to the room, they can inject attacks through the switches, routers, and patch panels that are housed within the closet. Two-factor authentication into the wiring closet should be considered the minimum for securing the area. Many organizations are not aware of the security issues surrounding the wiring closet, and it ends up becoming their coat closet and/or storage room. As a security professional, you must make sure this does not happen.

Intermediate distribution facilities can provide data and communication services to smaller buildings and offices. For example, a 20-person call center may have a room with a network switch and patch panel that provide voice services to the center, thereby acting as its intermediate distribution facility.

Whether distributing services to a large data center or a small office, it's important to secure the intermediate distribution facilities from man-made or natural disasters by keeping equipment safe. For instance, equipment may be kept high off the floor so that it would not be harmed by a flood. Further, make sure that the intermediate distribution facility only serves one purpose to avoid unnecessary hazards; for example, if the room doubles as a storage room, a user could accidentally knock off cables or break the patch panel while bringing other boxes and equipment inside.

Restricted and Work Area Security

When it comes to protecting the server rooms, data centers, media storage facilities, and evidence storage, physical security is not enough. Additional layers of defense, both technical and administrative, must be enacted to mitigate threats against an organization's valuable personally identifiable information or personal health records. Mitigating external threats is important, but so too is the consideration of internal threats and deployment of additional controls in areas such as evidence storage, media storage, and data centers.

Server Rooms and Data Centers

In all secure areas, make sure to store important assets in controls that can provide a high level of protection and security. For example, secure locking cabinets and safes can mitigate internal and external attacks. Acquire safes and secure cabinets that feature two-factor authentication to enhance security.

Also, when working in a data center shared by different organizations, servers can be placed in colocation cages so that staff only access systems that belong to their organization.

High-security servers should be air-gapped from the network. Air-gapped systems are those that are not connected to the internet, or any network, and may include additional security features such as validation for connected devices, armed guards that protect the facility, guard dogs that defend the grounds, and triple-barbed-wire fencing. Management controls include policies that mitigate unauthorized use, hosts, and storage devices.

Media Storage Facilities

Backup tapes, USB storage devices, and other storage media such as external hard drives can also be stored in a safe or vault. There are safes that are fireproof and waterproof. Also, there are safes that, depending on the value of the item, weigh heavier. Such safes are difficult for an attacker to lift, carry, and steal from the facility. For better security, acquire a safe with at least two-factor authentication, for example, one that requires a key and pin code to open.

Vaults allow humans to enter a protected environment and can protect against attacks such as drilling by sensing heat. Heat sensors also mitigate against fire. High-security air-gapped servers, especially those working as root certificate authorities, are good candidates for operating within a vault.

Physically securing data cables is critical to mitigating eavesdropping or DoS attacks within organizations. Security professionals must consider securing their communication and data cables. **Twisted pair cables** provide basic protection against eavesdropping by making it harder for attackers to intercept signals without specialized equipment by protecting against electromagnetic interference. **Shielded cables** offer enhanced protection against both eavesdropping and signal interception, as the shielding helps to contain and protect the data signals within the cable. You can use a **middle conduit**, which is a protective tube or channel to prevent unauthorized access and make it more difficult for attackers to manipulate or tap into the cables without detection. **Armored systems** protect cables by providing an extra layer of armor.

A Faraday cage prevents **electromagnetic fields** (**EMFs**) from entering or leaving the location and mitigates the risk of eavesdropping. Faraday cages follow the **transient electromagnetic pulse emanation standard** (**TEMPEST**) created by the United States Department of Defense, which lists best practices for shielding electromagnetic signals.

Evidence Storage

Computers and media removed from crime scenes must be labeled and bagged into tamper-evident, anti-static bags to mitigate data corruption. A chain of custody must also be attached to the evidence detailing how the evidence was involved in the crime. This includes **who** has handled the evidence, **what** the evidence is specifically, **when** and **where** the evidence was handled, and, if possible, **why** it was handled.

An evidence storage facility must provide access control and environmental control so that the data retains provenance in case the case goes to court proceedings.

Utilities and Heating, Ventilation, and Air Conditioning (HVAC)

In order for data centers to operate efficiently, they must maintain certain levels of temperature and humidity. Control systems installed to maintain the proper environment for data centers are known as HVAC systems.

HVAC systems make sure that the humidity is not too high as this could cause the corrosion of computer components, and not too low because this could cause static shock leading to a system fault. The proper humidity level should be between 40 and 50%. Temperature should be kept at about 65°F or 19°C. If the temperature gets too high, components can melt.

To ensure that temperature and humidity are at an optimum level, server rooms should be designed to make the environment operate under the best conditions. This can be done by implementing hot and cold aisles. Servers should be placed in such a way that their hot sides face only each other, and the air intake or cold sides face each other.

This way, the warm exhaust air rises into the cooling system and the cooled air from there is forced through the raised floor directly into the computers' intake. This causes less mixing of hot and cold air, resulting in a quality server room environment.

Environmental Issues

Make sure your cables are secured with cable ties and don't run across walkways, where people can trip on them or intruders can cut them. Instead, cables should run through either a raised floor or the ceiling, so they don't interact with humans. If you are running a cable through the ceiling, make sure the cable is fire retardant; such a cable is called a plenum-grade cable. A plenum-grade cable will not give off the toxic gas that other cabling gives off when on fire. Make sure to keep cables away from sources of interference such as fluorescent light and electromechanical devices such as fridges or industrial machinery. These emit electromagnetic interference and affect the integrity of data passing through the cable. Also, be careful of too many cables crossing each other because this is another source of electromagnetic Interference and could affect the integrity of the data.

Fire Prevention, Detection, and Suppression

Most US-based organizations and many others worldwide have to follow regulations that dictate the installation of different types of suppression and detection equipment to mitigate fires. In the US, this is known as **Occupational Safety and Health Administration (OSHA)** regulations.

Other countries have similar regulations. OSHA requires certain minimum safety standards for an organization when it comes to physical security. Some of these standards include the following:

- Marked exits

- Testing evacuation procedures at least annually

- Constructing buildings to resist fire (such as fireproof walls and doors)

- Employing smoke and fire detection controls

- Alarms to warn individuals that a fire has started

Beyond preparing for the threat of fire, thought should be put into preventing it. Fire suppression controls work based on the fire source triangle—the three elements that fire needs to ignite. These are heat, oxygen, and fuel to burn. Suppression controls remove at least one of the elements and aid in the suppression and prevention of fires. In the US, portable fire extinguishers come in different. Each class fights a specific type of fire. *Table 9.1* shows the fire extinguisher classes and the five types of fire:

Fire Extinguisher Class	Class of Fire
A	Wood/paper
B	Liquid/oils
C	Electric/computer
D	Metal
K	Kitchen

Table 9.1: Fire extinguisher types

The following are some systems that extinguish fires:

- **Wet pipe systems**: Wet pipe systems hold water under pressure, and when the sprinkler head solder melts because of fire, water is released immediately into the room. This system can extinguish the fire but damage computers.

- **Dry pipe systems**: Dry pipe systems contain no water and are often used in areas of freezing temperatures. Water enters once the sprinklers are triggered.

- **Pre-action systems**: Pre-action systems fill with water once the fire alarm is triggered, but the sprinkler system will not start unless the temperature in the room reaches a specific level. This allows individuals some time to manually contain the fire to save electronic equipment before the sprinkler triggers.

- **Deluge systems**: Deluge systems dump water in huge volumes in case of a fire. These systems are used in environments where fires can spread quickly, such as firework factories.

- **Halon**: Halon systems can put out fires without harming electronic equipment by interrupting the process needed for fire to sustain itself. However, halon has been found to be ozone-depleting, and further installations have been banned.

- **Clean agent**: A clean agent is an alternative to Halon because it is less damaging to the environment and is non-toxic to humans. Clean agent examples include INERGEN, a mix of argon, nitrogen, and carbon dioxide. Carbon dioxide can be used on its own but is not as safe for humans.

Backup and Redundant Power

Computer systems require clean power to run, but surges or voltage spikes can shut down computer systems and networks. Blackouts and brownouts can also cause system failures, so power management must be considered to maintain system uptime.

Backup power can be provided through dual power supplies, **uninterruptable power supplies** (**UPSs**), backup batteries, and backup generators. Hot pluggable **power supply units** (**PSUs**) provide failover capability without shutting the system down.

Power distribution units (**PDUs**) have features that include power cleaning, spike protection, surge protection, and brownout mitigation and can work in the same systems as UPSs. Advanced PDUs called managed PDUs can report load status and also have power-switching functionality.

Backup batteries are available for full systems or even components of systems, such as RAID arrays. They prevent data corruption or data loss during a power outage. UPSs can be acquired to provide power for 30 minutes, 30 hours, or even more. If power has not resumed before the battery dies, the UPS can send a graceful shutdown signal to the computer system to mitigate data corruption. UPSs consist of storage batteries and a DC-to-AC power inverter.

Backup power generators provide power to all critical assets in a building. Generators can use gasoline, diesel fuel, propane, or natural gas for fuel. Reserve backup power can also be provided by solar power, geothermal power, or wind power.

> **Note**
> Backup generators primarily fail because there is no fuel in the tank. This is either because fuel was never added or over several months, the fuel evaporated.

Summary

Without robust physical security, even the most advanced cybersecurity strategies can be undermined. Securing data not only requires technical and management solutions but also physical solutions that keep an attacker from walking into a building and carrying out attacks on physical infrastructure. This chapter highlighted the importance of securing both external and internal boundaries, using strategies such as perimeter fencing, surveillance systems, and controlled access points to prevent unauthorized access and protect against natural disasters.

We started the discussion with parking lot security, covering fencing, lighting, and cameras. Attackers seek vulnerabilities and notice when a facility is uncared for. Using principles of crime prevention through environmental design reduces the likelihood of attack because the facility appears to be tended to.

Then, the chapter discussed how to ensure building security with security guards and by logging who is entering and exiting the building and when. Alarm systems can alert security guards of intruders and security cameras can capture these events in action.

You covered secure storage, including knowing whether it's data on tapes or hard drives and whether it's evidence. Evidence storage requires stricter controls, including tamper-evident bags. Finally, you learned about maintaining the server room and ensuring clean electrical power. Hot and cold aisles help maintain a consistent environment in the server room, and redundant power controls and generators can provide electricity for some time if the power plant goes down.

For security professionals, these concepts are vital because they play a part in ensuring comprehensive protection for an organization's assets and help manage risks that could lead to significant data breaches or operational disruptions. In the next chapter, you will look at various networking technologies and how to secure them.

Further Reading

- *How DNA from family members helped solve the 'Golden State Killer' case: DA*: `https://packt.link/qIg4l`, ABC News, April 18, 2018.

- *Symmetric quantum fully homomorphic encryption with perfect security*: `https://packt.link/jY8Fg`, NASA Astrophysics Data System, Dec. 1, 2013.

- *PKI Fundamentals*, `https://packt.link/r45N0`, US Department of the Treasury, April 4, 2021.

Exam Readiness Drill – Chapter Review Questions

Apart from a solid understanding of key concepts, being able to think quickly under time pressure is a skill that will help you ace your certification exam. That is why working on these skills early on in your learning journey is key.

Chapter review questions are designed to improve your test-taking skills progressively with each chapter you learn and review your understanding of key concepts in the chapter at the same time. You'll find these at the end of each chapter.

> **How to Access These Materials**
>
> To learn how to access these resources, head over to the chapter titled *Chapter 24, Accessing the Online Resources*.

To open the Chapter Review Questions for this chapter, perform the following steps:

1. Click the link – https://packt.link/chapter09.

    Alternatively, you can scan the following **QR code** (*Figure 9.10*):

Figure 9.10: QR code that opens Chapter Review Questions for logged-in users

2. Once you log in, you'll see a page similar to the one shown in *Figure 9.11*:

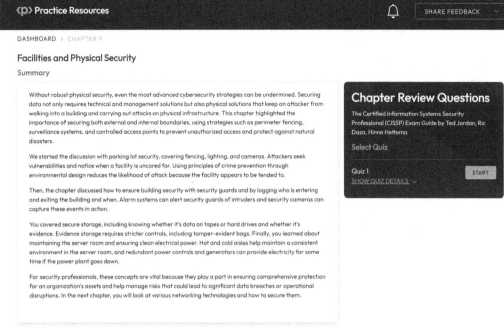

Figure 9.11: Chapter Review Questions for Chapter 9

3. Once ready, start the following practice drills, re-attempting the quiz multiple times.

Exam Readiness Drill

For the first three attempts, don't worry about the time limit.

ATTEMPT 1

The first time, aim for at least **40%**. Look at the answers you got wrong and read the relevant sections in the chapter again to fix your learning gaps.

ATTEMPT 2

The second time, aim for at least **60%**. Look at the answers you got wrong and read the relevant sections in the chapter again to fix any remaining learning gaps.

ATTEMPT 3

The third time, aim for at least **75%**. Once you score 75% or more, you start working on your timing.

> Tip
>
> You may take more than **three** attempts to reach 75%. That's okay. Just review the relevant sections in the chapter till you get there.

Working On Timing

Target: Your aim is to keep the score the same while trying to answer these questions as quickly as possible. Here's an example of how your next attempts should look like:

Attempt	Score	Time Taken
Attempt 5	77%	21 mins 30 seconds
Attempt 6	78%	18 mins 34 seconds
Attempt 7	76%	14 mins 44 seconds

Table 9.2: Sample timing practice drills on the online platform

> Note
>
> The time limits shown in the above table are just examples. Set your own time limits with each attempt based on the time limit of the quiz on the website.

With each new attempt, your score should stay above **75%** while your "time taken" to complete should "decrease". Repeat as many attempts as you want till you feel confident dealing with the time pressure.

10
Network Architecture Security

Ensuring efficient network communication is a core objective of any network engineer. Yet modern IT networks are complex, and as we have seen, threats and vulnerabilities are numerous, so making this communication, and the networks, secure demands careful analysis and planning. The **Open System Interconnection (OSI)** and **Transmission Control Protocol/Internet Protocol (TCP/IP)** models are central to this discussion, as they provide structured frameworks for securing different layers of network communication.

The OSI model, with its seven layers, provides a comprehensive approach to ensure that security measures are applied consistently across all aspects of a network. The TCP/IP model, although more practical with its four layers, similarly aids in the implementation of security protocols. This chapter will cover a part of *Domain 4 – Communication and Network Security*. The focus of this domain is on two main elements: understanding various networking technologies and how to secure them.

The chapter also looks at advanced networking concepts, such as **Fiber Channel over Ethernet (FCoE)**, **Internet Small Computer Systems Interface (iSCSI)**, and **Voice over IP (VoIP)**. These converged protocols operate over IP networks, raising significant security considerations. Understanding these protocols helps in predicting and mitigating potential security threats, a key skill for any security professional. The chapter also touches on micro-segmentation and virtualization-aware technologies such as **Software-Defined Networks (SDN)** and **Software-Defined Wide Area Networks (SD-WANs)**, which are increasingly relevant in securing modern, complex network environments.

Without that critical foundational knowledge of the concepts in this chapter, it would be difficult to understand *Chapter 11, Securing Communication Channels*. It's challenging enough to secure network communications and will only be more so if you don't thoroughly understand them.

By the end of this chapter, you will be able to answer questions on the following:

- Applying secure design principles in network architectures
- The seven layers of the OSI
- The TCP/IP model
- The implications of multiplayer protocols

We will start by looking at network architectures.

Secure Design Principles in Network Architectures

When building and assessing IT systems for security, it is useful to have principles to guide you. Modern architectures are large and complex, so to fully understand them, it helps to break them down into key concepts, or layers. The CISSP exam expects you to understand two key models, the OSI and TCP/IP models. The OSI model helps with secure design principles by providing a structured framework that can be used to ensure security measures are applied consistently and comprehensively across all aspects of a network. Each layer of the OSI model addresses a different aspect of network communication, and by considering security at each layer, a more robust and secure system can be designed. TCP/IP similarly provides a structured approach to implementing security measures across different layers of network communication, but with four layers instead of seven.

The following sections will first take you through the multi-layer protocols of the OSI model. You will also look at other protocols, such as FCoE, iSCSI, and VoIP. Finally, you will read about other more advanced networking concepts such as micro-segmentation, wireless and cellular networks, and content distribution networks.

The following diagram shows the layers of the OSI model and the TCP/IP model, with the seven layers of the OSI model on the left and the five layers of the TCP/IP model on the right. They are essentially describing the same thing, but the TCP/IP model collapses the top three layers of the OSI model (application, presentation, and session) into one application layer.

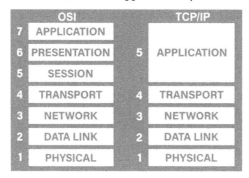

Figure 10.1: The layers of the OSI and TCP/IP models

The Seven Layers of OSI

The seven layers of the OSI model, from the bottom up, starts with the transmission media, that is, the electrical signals sent over a wire or cable, radio signals for wireless, and light pulses sent over fiber cable for optical, and concludes at the top with computer interface with the network (typically, some form of application such as a web browser, or an email client).

Two key concepts explain the need for the OSI model: interoperability and troubleshooting. Imagine you have two or more computers you would like to link together to communicate. Where do you start? First, you need a cable, then a network card, and finally, the protocol that spells out how each component would communicate with the others. If you were going to buy each one of these components, which brand would you buy to make sure the two machines could communicate? Interoperability refers to the ability of different systems, devices, or applications to work together and communicate effectively, regardless of their underlying technologies or manufacturers. The OSI model ensures this by creating layers that are standardized. Similarly, this structured layer approach helps with troubleshooting by giving a clear structure, allowing a systematic approach.

Before we look at the OSI layer by layer, it's worth noting the comparison to the TCP/IP model. You will notice many similarities between the TCP/IP model and the OSI model. They serve the same function but organize the information differently. You can think of the OSI model as theoretical and the TCP/IP model as practical. For clarity, there are references to the TCP/IP model when describing the OSI model.

> **Note**
> Because some of the names are interchangeable, it is always good to clarify which model you are talking about. The TCP/IP model will be discussed in more detail after you have seen all seven OSI layers.

OSI Layer 1 – Physical

The first is the physical layer, which is so named because it is something physical that you can touch, although with the advent of wireless networks, it's no longer completely accurate. The physical layer consists of the network cable, fiber cable, or wireless radio signal (not so physical...). The most common network cables are coaxial or pair cables. Common types are 10Base2 or 10BaseT, or category 5(CAT-5) or category 6(CAT-6) cables. A coaxial cable is also known as **coax** (see *Figure 10.2*).

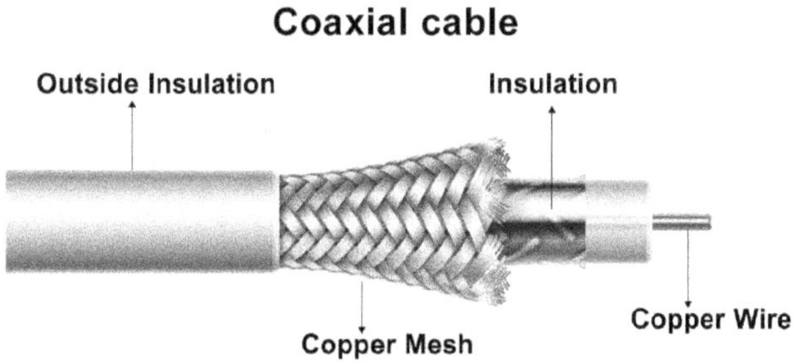

Figure 10.2: Coaxial cable

You may recognize this type of cable from the back of your cable TV box, but it can also be used as a networking cable and was one of the first **local area network** (**LAN**) cables used in the 1980s when building LANs became popular. This cable relied on the standard 10BASE2, which is a 10 **megabit per second** (**Mbit/s**) Ethernet standard using a coaxial cable. These standards range all the way up to 10GBASE-T, which is a standard to provide 10 Gbit/s connections over unshielded or shielded twisted pair cables, over distances up to 100 meters. The standards mentioned here are merely the tip of the iceberg; there are different standards for wireless, optical, and copper connections.

OSI Layer 2 – Data Link

Now that you have an idea of the medium used to send your bits and bytes, you can start putting those bits into what is called data link frames. A data link frame contains the payload (the actual data being transmitted, such as a text string) along with additional information for error checking and control. This additional data ensures that the frame can be properly routed and that any errors that occur during transmission can be detected and potentially corrected.

There are many types of frames to choose from such as 802.3 (Ethernet), 802.5 (token ring), 802.11 (wireless), Frame-Relay, and ATM. These frame choices are usually dictated by the networking equipment they are attaching to; for example, if you're plugging into an ethernet switch, you have to use 802.3. If you are plugging into a token ring switch, then you have to use 802.5 frames and must also have token ring **network interface cards** (**NICs**) in your computer.

The data link layer is the layer where the **media access control** (**MAC**) address is used. The MAC address is a unique identifier of the machine; it is typically the serial number burned into the NIC by the manufacturer.

When discussing network architectures, devices are often described by the OSI model layer at which they operate. For example, a network switch is typically referred to as a layer 2 (L2) device. This means it operates at the data link layer of the OSI model, handling the switching of data packets based on MAC addresses to direct them to the correct destination within a local network. In contrast, a network hub operates at layer 1 (L1), the physical layer, where it functions essentially as a repeater, amplifying and forwarding signals without any intelligent processing.

A layer 2 network is also known as a broadcast domain, where devices can directly communicate with each other using MAC addresses. Another L2 device is a bridge, which connects and filters traffic between two or more broadcast domains, helping to segment networks and reduce collisions.

OSI Layer 3 – Network

MAC addresses are only locally significant, meaning they can be used for two or more machines to communicate with each other on the same broadcast domain—that is, the same layer 2 network. It would not be very efficient to have one global broadcast domain, as there would be congestion as the network expanded to include more MAC addresses. It would also create security issues if all machines had access to all the data in the system.

To prevent this, networks are divided into small broadcast domains using devices such as routers and bridges. This is known as the network layer, or layer 3, and it links the layer 2 networks together and facilitates their communication.

When communication needs to happen, the first step is to determine whether the target is in the same broadcast domain or in a remote one. This determination is done by performing a bitwise AND operation (binary ANDing) between the sender's IP address and subnet mask to identify the network portion of the IP address. If the destination machine is on the same broadcast network, an **Address Resolution Protocol (ARP)** message asking the destination machine for its MAC address is sent to everyone on the network.

If the binary ANDing process determines that the destination is not on the same network, then all traffic for that destination is sent to the default gateway, which is the router that handles traffic for which the communicating node does not have a defined route. The router will receive the traffic and perform its own ANDing process on the first packet it receives to determine where to send it next. In TCP/IP, the subnet mask is used to determine which part of the address designates the network and which part of the address designates the host. The router only deals with the network portion of the address, using that to look up the network in its routing table. If the router has an entry for the network in its routing table, that entry will have an associated interface for that route. This interface allows the router to know which interface to use for sending all traffic to that network, and then the router can forward the traffic through that interface.

If the next router has an interface that shares the same layer 2 network as the destination, the router will perform an ARP lookup of its own to determine the MAC address of the destination host. This process of determining whether to send it to the next router along the way or send it to the actual destination host is performed at every step along the path by every router in the path. This is basically how global routing is achieved, whether your traffic is going from one broadcast domain to another on the other side of the router or the other side of the world. Routers share information about the networks they know about using various routing protocols such as **Routing Information Protocol (RIP)**, **Open Shortest Path First (OSPF)**, **Enhanced Interior Gateway Routing Protocol (EIGRP)**, and **Border Gateway Protocol (BGP)**, among others.

OSI Layer 4 – Transport

While layer 3 provides the connection between hosts, there are many applications and services that wish to connect with each other. Each of them is represented by a port number, which you can think of as a unique identifier within the host of applications and services – kind of like a mailbox on a host. Imagine the host as an apartment building. The IP address is the address of the apartment building, and the port number is the mailbox within the apartment building that belongs to the application or service living in that building. The following example will make this a little easier to understand.

Suppose you're trying to connect to a webpage on the internet. The first thing you need is the address of that site. IP addresses are very cumbersome to remember, which is why there are domain names such as `Google.com`. However, networking protocols do not communicate by name, so the name must be converted to an IP address.

This is what **domain name service (DNS)** does. After you put a **uniform resource locator (URL)** in your browser, your machine asks DNS for the IP address belonging to that name. In the case of this example, `Google.com` has an IP address of `142.250.189.142`.

Now that your browser knows the destination, the next step is to figure out the port. This is where the other part of the URL comes into play – `http`. **HTTP** stands for **hypertext transport protocol**, which, among other things, tells your browser to connect to destination port `80`. A URL often has `https` at the beginning, which stands for **hypertext transport protocol secure** and denotes port `443`. The port numbers `0` to `1024` are reserved for commonly used applications and services. This list of port numbers is discussed further in RFC 1700.

At the same time as the port number is being determined, so is the kind of transportation to use – reliable or unreliable. In the case of TCP/IP (the protocol, not the model, which we have not yet discussed), this will be **transmission control protocol (TCP)** and **user datagram protocol (UDP)**. TCP is the transport to use if you want reliable connectivity – that is, assurances that your traffic is reaching the destination. These assurances are provided through the use of several mechanisms, such as sequence numbers on messages, acknowledgment from the receiver of all numbers sent, and retransmission of missing messages. TCP is slower and computationally expensive but reliable. If you want speed, you go with UDP. UDP does not focus on whether the traffic gets there or not. In fact, it doesn't even check to see that the host is up and running. It simply sends the traffic; this is why it's typically referred to as best-effort.

OSI Layer 5 – Session

The session layer monitors all the network sessions and provides easy access to establish sessions for applications. When the application wants to access a remote resource, it just chooses a session and is not concerned with how that session is created or maintained. In this layer, you'll find the **remote procedure call (RPC)**, **session control protocol (SCP)**, **Network Basic Input Output System (NetBIOS)**, or some commonly used security protocols, such as **Password Authentication Protocol (PAP)**.

OSI Layer 6 – Presentation

The second of the aforementioned application layers, the presentation layer is also aptly named, as it packages the information to be transmitted across the network. An example is the **American Standard Code for Information Interchange (ASCII)** format used by many programs, such as telnet, FTP, SCP, SSH, and many others. Another example you are probably very familiar with is **hypertext markup language (HTML)**, which is the file format used to present documents in your web browser.

OSI Layer 7 – Application

The application layer, as its name implies, is where applications live, particularly applications with the ability to access a network. This is also the layer where humans interact with the network, usually through an application such as a terminal window, web browser, or Microsoft Word. This is also where protocols such as HTTP, SMTP, and FTP can be found. These protocols represent the **application programming interface (API)**, which the aforementioned applications use to access the network. HTTP is the protocol for your web browser, SMTP for your email application, and NetBIOS for Microsoft Word or Windows File Manager.

TCP/IP Model

The network protocol used on the Internet is named TCP/IP after two of its components: TCP and IP. The TCP/IP model uses a five-layer structure (formerly a four-layer model) in its protocol suite. In the TCP/IP model, the layers have names rather than numbers.

Although sometimes people still refer to TCP/IP layers by their number, such as *layer 1* or *layer 4*, this kind of reference is more appropriate for the OSI reference model. Layers in the TCP/IP model are not broken down from the bottom (physical) to the top (application), like the OSI model. Instead, they are grouped on the basis of functions.

The following diagram shows the layers of the TCP/IP model with some corresponding protocols and devices at each layer.

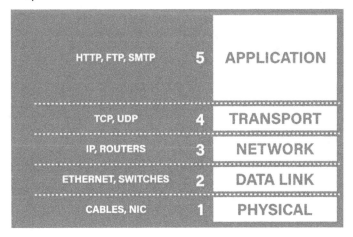

Figure 10.3: TCP/IP model with protocols

Reading from the bottom to the top, starting with the physical layer, you will notice that the layers map exactly onto the bottom four layers of the OSI model of the same name. The fifth layer, application, groups the top three layers of the OSI model – session, presentation, and application.

Now that you have gone through the similarities between the two models, you can focus on the differences between them. One of the things that is unique about the TCP/IP model is its naming convention for information as it moves through the various layers of its model. The following diagram shows the layers of the TCP/IP model with the corresponding name of the communication at that layer.

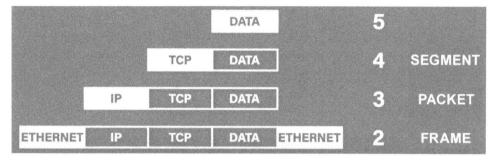

Figure 10.4: Name for each data layer of the TCP/IP model

At the physical layer, things are virtually indistinguishable between the TCP/IP model and the OSI model. Once information has moved from the physical layer to the data link layer, that information is referred to as a frame. Moving higher up the model to the network layer, the data link layer information is stripped away and is now referred to as a packet. Moving up to the next layer above, the transport layer, the IP information is removed from the packet and that information now becomes a segment. As the information is passed to the final layer, the application layer, the information is simply referred to as data.

Now that you understand how protocols communicate, you can take a closer look at the implications of multilayer protocols.

Implications of Multilayer Protocols

With IP networks becoming ubiquitous and inexpensive, the list of all the things that cannot be run over a modern IP network would be shorter than all the things that are now run over an IP network. As was mentioned earlier, the OSI model makes possible the transmission of just about anything from voice, video, and block-level traffic to **supervisory control and data acquisition** (**SCADA**) traffic over a regular IP network. The disadvantage of these applications was the requirement for their own network and equipment and the associated costs. The advantage, however, was that these networks were all separated instead of being interconnected into one giant **Internet of Things** (**IoT**), which is what is currently used.

Take a moment to consider the security ramifications of this cost-effective convergence. Networks and networking have many advantages, but they have one key disadvantage, which is vulnerability. That is why we need for information security.

One of the security techniques that has generally been considered the most effective is the air gap, which is a physical separation between the computer and everything else, removing it from the network. Some highly sensitive systems with high-level security implications, such as government security, are air-gapped. But if you want to be in a network that is accessible to everyone and everything, then remaining secure is difficult. There are at least two key challenges to securing multi-layer protocol networks. The first is to understand all of the protocols/applications operating on that network. The second is to have sufficient imagination to see how any of these protocols/applications can be a desired target or an attack vector. The next section discusses a few of the manifold converged protocols/applications that can be run over an IP network.

Converged Protocols

Some important communications functions, such as telephony, have historically been dealt with by independent networks, such as **public switched telephone network** (**PSTN**). However, in modern IT infrastructure, these functions are increasingly being handled by the standard network. Essentially, standard analog phone lines are being supplanted by VoIP. This is known as a converged protocol.

This section will discuss some of the many converged protocols that previously had to run on their own dedicated network and hardware and now can be run over an Ethernet or IP network. Understanding these protocols will help you to predict and mitigate attacks and other security incidents.

FCoE is a technology that allows the transport of fiber channel frames over high-speed Ethernet networks – high speed here refers to 10 Gbps or faster. Before FCoE, if you had a server or servers in your data center and you wanted to connect them to a **storage area network** (**SAN**), you would have had to install a **host-based adapter** (**HBA**). This meant purchasing and building an entirely separate network infrastructure just to transport fiber channel frames between your servers and your SAN, servers that no doubt already had a network adapter to provide network connectivity. FCoE allows you to use 10 Gbps Ethernet adapters in your SAN and leverage the same 10 Gbps adapters you already have in your servers, facilitating access to your SAN and IP networks simultaneously.

A similar technology used to connect to hard drives and other peripherals from the 1980s is **Small Computer Systems Interface** (**SCSI**), which, like FCoE, has also been modified to operate over networks. While SCSI is rarely found in modern computers, back in the 80s, it was one of the fastest technologies for connecting to peripherals and transferring data. The network-enabled version of SCSI is known as **Internet Small Computer Systems Interface** (**iSCSI**) and has comparable functionality and performance to FCoE.

Finally, the best-known conversion protocol is VoIP. Traditional phone systems ran on what is called **Time Division Multiplexing** (**TDM**) technology. This is an old technology that you may have seen in movies, with an operator sitting in front of a switchboard plugging and unplugging cables to connect calls. The technology did start out that way and is still powering many large enterprises' phone systems, known as a **Private Branch Exchange** (**PBX**), and our private home phone systems, also known as PSTN, or **plain old telephone service** (**POTS**).

Traditional voice technology and modern voiceover IP technology are analogous to vinyl records and **compact discs** (**CDs**) in that the old technology is analog and the new technology is digital. The steps required to make a VoIP telephone call are the same as traditional phone service – both requiring signaling, channel setup, digitization of analog voice signals, and encoding. The primary difference between the two is that traditional voice calls are transmitted over a circuit-switched network, while VoIP calls (also known as IP telephony) are packetized and sent over a packet-switched network such as the internet.

Micro-Segmentation

One of the many important trends in information security and networking has been micro-segmentation, which is the segmentation of things or objects using software. Those things or objects can be servers, applications, data, networks, or just about anything that can be accessed. One purpose of segmentation is to group things by attributes such as common security risks or access requirements. Another purpose of segmentation is compartmentalization – access to one compartment does not necessarily grant you access to another. This helps minimize the lateral movement of any hackers in your environment. This section will explore more carefully what micro-segmentation is, what technologies are necessary to achieve segmentation, and what benefits it provides to security.

Traditionally, segmentation was achieved through the use of technologies such as VLANs, firewall appliances, **access control lists** (**ACLs**), and security groups. However, these tools are very cumbersome to implement in a way that achieves the desired effect, and they are very error-prone. Modern networks are highly virtualized, at the network layer, at the host layer, and at the software layer. So, modern segmentation tools must also be virtualization-aware. The following subsections will discuss a few of such virtualization-aware technologies that are the backbone of modern networks. Understanding how they operate and how to use them is core to securing the modern enterprise.

Software Defined Networks and Software-Defined Wide Area Networks

SDNs and SD-WANs are both extensions of virtualization technology applied to networking – that is, the separation of the software from the hardware that runs it. Traditional networking uses integrated hardware and software that are individually configured to work with other pieces of networking equipment to forward traffic. In contrast, SDN was originally conceived to separate the control plane (where networks are managed) from the data plane (where traffic flows), to create more agile, flexible, and manageable networks.

SDN achieves this separation through the use of a smart controller that configures and manages the network. The smart controller pushes the necessary commands to the data plane. The advantage of this architecture is that it allows a single place to configure an entire network without having to deal with hundreds of routers and switches.

Imagine a large traditional networking environment (i.e., without an SDN) with hundreds of switches and routers, and you want to deploy VoIP. That would normally require manually handling each router and switch to replicate a virtually identical configuration. In a similarly sized SDN environment, a VoIP policy would be created on the smart controller, and that policy would be propagated automatically to all the routers and switches in the network.

SD-WAN is the same concept as SDN but is applied to the **wide area network (WAN)** arena. Imagine a business with hundreds of remote branch offices that have hundreds of routers, each with at least two connections – one to the internet and one to some kind of a WAN, such as **multi-protocol label switching (MPLS)**. Suppose management wants to configure each branch office to be able to use its internet connection as a failover for the WAN connection, using a **dynamic multipoint virtual private network (DMVPN)**. Achieving the goal would require connecting to each of the individual branch offices' routers and configuring them for this solution. With the SD-WAN setup, this configuration will again be done automatically from a smart controller.

Virtual Extensible Local Area Network Encapsulation

Virtual eXtensible Local Area Network (VXLAN), simply put, is a way to extend/connect layer 2 networks over a layer 3 network. This is done by tunneling or encapsulating layer 2 Ethernet frames inside layer 3 IP packets. VXLANs were created to solve the following three problems/limitations related to traditional VLANs:

- **Spanning tree**: Spanning tree is an old protocol whose sole purpose is to prevent loops in a layer 2 topology. It prevents loops by blocking redundant links. However, this process is very inefficient from a bandwidth perspective.

- **Limited number of VLANs**: Traditional VLANs have a 12-bit VLAN ID field, which means they can support between 0 and 4,095 VLAN ID numbers. This may sound like a lot, but not when you consider the typical service provider architecture that may have extensive segmentation and large numbers of VLANs. VXLANs, on the other hand, have a 24-bit VLAN ID field that provides about 16 million IDs.

- **Large MAC address tables**: This problem is created by an increase in server virtualization. Traditionally, a switch had to only know one MAC address per physical port. Now, one physical machine can have 2 to hundreds of virtual machines/containers, each with its own virtual MAC address. This significantly expands the number of MAC addresses a switch must handle, leading to scalability and security issues.

Wireless Networks

Wired networking technology is pretty much the sole purview of server computers today in that laptops and mobile devices predominantly rely on wireless technology for their connectivity. There are several different options for wireless networking technology, and a few of the more prevalent options will be discussed here.

The most commonly used and well-known technology is **Wireless Fidelity (Wi-Fi)**. Wi-Fi refers to a collection of wireless network protocols based on the IEEE standard 802.11. These are used for short-range LAN networking of devices, making it possible for these devices to exchange data via radio signals.

Li-Fi (which stands for **light fidelity**) is a wireless networking technology that utilizes light to transmit data. Li-Fi is capable of transmitting data at high speeds over the visible light spectrum. While Li-Fi is similar to Wi-Fi, there are some pros and cons to its use. For example, Li-Fi is not ideally suited to devices that move (i.e., non-stationary devices) as it does not handle handoffs from one receiver to another well. It is much more suitable for environments that are sensitive to electromagnetic interference, but where receivers are stationary relative to the transmitter, such as airplane cabins, hospitals, and nuclear power plants.

Zigbee is another wireless communication protocol and is ideal for short-range communication and **personal area networks (PANs)**. While Zigbee is not a particularly fast protocol, it does have the advantage of low power consumption, which is ideal for battery-operated devices. Zigbee has a range of about 10 to 100 meters of line-of-sight, which is ideal for many IoT devices, such as smart bulbs or thermostats. Finally, there is satellite communication, which is very old but is becoming popular again as **Internet service providers (ISPs)** strive to provide service around the world to less densely populated areas.

Cellular Networks

Cellular networking technologies predominantly provide internet connectivity to mobile devices, such as smartphones and laptops, that are connected to the internet and telephone network by radio signals through a local cellular tower.

There are many different cellular technologies, but the primary ones in use are 4G and 5G. **4G** or **fourth generation**, also known as **long-term evolution (LTE)**, is currently the most widely deployed cellular standard, as 5G is slowly being rolled out all over the planet. 4G's primary benefit over previous generations of cellular technology is increased capacity, a simpler network using an IP packet-based architecture, and reduced latency.

5G networks have higher download speeds than earlier generations of cellular technology; eventually, up to 10 GB/s could be possible. With speed like that, 5G could eventually supplant traditional ISPs, particularly in population-dense areas.

Content Distribution Networks

Content Distribution Networks (CDNs) and edge computing are architectures that strive to move data or computing power close to the requester, as opposed to a distant data center. Modern technologies such as IoT, voice-driven home automation, and autonomous driving vehicles require fast response times to be usable. Originally, CDNs cached static content closer to the requester in order to speed up the response time.

However, with IoT, it's less about serving content quickly and more about processing requests from IoT devices, such as sensors and monitoring equipment, that need rapid compute response time. Edge computing will not replace the backend data center for complex computational work but, rather, augment the architecture to allow simple time-sensitive computing to be performed closer to the endpoint.

Summary

This chapter covered various networking technologies, both current and emerging. Beginning with a discussion on how networks function through an overview of the OSI model and the TCP/IP model, this chapter then took you through an overview of multi-layer protocols and converged protocols. You also examined modern concepts such as micro-segmentation and various wireless networking technologies.

Now that you understand how protocols communicate and the challenges provided by multilayer protocols, you are ready to explore the next chapter, which focuses on the security of network components and their protocols.

Further Reading

- Saxena, Piyush. *OSI reference model–a seven layered architecture of OSI model.* International Journal of Research 1, no. 10 (2014): 1145-1156: `https://packt.link/NIL3C`

- Nath, Pranab Bandhu, and Md Mofiz Uddin. *TCP-IP model in data communication and networking.* American Journal of Engineering Research 4, no. 10 (2015): 102-107. `https://packt.link/Wbf2x`

- Keromytis, Angelos D. *Voice-over-IP security: Research and practice.* IEEE Security & Privacy 8, no. 2 (2010): 76-78: `https://packt.link/IWrsT`

- Mämmelä, Olli, Jouni Hiltunen, Jani Suomalainen, Kimmo Ahola, Petteri Mannersalo, and Janne Vehkaperä. *Towards micro-segmentation in 5G network security.* In European Conference on Networks and Communications (EuCNC 2016) Workshop on Network Management, Quality of Service and Security for 5G Networks. 2016: `https://packt.link/Cy7SX`

Exam Readiness Drill – Chapter Review Questions

Apart from a solid understanding of key concepts, being able to think quickly under time pressure is a skill that will help you ace your certification exam. That is why working on these skills early on in your learning journey is key.

Chapter review questions are designed to improve your test-taking skills progressively with each chapter you learn and review your understanding of key concepts in the chapter at the same time. You'll find these at the end of each chapter.

> **How to Access These Materials**
>
> To learn how to access these resources, head over to the chapter titled *Chapter 24, Accessing the Online Resources*.

To open the Chapter Review Questions for this chapter, perform the following steps:

1. Click the link – `https://packt.link/chapter10`.

   Alternatively, you can scan the following **QR code** (*Figure 10.5*):

Figure 10.5: QR code that opens Chapter Review Questions for logged-in users

2. Once you log in, you'll see a page similar to the one shown in *Figure 10.6*:

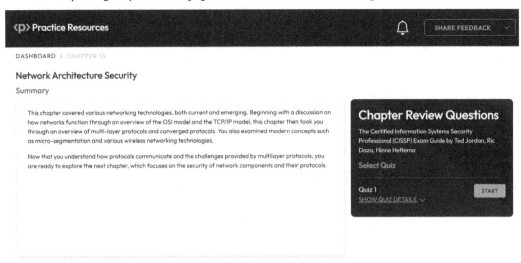

Figure 10.6: Chapter Review Questions for Chapter 10

3. Once ready, start the following practice drills, re-attempting the quiz multiple times.

Exam Readiness Drill

For the first three attempts, don't worry about the time limit.

ATTEMPT 1

The first time, aim for at least **40%**. Look at the answers you got wrong and read the relevant sections in the chapter again to fix your learning gaps.

ATTEMPT 2

The second time, aim for at least **60%**. Look at the answers you got wrong and read the relevant sections in the chapter again to fix any remaining learning gaps.

ATTEMPT 3

The third time, aim for at least **75%**. Once you score 75% or more, you start working on your timing.

Tip

You may take more than **three** attempts to reach 75%. That's okay. Just review the relevant sections in the chapter till you get there.

Working On Timing

Target: Your aim is to keep the score the same while trying to answer these questions as quickly as possible. Here's an example of how your next attempts should look like:

Attempt	Score	Time Taken
Attempt 5	77%	21 mins 30 seconds
Attempt 6	78%	18 mins 34 seconds
Attempt 7	76%	14 mins 44 seconds

Table 10.1: Sample timing practice drills on the online platform

---

Note

The time limits shown in the above table are just examples. Set your own time limits with each attempt based on the time limit of the quiz on the website.

With each new attempt, your score should stay above **75%** while your "time taken" to complete should "decrease". Repeat as many attempts as you want till you feel confident dealing with the time pressure.

11

Securing Communication Channels

Network security has traditionally focused primarily on ensuring availability. However, as we saw in *Chapter 10*, there is a lot of critical information now being transferred over networks that need to be secured. This is a multifaceted task that requires a thorough understanding of both the physical and virtual elements within an IT architecture. This chapter delves into the security considerations across various layers and components of a network, ensuring robust protection against vulnerabilities and potential breaches. You will also review the other core security concepts of confidentiality, integrity, availability, authenticity, and non-repudiation that will all come into play for networks.

Communication channels within a network, such as voice, multimedia collaboration, and remote access, must also be secured to protect data integrity and confidentiality. In this chapter, you will see what this means for VoIP systems, multimedia collaboration, and remote access. This chapter focuses on two main concepts: securing the components that make up networks and securing the communications carried over those networks.

By the end of this chapter, you will be able to answer questions on the following:

- Secure network components
- The operation of hardware
- Transmission media, network access control devices, and endpoint security
- Implementing secure communication channels according to the design

You will start by looking at secure network components.

Secure Network Components

In *Chapter 10*, we saw how the **OSI** and **TCP/IP** models can be used to break down IT architecture into different layers. The first of these layers is the physical layer—essentially, the equipment, transmission media, and endpoints that connect to a network. Before you examine the ways in which different network components can be secured, it is crucial to understand **high availability** (**HA**) and **fault tolerance**. As we covered in *Chapter 1*, availability is one of the five core security concepts. HA and fault tolerance are how you provide availability in your networks.

A fault-tolerant environment has zero service interruption but comes at a much higher cost, whereas a highly available environment has tolerance for minor service interruptions and is cheaper to set up. At their core, these two concepts have the same objective (to provide the required level of availability). The only difference between them is the cost associated with achieving that objective.

Fault tolerance is usually built using specialized hardware that can detect a hardware-level failure and instantly switch to a redundant component. This transition is designed to be seamless and provide uninterrupted service, though there is usually a high cost in both hardware and performance due to duplicate components that remain idle until called into service. However, one of the downsides of the fault-tolerant model (besides its obvious high cost) is that it is not meant to deal with software failures, which are by far the most common cause of downtimes.

HA, on the other hand, strives to achieve availability not by replicating physical components but by treating a system as a holistic collection of components that work together to provide a service. HA combines software and hardware to rapidly restore critical services when a system, component, or application fails, thereby minimizing downtime. While HA recovery is not as seamless as a fault-tolerant model, the restoration of services is still quick, often taking only a minute or two.

This concept of balance between outage time and recovery time will be covered in more detail in *Chapter 18*, *Disaster Recovery*. To achieve HA, it is important to consider even the most basic components, as will be covered next.

Operation of Hardware

When considering how robust your architecture is, it's easy to overlook its most basic component—electricity. No piece of technology can work without it; even a mobile device has a battery. If you want to ensure availability for systems that rely on electricity, you must provide redundancy for power to counter power outages. This is typically done through the use of various forms of **uninterruptible power sources** (**UPSs**), which are electrical devices that deliver emergency power to one or more computers when the main power source fails. A UPS may come in various sizes, from small ones that can provide backup power to a single computer for a few minutes to large ones that can power an entire data center for days at a time. UPSs are the first line of defense against a power interruption, but they usually only last for as long as the batteries contained within the UPS can provide power. For an extended outage, another form of backup power (such as a generator) is required.

As alluded to earlier, most things in cybersecurity are a function of cost and tolerance. Not every component of a system requires 100% availability. In the event of a failure, some components can be replaced within hours or days with an acceptable impact on the business. This is where a component's warranty and support coverage come into play.

As was discussed in *Chapter 3*, *Security Policies and Business Continuity*, under **business impact analysis (BIA)**, once you understand the manifold systems and components your organization relies upon and their respective dependencies, **Maximum Tolerable Downtime (MTD)** and **Recovery Time Objective (RTO)**, you will be prepared to select the appropriate level of warranty and support coverage as commensurate with the criticality of those assets. The level of coverage primarily refers to how quickly the inoperative component is repaired or replaced. Typically, the faster the corrective response, the more expensive the item's original cost and ongoing support cost.

Transmission Media

One of the often overlooked components of physical security is the networking cable plant. This refers to the transmission media that makes up the **local area network (LAN)**. If a would-be attacker can see all of a company's LAN traffic via a network tap or network port span, they can gather quite a bit of information to enable further compromise. The transmission media to be protected does not just include copper, coaxial, and fiber cabling; it also includes wireless transmission media such as Wi-Fi and microwave transmissions.

A detrimental compromise of the company's networking cable plant could be through the **intermediate distribution frame (IDF)** or a **main distribution frame (MDF)** closet. Since these rooms typically have limited access, they are hard for attackers to get into. However, for the same reason, it might also be very difficult to detect a network tap in these rooms as they can go months or even years without a visitor. For a more complete discussion on how to protect transmission media, see *Chapter 9*, *Facilities and Physical Security*, and *Chapter 19*, *Business Continuity, Personnel, and Physical Security*.

Network Access Control Devices

When a new component, such as a computer or an IoT device, connects to a network, it has the potential to introduce security vulnerabilities. A computer that is not robustly protected could introduce a virus, or even allow an attacker into a network. **Network access control (NAC)** is a solution that uses technologies and protocols to ensure that when new nodes are added to a network, they conform to its security needs.

Initially, NAC began with the creation of **IEEE standard 802.1X**, but it has since been extended by various vendors to increase its functionality. At a high level, NAC begins working from the moment a device connects to a network. The network device interrogates the newly joined endpoint and requires authentication credentials to allow access to the network. The network device in this case can be a router, a switch, or a wireless access point. The authentication credentials can be a number of things, such as a MAC address, a digital certificate for the device, a username and password attached to the user who is using the device, or the security posture of the device.

At a minimum, NAC requires three main components: a supplicant, an authenticator, and an authentication server. A supplicant refers to a client, an endpoint, or any entity that wishes to be authenticated and granted access to a resource—in this case, the network. The role of a supplicant can also be played by installed software on the client device that can gather the security posture of the endpoint. Supplicant software is pre-installed on most common operating systems and only needs to be enabled and configured.

The authenticator (usually a network device) is the gatekeeper to the network. It can prevent access to the network by ensuring that only authenticated users connect. It can also move users to different VLANs based on role or authentication status. By using and modifying **access control lists (ACLs)**, it can change the access for different users based on security policies and roles.

The role of the authentication server is typically played by an **authentication, authorization, and accounting (AAA)**, **Remote Authentication Dial-In User Service (RADIUS)** or **Diameter** server. The authentication credentials and/or security posture of the device collected by the authenticator and forwarded to the authentication server determines what network to place the device in and what access restrictions to place on that device. This "policy" is then passed back to the authenticator to be enforced on the supplicant device.

Figure 11.1 shows the three main components (a supplicant, an authenticator, and an authentication server) and how they relate to each other.

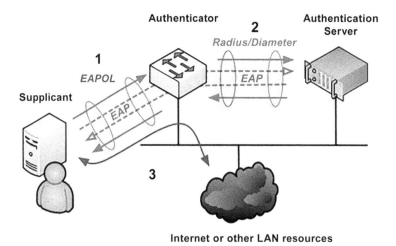

Figure 11.1: NAC architecture

Access to the network via NAC can be restricted on a number of criteria, such as user or device identity, device security posture, or user or device role. User identity can be established with a username and password combination or a digital certificate. Device identity can be established using the device's **media access control (MAC)** address.

Some examples of user roles that could be restricted via NAC are the common guest access role, a contract employee role, or an employee's job role in the organization such as finance or HR. NAC networks can be configured to enforce access to only resources that a particular user's job role requires and no one else. Similarly, devices can be restricted based on their role in the network, such as printers, IP phones, IP cameras, or other IoT devices. Access to these devices can be limited to just the networks and resources they require to fulfill their function and no more.

Additionally, NAC can be configured to verify that a security posture is being upheld before granting access to the network. A few ways in which the posture status can be determined are by the version of the antivirus scan engine ID, when the last full computer antivirus scan was performed, and whether the device was found to be infected. A connecting device, when found to be out of posture, can be handled in a number of ways, with the most obvious being to block all access.

Organizations, however, tend to opt for a more graceful approach, which is to grant what they call "quarantine" access where the user can self-remediate their posture and then re-authenticate and be granted full access. Quarantine access allows limited access to the network to reach resources that would allow the user to update their antivirus scan engine, perform a new full disk scan, or do whatever is necessary to bring their device into compliance.

There are several popular use cases for NAC; generally, they fit into two categories: role-based restriction and incident response. We have already touched on a few examples of role-based use cases such as guest access restriction, device-based restriction, and job role-based restriction. Another popular role-based use case that we have not discussed yet is the **bring-your-own-device** (**BYOD**) access. This use case involves users bringing their own devices and accessing the network or some parts of the network without having to be on an authorized device.

One of the newest use cases for NAC is incident response quarantining in real time using the network. In this scenario, if malware is detected on an endpoint, that endpoint and any others that have interacted with it can be quarantined before the malware spreads to the entire network. This is exciting because it extends a security professional's ability to control a possibly risky endpoint.

Endpoint Security

An endpoint is any device that accesses the network but is not part of delivering network services. Rather, it is only a client of the network services. Some examples of endpoints include desktops, laptops, smartphones, tablets, workstations, or any IoT device. Endpoints represent a major threat vector for incoming attacks. You can consider them the soft underbelly of an organization's network, which makes endpoint protection a critical aspect of security.

The endpoint security space has changed since the 2010s, evolving from basic antivirus software into more comprehensive defense systems. While basic antivirus marks the initial stage of endpoint protection, a more comprehensive defense system is necessary because viruses are not the only threats that endpoints encounter. Such a defense system includes next-generation antivirus, an **endpoint protection platform** (**EPP**), **endpoint detection and response** (**EDR**), **extended detection and response** (**XDR**), **mobile device management** (**MDM**), **data loss prevention** (**DLP**), and other considerations to face evolving threats.

EPPs are software installed on endpoint devices to protect endpoints against file-based malware attacks. They work by detecting, investigating, and remediating malicious activity and responding to dynamic security incidents and alerts. The terms EPP and EDR can be used interchangeably, but most vendors will prefer one over the other. The main difference between these two and XDR is the use of artificial intelligence and machine learning to predict and detect attacks and minimize false positives. XDR is discussed in further detail along with EDR in *Chapter 16, Planning for Security Operations*.

As you have seen here, there are some basic, though sometimes easy to overlook, concepts that you need to get right to ensure your IT architecture is robust. It is important to keep in mind how to maintain reliable power supplies and look after cabling. It is also important to recognize and mitigate security issues caused by devices joining the network using NAC and endpoint security concepts. With this secure base, it's now time to consider communication channels.

Secure Communication Channels

In the previous chapter, you looked at network architecture and how network protocols communicate, but if you are tasked with securing a network, you also need to know the risks that they introduce and how to mitigate them. This section focuses on some of the critical design aspects of building secure communication channels such as voice, multimedia collaboration, remote access, and virtualizing networks. You will look at remote access connectivity and the security ramifications of multimedia collaboration and third-party connectivity.

Voice

The most important concerns with regard to securing voice communications over a network are confidentiality and availability. As with traditional voice calls, the chief threat to the confidentiality of **voice over IP** (**VoIP**) calls is eavesdropping. While traditional voice communication requires access to the cable that carries the voice signal, **VoIP** only requires the ability to intercept the traffic that passes over a network. Eavesdropping on a traditional voice call requires physically tapping into the copper wire carrying the signal, whereas eavesdropping on a VoIP call can be done from across the world by means of a person- (or man-) in-the-middle attack.

The second area of concern for voice security is the availability of the system. This is usually undermined by attacking the infrastructure that sets up and delivers the calls—both the signaling traffic (usually **Session Initiation Protocol**, or **SIP**) and the bearer traffic (**Real Time Protocol**, or **RTP**). One of the most common ways to affect the availability of a VoIP network is by means of a **denial of service** (**DoS**) attack. This attack is easy to carry out since voice traffic is notoriously sensitive to latency, delay, and jitter. If your voice network is starved of resources, it may begin to drop calls and interrupt service.

The primary countermeasure for protecting VoIP against eavesdropping is encryption. Another important countermeasure is network segmentation, which protects against DoS attacks and helps protect availability. This entails separating voice VLANs from data VLANs and controlling which devices can join them, which makes it more difficult for an attacker to join a VLAN where voice traffic is found in order to execute an attack.

If your budget allows for it, designing networks without **single points of failure** (**SPoFs**) is also an effective method to protect VoIP availability. One of the ways to affect the DoS attack is to identify an SPoF and neutralize that point—for instance, by introducing redundant components such as switches, or increasing the diversity of network paths. If a component fails or a path is attacked, the redundancy can kick in.

In the context of VoIP, where uninterrupted communication is crucial, designing without SPoF helps maintain service continuity. Even if one component or link fails, redundant systems ensure that voice calls can continue without interruption or minimal downtime.

Multimedia Collaboration

Multimedia collaboration encompasses the applications that professionals use to work and collaborate with others. These include applications that allow real-time voice and video communication as well as screen sharing and remote control of desktops. Some examples are *Zoom*, *WebEx*, *Google Meet*, and *Microsoft Teams*. In addition to remote meeting applications, there are also instant messaging applications such as *Slack* and *Skype*.

These applications have their own security challenges that require constant vigilance. A common vulnerability is inadvertently allowing uninvited members to join and participate. On the surface, this may seem fairly innocuous but can prove to be quite dangerous in modern-day meetings, such as board of directors meetings that discuss sensitive financial information or public safety/policy meetings that have topics that are not fit for broad public consumption. Keep in mind the kind of information that is shared during these meetings and remember your duty to protect the confidentiality, integrity, and availability of your organization's information. The primary protection against unwanted information disclosure during multimedia collaboration meetings is awareness and training.

Remote Access

Remote access technologies allow users to connect to networks or machines on other networks over the internet. In most cases, a secure connection is vital to protect data flows and stop attacks, so using the right technologies is important when considering remote access in your architecture. The most commonly known access technology is **virtual private networks** (**VPNs,**) which cover several different technologies, including various forms of desktop virtualization such as Citrix and **Remote Desktop Protocol** (**RDP**) and old remote access technologies such as **Secure Shell** (**SSH**).

VPNs facilitate secure communication between parties over insecure networks such as the internet. They use encryption and are typically set up through **Internet Protocol Security** (**IPSec**). VPNs can also be built over **Secure Sockets Layer** (**SSL**) tunnels, **Point-to-Point Tunneling Protocol** (**PPTP**), and **Layer 2 Tunneling Protocol** (**L2TP**). VPN tunnels can be built between two network devices, such as routers or firewalls, but most remote access VPNs are built between a network device, such as a VPN concentrator or firewall, and client software installed on an endpoint. Authentication and authorization of remote access VPNs are typically handled by a **RADIUS** server, sometimes using an external authentication store such as **Active Directory** (**AD**), or a **Lightweight Directory Access Protocol** (**LDAP**) server.

Desktop virtualization technologies encompass any technology that allows you to view and manipulate the user interface of a remote desktop from your own machine. Desktop virtualization is not a new concept; it was done using X Window System forwarded over SSH tunnels many years ago and is now done using many other secure protocols. All desktop virtualization technologies transmit screen changes over a secure channel (i.e., some encrypted protocol).

Desktop virtualization technologies can be separated into two distinct categories: those that connect to an actual desktop and the simulation of a virtual desktop workspace. **Virtual desktop infrastructure** (**VDI**) involves simulating desktop environments within virtual machines on a central server, which users access over a network. In contrast, connecting to an actual desktop typically involves using protocols such as RDP or **virtual network computing** (**VNC**). However, with the advancements in virtual machine technology, the difference between these approaches is often more about terminology than functionality.

Data Communications

This section will dig a little deeper into data communications: **network sockets**, **remote procedure calls** (**RPCs**), and backhaul networks including satellite networks. These two forms of data communication (network sockets and RPC) apply to the entirety of the telecommunications hierarchy, from the core or "backbone" network (global **internet service providers**, or **ISPs**) to backhaul networks.

The backhaul aspect of the telecommunications hierarchy refers to the intermediary connections between the backbone network and the smaller subnets at the *edge* of the network (i.e., private networks, smaller regional ISPs, etc.). A backhaul might include wired, fiber optic, and wireless components including satellite networks. The decision of which backhaul technology is used must consider such factors as capacity, cost, reach, and the need for resources such as frequency spectrum, optical fiber, wiring, or rights of way. In short, the geography between the two points to be connected in the options at hand will ultimately dictate how the backhaul is connected.

As you may remember from *Chapter 10, Network Architecture Security*, network sockets are an OSI layer 4 (transport layer) concept. They are composed of four components: a source address, source port, destination address, and destination port.

When a user initiates a connection to a remote server, the machine generally knows the destination IP address and the destination port associated with the application desired, such as port 80 or 443 for web servers. Your machine knows the source address, which is its own, and for an outbound connection, it randomly chooses an unused source port from the range of 1024 to 65535. The established point-to-point connection between your machine and the remote web server is called a network socket.

You can see these connections on your machine using the netstat command, which comes preinstalled on **Windows, Linux,** and **macOS.** Commands such as netstat are invaluable for cybersecurity professionals to monitor what connections are coming into or leaving a given host for two reasons: first, to validate that the connections on a host are meant to be there and don't represent a non-authorized back channel from an attacker, and second, to validate that those connections are all secure and encrypted if applicable.

A **remote procedure call (RPC)** is an **OSI layer 5 (session layer)** concept. RPC enables distributed computing, meaning that it allows you to do part of the work on your local machine and distribute part of the work to a remote machine by sending instructions over an RPC communication. For example, if an organization has a server running an application that performs accounting functions but does not perform the tax calculation locally, that calculation portion is sent to a server that has a complete database of local tax rates. In that situation, when the local server needs to calculate taxes for a specific location, it sends the specific geographic location and the transaction amounts via RPC to the remote server. The remote server sends the calculated output back to the local server over the same RPC channel. RPC communication is typically used in a private network environment; if transport over a public network is necessary, it is usually tunneled over some type of encrypted protocol such as IPSec or SSL.

Virtualized Networks

Virtualizing networks involves the separation of hardware and its functions. Specifically, a virtualized network implies that the functions of networking hardware (a switch, a router, a firewall, or a load balancer) are being performed in a virtual environment by software. The virtualization of these networks is referred to as **network function virtualization** (**NFV**), which is closely related to **software-defined networking** (**SDN**). Nowadays, enterprise networks are increasingly being virtualized and/or placed in the cloud, so it is crucial that cybersecurity professionals have a clear understanding of the process of virtualization.

Virtualizing network functions is not just reserved for cloud operations; it can also be done in the enterprise network. As long as you understand a physical device and how it operates, you will be able to understand how its virtualized version operates. A disadvantage of a virtual device is that you don't have the benefit of a physical representation of the device through a visible interface, such as a physical firewall where you can check that the right leads are connected to the right ports just by looking at it. This means that it's easier to make a configuration error that leaves a security vulnerability open.

Like virtual machines, the software for virtual network devices runs in a **hypervisor**, which is a program used to run and manage one or more virtual machines on a computer, as covered in *Chapter 8, Architecture Vulnerabilities and Cryptography*. The hypervisor enables control over these virtual network devices and is thus often the focus of attackers. Therefore, it is also the chief priority of the cybersecurity professional to thoroughly understand how hypervisors work and how to secure them. One of the greatest vulnerabilities in hypervisors is software updates not having been applied and thus not including the latest security patches. Accidental misconfiguration can also create issues if it doesn't adequately control access.

Third-Party Connectivity

The modern-day organization is no longer a standalone entity but rather an ecosystem of business and security relationships. History is replete with examples of vulnerabilities exposed through third-party relationships, not the least of which is the Target breach. In 2013, cybercriminals compromised a third-party vendor's credentials to access Target's network. The criminals then installed malware on Target's **point-of-sale** (**POS**) systems to capture credit card data, leading to significant financial loss and legal implications for Target.

Utilizing third-party vendors can be an important part of a business's operations, as they bring expertise that might be too expensive to have in-house. However, allowing a third party to have access to your systems can be risky as you do not have the same knowledge of their information architecture or security practices and could be introducing a vulnerability, such as the Target example.

You can mitigate the risk posed by third-party relationships through the time-tested security tenets of **least privilege** and **separation of duties (SoD)**. These tenets aim to limit the damage that can be caused by a single user. The principle of least privilege limits damage by ensuring that a user only has the privileges necessary to do their job and no more. Similarly, SoD entails making sure that no single person can perform critical job functions without the oversight provided by another person. For instance, a person who writes or prints checks should not also be able to authorize and/or sign them.

To ensure security, SoD can be combined with NAC. For instance, if you have to allow third-party business relationships access to your network, grant them access to only what they need to perform their function. This setup can further be enhanced by creating networks that contain only the resources they need access to while the other parts of the network remain inaccessible. NAC can assign third-party users to VLANs based on their role and identify and limit their access through firewall policies.

Last but certainly not least, remember the concept of threat modeling from *Chapter 4*. Threat modeling is the act of walking through every possible negative scenario that could occur in a virtual environment as a result of third-party connectivity. Then, the security professional can model the ways and countermeasures that would mitigate the identified risks. This type of effort should become second nature to experienced cybersecurity professionals whenever they are presented with a new situation. It should become automatic to consider the risks and, simultaneously, the mitigation of those risks.

Summary

This chapter reinforced the core security concepts of confidentiality, integrity, availability, authenticity, and non-repudiation and how they come into play when securing communications channels. It emphasized the importance of securing various network elements, focusing on HA and fault tolerance, NAC, and endpoint security. You looked at the difference between HA and fault tolerance, and how they both aim to ensure network availability but differ in cost and implementation. Fault tolerance offers zero service interruption using redundant hardware, while HA uses a holistic approach combining hardware and software to restore services quickly after a failure, making it more cost-effective. The chapter also highlighted that power redundancy, warranties, and support are crucial in maintaining robust network architecture, with components such as UPSs providing backup during power outages.

You also looked at network security components, including transmission media, NAC devices, and endpoint security. NAC was presented as a critical tool in controlling the security posture of devices joining a network, employing methods such as IEEE 802.1X standards, authenticators, and authentication servers to enforce access control. Endpoint security, a vital aspect of network defense, has also evolved beyond basic antivirus solutions to include comprehensive systems such as EPPs and XDR.

Lastly, the chapter explored secure communication channels, particularly for voice, multimedia collaboration, and remote access. It highlighted the security challenges associated with VoIP systems, such as eavesdropping and DoS attacks, and suggested encryption and network segmentation as countermeasures. Multimedia collaboration tools such as Zoom and Microsoft Teams require vigilance to prevent unauthorized access, while remote access technologies such as VPNs and desktop virtualization ensure secure connectivity over the internet.

In the next chapter, you will look at how to secure agents in your system by looking at identity, access management, and federation.

Further Reading

- Mesic, M., and M. Golub. *An Overview of Port-Based Network Access Control*, Proceedings of the Information Systems Security, MIPRO 2006 (2006).

- Slate, Sam. *Endpoint Security: An Overview and a Look into the Future*, Lat. Am. Polit. Hist (2018): 9780429499340-15.

- Keromytis, *Angelos D. A comprehensive survey of voice over IP security research*, IEEE Communications Surveys & Tutorials 14, no. 2 (2011): 514-537.

- Cziva, Richard, Simon Jouet, Kyle JS White, and Dimitrios P. Pezaros. *Container-based network function virtualization for software-defined networks*. In 2015 IEEE Symposium on Computers and Communication (ISCC), pp. 415-420. IEEE, 2015.

- Keskin, Omer F., Kevin Matthe Caramancion, Irem Tatar, Owais Raza, and Unal Tatar. *Cyber third-party risk management: A comparison of non-intrusive risk scoring reports*. Electronics 10, no. 10 (2021): 1168.

Exam Readiness Drill – Chapter Review Questions

Apart from a solid understanding of key concepts, being able to think quickly under time pressure is a skill that will help you ace your certification exam. That is why working on these skills early on in your learning journey is key.

Chapter review questions are designed to improve your test-taking skills progressively with each chapter you learn and review your understanding of key concepts in the chapter at the same time. You'll find these at the end of each chapter.

> **How to Access These Materials**
>
> To learn how to access these resources, head over to the chapter titled *Chapter 24, Accessing the Online Resources*.

To open the Chapter Review Questions for this chapter, perform the following steps:

1. Click the link – `https://packt.link/chapter11`.

   Alternatively, you can scan the following **QR code** (*Figure 11.2*):

Figure 11.2: QR code that opens Chapter Review Questions for logged-in users

2. Once you log in, you'll see a page similar to the one shown in *Figure 11.3*:

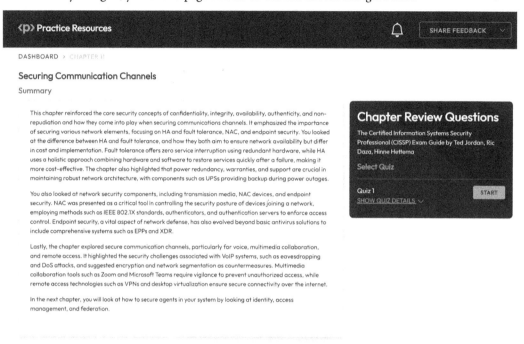

Figure 11.3: Chapter Review Questions for Chapter 11

3. Once ready, start the following practice drills, re-attempting the quiz multiple times.

Exam Readiness Drill

For the first three attempts, don't worry about the time limit.

ATTEMPT 1

The first time, aim for at least **40%**. Look at the answers you got wrong and read the relevant sections in the chapter again to fix your learning gaps.

ATTEMPT 2

The second time, aim for at least **60%**. Look at the answers you got wrong and read the relevant sections in the chapter again to fix any remaining learning gaps.

ATTEMPT 3

The third time, aim for at least **75%**. Once you score 75% or more, you start working on your timing.

> **Tip**
>
> You may take more than **three** attempts to reach 75%. That's okay. Just review the relevant sections in the chapter till you get there.

Working On Timing

Target: Your aim is to keep the score the same while trying to answer these questions as quickly as possible. Here's an example of how your next attempts should look like:

Attempt	Score	Time Taken
Attempt 5	77%	21 mins 30 seconds
Attempt 6	78%	18 mins 34 seconds
Attempt 7	76%	14 mins 44 seconds

Table 11.1: Sample timing practice drills on the online platform

> **Note**
>
> The time limits shown in the above table are just examples. Set your own time limits with each attempt based on the time limit of the quiz on the website.

With each new attempt, your score should stay above **75%** while your "time taken" to complete should "decrease". Repeat as many attempts as you want till you feel confident dealing with the time pressure.

12
Identity, Access Management, and Federation

One of the most crucial things to control to keep an IT system safe is who has access to it. Without strict controls on who can access a system, all other security measures are almost useless. If just anyone can access your data, nothing is secure. At the same time, for IT systems to be effective, people and other agents, such as apps, need to be able to efficiently access the data needed to fulfill their functions. Striking the balance between the security of data and the availability of data is the core of identity and access management.

Controlling access means having barriers that keep the bad agents out and the good agents in, both for physical and digital systems. Controlling identity means being confidently able to tell the two apart. Once people have gained access to systems, they should be able to efficiently access the right data and systems. Because people (and other agents) might have to access more than one system, often, while carrying out one task, they use systems of shared identities and access; this means they only have to log in once. This process is known as federation.

By the end of this chapter, you will be able to answer questions on the following:

- Critical physical and technical controls, and how to secure them
- Forming security practices depending on the corporation's environment
- Performing accounting on standard user and system access
- Provisioning and deprovisioning identities securely

The general term for a user is a *subject*, and the general term for a file or record is an *object*. In the next section, you will learn about securing the subject's access to their objects or data.

Securing Access Control

A major role of an organization's CISO, CSO, and CISSP is to protect the organization's assets. This section will discuss important concepts for protecting object access, and where to apply these access controls for the best cybersecurity protection.

Securing Data and Information

Securing data starts with user authorization and identity protection. This data, or assets, can be accessed by individuals through either technical or physical systems. Examples of physical systems include doorways, and displaying an identification badge to a guard before parking your car or entering a building.

Technical systems, on the other hand, are those that require a login name and password such as a computer, network-based router, or application. Access control systems are managed through either a centralized system such as **Lightweight Directory Access Protocol (LDAP)** or active directory, or a decentralized system where owners manage their own systems.

The downside of decentralized systems is that the users' login names and passwords can differ from system to system. For example, a single user might set their Facebook username to `Oscar` and password to `PW1` but their Twitter/X username to `Oscar123` and password to `PW2`, and so on for possibly tens of different usernames and passwords that this user needs to manage. Identity becomes very cumbersome, and so one can see the advantage of a centralized system in which a single administrator manages all usernames and passwords.

However, a major disadvantage of a centralized system is the **insider threat**. When an insider turns rogue, thousands of users in a huge organization may not trust the integrity of their data. If all members of the company have access to all systems, this increases the risk of insider threats, so access management becomes key. Even if access is restricted, there is still the risk that employees with access to critical information, such as company accounts or payroll, could go rogue. To reduce the risks of incidents such as embezzlement and corporate espionage, some organizations will employ policies such as job rotations, separation of duties, and mandatory vacations to ensure that all employees can have their work audited.

Most organizations operate a hybrid approach and manage and secure both decentralized and centralized systems depending upon the circumstance. In general, decentralized systems are non-production (i.e., not used in production), whereas centralized systems are production systems set up for day-to-day use. For example, systems in a test/sandbox environment are decentralized because these are not involved in day-to-day operations (i.e., non-production).

Critical company applications used by most of the staff (such as email, file sharing, web browsing, etc.) are examples of centralized systems.

Access Controls

An organization's assets not only include physical assets such as cash or machinery but also intangibles such as sales data, design blueprints, or client tax identification numbers. The first line of protection against attacks on these assets is access control. Management determines the value of assets, and the CISO/CISSP makes recommendations to protect them. For example, if the management team decides to protect customer tax identification numbers, the CISO/CISSP will recommend implementing usernames and passwords and a PIN to unlock the door to the server room.

Depending on whether the organization follows NIST standards or ISO standards, access controls are called either technical or physical (ISO calls the technical standards logical standards). In general, the main difference is that a physical system is one that a person can touch, such as a lock or a key, whereas a logical system usually requires a username and password or PIN and some kind of electronic data processing executed in the background. Examples of each are shown in *Table 12.1*:

Logical Access Controls	Physical Access Controls
Firewall	Standard door lock and key
One-time password	Turnstile
Logon and password	Mantrap
Retina scanner	Gate

Table 12.1: Examples of logical and physical access controls

Device Types

There are varying types of hardware and software devices available to users to allow them to access buildings or data. Some logical devices include tokens or PINs combined with biological information, such as a fingerprint or retinal scan to enhance security.

Many organizations also provide visitors with **personal identity verification** (**PIV**) cards to grant access to permitted areas of the building or campus. Processing the access of such visitors is called **visitor management** and involves vehicle permits, which require visitors to obtain a parking pass or even a vehicle pass to drive on the campus. Larger organizations even handle traffic management, employing their own police force that issues drivers parking and speeding tickets if they violate the rules. To obtain the automobile pass, the visitor must provide their name, license plate, and insurance information. Visitor management then provides a pass with an expiration date.

There is also **identification management**, which is handled by visitor management. The visitor provides their name and signature, who they are visiting, their arrival time, and their license plate number on a sign-in sheet so that the organization has those details on record. The departure time is also recorded when the guest leaves the facility.

Finally, intrusion detection devices such as alarms and video cameras are installed in the facility to manage visitor ingress and egress. These alarms are tripped anytime a visitor's PIV card or PIN fails, or if an attacker attempts forced entry, as they are disallowed. A record of failures is recorded in a log and managed by on-site security staff.

Identity Provisioning

Though denying access is an effective way of protecting an asset, security cannot be so restrictive that employees and contractors are unable to do their jobs. The process of granting, managing, and deactivating access to systems is known as **identity provisioning**. For the best balance of security and production, the CISSP, CISO, or CSO will recommend that users have exactly the privileges required to perform their job role, and only those. This concept is called **least privilege**. Users with higher privileges (also known as superusers) require access to restricted areas to do their jobs. The best security option for either system is to provide the least privilege.

Finally, as enrollment of new staff and contractors occurs, access to systems they require needs to be granted, and when that person leaves the company, their access should be revoked. The next section will discuss how to account for these users and superusers and ensure that the users are provisioned and deprovisioned from the organization properly.

Standard User and Superuser Accounts

Normal or standard account access reviews must be done within organizations to ensure users only access files necessary for their jobs. Reviews should either be scheduled or randomly performed within the user life cycle to help discover security vulnerabilities such as poor passwords, improper access to departments due to privilege/authorization creep, and privileges that exceed current rights. When such vulnerabilities are found, they can either be mitigated or removed. If the status of the employee has changed, there are some standard account actions that should be taken to ensure the least privilege, as follows:

- **Delete**: Delete the user's account if the employee has already left the company
- **Modify**: Modify the user account of a temporary worker whose contract has changed
- **Restrict**: Restrict the user's account if the user has changed jobs, resulting in unneeded privileges in past departments

Superusers or system accounts include names such as Administrator or Root and are very advantageous for hackers because these accounts have unrestrictive privileges. These accounts can do the following:

- Create, update, or delete any file

- Insert or delete firewall rules

- Start or stop any job

- Create or delete user accounts

Attackers might target a superuser's accounts not only because they have more access but also because it allows them to actually control access. For example, it allows them to modify firewall rules that grant them further access to the system and start background processes to download top-secret files. Then, when the attacker completes their attack, they can even change log files to delete records of their nefarious activity.

> **Note**
>
> Attackers target the administrator account on Microsoft Windows systems, which is userid 500, or the domain admins group account of userid 512. On Linux and Unix systems, attackers target the root account, which is userid 0.

For security, many systems administrators rename the administrator and root accounts to make it harder for hackers to exploit these accounts. Another way to ensure security is to make sure that system administrators only use superuser accounts for what they are intended to do. For instance, if an administrator needs to add a new user or change a password, they temporarily switch from a standard account to an administrator account to perform that specific function. Once complete, they immediately exit, returning to the account of a standard user. This also helps prevent mistakes with a significant impact, such as shutting down a system while 100 users are working on it when they only intended to shut down a single virtual machine instance. When administrators operate with standard accounts for routine tasks, the chances of such critical errors are minimized.

In addition to system administrator accounts, there are service accounts that have extended privileges as well, and these need to be monitored. Service accounts as displayed on a Linux system are shown in *Figure 12.1*:

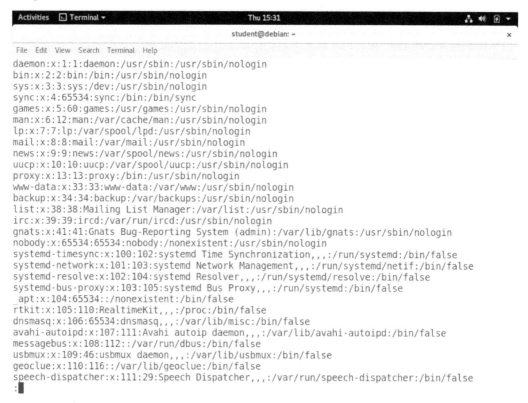

Figure 12.1: Service account listing on a Linux system

Each line consists of seven columns separated by colons. The first column is the service name, and the third column lists the user IDs. Service accounts manage printers, video, email, games, backups, and so on as background jobs. On modern Linux systems, service account user IDs go from 1 to 999. This file is crucial for user management and system administration, providing essential information about each user account on the system.

User Account Setup and Removal

The setting up and destruction of user accounts is called provisioning and deprovisioning, and this process determines which objects users have access to. The life cycle of user accounts from when an employee or contractor is given privileges to the system up until they leave the organization is controlled by **identity and access management (IAM)**.

When a new user is provisioned, they obtain their user ID and required privileges to do their jobs. Those privileges can also be modified later if their role changes. For example, if they get promoted, they may need additional rights, or if the user leaves a specific department, privileges received from that department will need to be removed. Finally, an account is deprovisioned when the user leaves the organization and their user ID is deleted. The person taking over their job will then be given access to those objects.

The CISSP, working with HR, recommends a specific process for creating (provisioning) and removing (deprovisioning) user accounts to ensure optimal security. This process might include an identity manager, which is a role or a system responsible for managing user identities and their access rights within an organization.

For example, the steps for provisioning a new user account or modifying an existing account could be as follows:

1. The user completes a request form to access the objects for their role.
2. The identity manager validates the specific access needs of the user.
3. The systems administrator creates the account for the user.
4. The systems administrator notifies the user that the account has been created.
5. When a user needs to modify their account, the user completes a request for account modification and is notified once the change is complete.
6. Finally, to deprovision an account, the systems administrator is informed that the user no longer requires access and then removes that user's account.

Identity and Authentication

Having access policies and controlling user accounts is only a part of the picture. There are several hardware and software technologies designed to grant authentication and authorization for users, networks, systems, and services. The CISSP must understand the technologies available for their organization and how to secure them to mitigate threats.

Components of Identity Management

The four main components of the **identity management** (**IdM**) process are identification, authentication, authorization, and accountability. Identification could be as simple as the user ID, which defines the person. For the person to prove who they are, systems add authentication such as passwords or PINs. Authorizations are the rights granted to the user, for instance, read or update permissions. Finally, accountability is the process of recording the activities of the user in logs, also known as **logging**. *Figure 12.2* shows the four stages of identity management:

Figure 12.2: Components of identity management

Authentication Methods

Authentication is the process of verifying the identity of a user or device before granting access to a system. There are various methods of authentication, each providing different levels of security. In the past, authentication methods have relied on a single factor, but combining multiple factors can enhance security.

Older authentication methods are still used today and include systems that ask the user to prove their identity by entering a password or PIN. This type of authentication is called *something-you-know* authentication. Other examples of *something you know* include your mother's maiden name, birth date, where you met your spouse, and the name of your first pet.

With this authentication, the user inputs what they know; a password, PIN, mother's maiden name, or first dog's name are all considered *something-you-know* factors. In general, this information gets saved as a hash and not as the word itself. This way, the systems storing this information do not know the exact word either. They just know the hash.

For example, when the user inputs a password, what actually gets saved is something such as 1@2#3$4%5^6. The system validates the user because, when they take that user's password and put it through the hashing algorithm, the same hash appears (1@2#3$4%5^6, in this example). When the hash matches, the system knows that the user has been properly identified. If a system is set up properly, other information the user knows (such as their mother's maiden name or the last four digits of their tax ID number) is also hashed.

Authenticating with a device that you hold or carry, such as a smart card or a one-time password, is known as *something-you-have* authentication. So, a user that has a card to withdraw money from an ATM combines this with *something-they-know* authentication to validate their identity.

Another example of *something-you-have* authentication is the **one-time password** (**OTP**) device or phone app. An OTP token, shown in *Figure 12.3*, generates a different 6-digit value every 30 seconds; this is algorithmically generated based on a shared secret with a system or website. The value is entered into the system the user is trying to access and, if correct, is validated:

Figure 12.3: OTP token

Other examples include USB security keys with saved passwords, shown in *Figure 12.4*, and **Common Access Card** (**CAC**) smart cards, which contain a user's credentials, fingerprints, and access certificates. Even a key to a building is considered a *something-you-have* type of device.

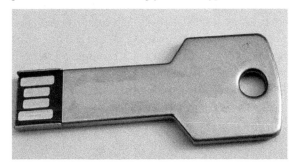

Figure 12.4: USB security key

Finally, when a user logs in using their fingerprint or facial recognition, this type of authentication is called *something-you-are* authentication. Many modern computers and phones can use fingerprints and facial recognition instead of passcodes. *Figure 12.5* shows a fingerprint scanner. Other *something-you-are* devices include iris pattern scanners, facial image technology, hand geometry, voice recognition, eye readers, retinal scanners, signature dynamics, and palm vascular patterns. This could even include *something* performance-based, such as typing speed or writing dynamics.

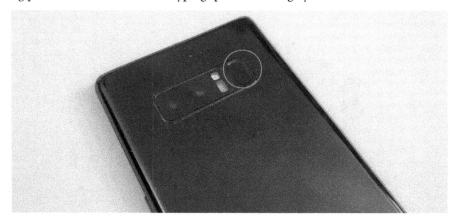

Figure 12.5: Fingerprint scanner

Single-factor authentication uses just one of the aforementioned methods to authenticate and is relatively secure. But what if someone steals an identity card? What if a hacker learns the user's password, or an attacker gets a copy of someone's fingerprint image? Then, the attacker can authenticate as the victim. To address the risk of the primary authentication being compromised, the user needs to combine different authentication methods. For example, an ATM card can be combined with a password or PIN, or the key to an office can be combined with a facial scanning system. This makes it much more difficult for an attacker to gain access to secured information. Combining two of these authentication factors is known as dual-factor or **two-factor authentication** (**2FA**). Combining *something you know* with *something you have* (as in the previous ATM example) or *something you have* with *something you are* (as in the office key example) are two examples of 2FA.

> **Note**
>
> For even better identity security, combine *something you know* with *something you have* and *something you are*. This is known as **three-factor authentication** (**3FA**), and, in general, using two, three, or more modes of authentication is known as **multi-factor authentication** (**MFA**). Next, you will look at the different types of authentication in more detail.

Combining multiple methods from the same factor is not MFA. For example, a PIN, password, birthday, and mother's maiden name are still considered single-factor authentication because even though the utility is requesting four different ways to authenticate, because they are from the same category (in this case, *something you know*), this is still considered single-factor authentication.

Authorization

Authorization is used to determine what kind of access a user has. For example, say you have a user named Shivani who has access to a shared document with other people on the team. The authorization Shivani has will allow (or deny) Shivani to modify, read, or even delete the file.

Now, if the system is set up properly, the user is assigned least privilege access, which grants them only enough access to do their work. There is another principle known as **need to know**. For example, user Arhant and user Shivani are both working in the Navy and have Top Secret clearance. Shivani works with submarines, and Arhant works with surface ships. However, top-secret projects for submarines are different from top-secret projects for surface ships. Arhant's top-secret privilege is for surface ships only and not for access to the documents for submarines. Similarly, Shivani does not have top-secret access to the documents for surface ships. This concept is known as *need to know*. They both have top-secret access but neither of them needs to know information for ships outside the scope of their responsibility.

As staff are transferred within organizations from department to department, they might end up having authorization for departments they no longer work for. This is known as **authorization creep**, and management should have systems in place to mitigate this, such as holding annual rights and permission account access reviews.

Session Management

To help protect organizations from man-in-the-middle attacks, it is important to implement cookies or session keys to mitigate session hijacking by a hacker. One such system comes from NIST SP 800-63-3, which gives guidelines for identity proofing, including non-repudiation. In the system, there are three guides: **identity Assurance Level (IAL)**, **authenticator assurance level (AAL)**, and **federation assurance level (FAL)**.

 IALs describe the level of assurance that an individual is who they say they are. For example, you need your birth certificate to obtain a driver's license or passport; without this evidence, the credential will not be granted. This is IAL 2.

IALs go from 1 to 3, the lowest to highest levels of assurance. IAL 1 is a self-asserted identity such as a Gmail address the individual created for themselves. IAL 3 includes biometric collection for non-repudiation, such as fingerprints.

Similarly, **AAL** reflects the level of assurance in authentication; 1 could be a password and 3 could be MFA. **FAL** refers to how secure the communication channel is. Session management is described by listing the three levels; for example, a top-level official such as a president who is sending a secure message to their team requires a high level of confidence. So, high identity assurance would be IAL3. AAL3.FAL3 – IAL3 for best identity proofing, AAL3 because the leader used 2FA including a biometric, and FAL3 because the conversation came via a secure channel.

Another example is a whistleblower working with the press that wants to protect their identity. A session would be described as IAL1.AAL3.FAL3 because IAL1 provides a pseudonym, which means the accuser can use a fake name to protect themselves from retribution in case their employer wants to fire them.

Cloud Authentication

Using cloud authentication, users can access cloud services securely from almost anywhere. Organizations can utilize a credential management system in which a binding is collected by the cloud service provider to increase confidence in the authentication process, by binding, or associating, the user to the authorized credential. This process is made up of five levels: sponsorship, enrollment, credential production, issuance, and credential life cycle management. **Sponsorship** means that an authorized organization owns the credential. **Enrollment** is the process of verifying the individual's identity. **Credential production** is the process through which forms of user credentials such as smart cards and certificates are created. **Issuance** means providing the user with the credential, and **credential life cycle management** ensures proper deprovisioning when a credential expires, or the user no longer works for the organization.

For example, if Jake gets a three-month contract job at MEMEco, the identity manager at MEMEco authenticates Jake into the credential management system in the **sponsorship** phase. During **enrollment**, the identity manager reviews Jake's driver's license or passport to verify his identity. The identity manager creates Jake's company identification card during **credential production** and provides it to Jake as part of **issuance**. When Jake's three-month contract ends, the **credential life cycle management** process deprovisions Jake by collecting his company identification card and closing his accounts. Deprovisioning is a very important step to protect the organization; otherwise, an unauthorized person could gain access to premises unquestioned because they still have their old credentials.

Authentication is essentially proof of identity. This can be something as simple as the password, or include biometrics, electronic credentials, or, very often, combinations of different authentication types. Care should be taken to balance security with practicality.

Summary

This chapter covered user access controls and management and reviewed some measures you can take to protect your organization from attackers using either physical or technical controls to breach the firm. A major role of the CSO and CISSP is to secure identity access with centralized utilities such as single sign-on so that users can do their jobs. Users enhance security by using strong passwords.

Next, you learned the differences between logical and physical controls and the security frameworks that organizations need to use, whether it be NIST, ISO, or another, to ensure they understand the risks of granting user access. Several devices can be used to identify a user for access to an organization's systems. When setting up new users, make sure they are provisioned properly and given only superuser or administrator access if required. Otherwise, they should only be given the privileges to do their job. This is called least privilege. Also, make sure that when an employee or contractor leaves a company, their access to the organization is removed through the deprovisioning process.

Identity can be managed with multiple methods. These include *something-you-know*, *something-you-have*, or *something-you-are* authentication factors. Identity security is enhanced by combining these methods as either two-factor or multi-factor authentication. Although two users might have the same level of security or access, such as Top Secret, that does not automatically give them top-secret access to every project. This is known as *need-to-know* access. As users move through different positions in their organization, system administrators must protect the organization by removing past privileges and granting the required rights to do their jobs; otherwise, users will have more rights than needed. This is known as authorization creep.

The chapter ended with a discussion on the credential life cycle, which is the process of provisioning, securing, and deprovisioning users that helps administrators establish that users are authorized properly, and ensures their credentials are being used appropriately within LAN and cloud systems. The next chapter takes you through identity management implementation.

Further Reading

- *NIST Special Publication 800-63 Digital Identity Guidelines*, https://packt.link/O1IgC, National Institute of Standards and Technology, March 17, 2023

- *FedRamp Digital Identity Requirements* https://packt.link/sXJAF, US General Services Administration, January 31, 2018

- *Frequently Asked Questions (FAQs) for Credentialing Standards Procedures for Issuing Personnel Identity Verification Cards under HSPD-12 and New Requirements for Suspension or Revocation of Eligibility for PIV Credentials*, https://packt.link/5GK2X, United States Office of Personnel Management, December 2020

Exam Readiness Drill – Chapter Review Questions

Apart from a solid understanding of key concepts, being able to think quickly under time pressure is a skill that will help you ace your certification exam. That is why working on these skills early on in your learning journey is key.

Chapter review questions are designed to improve your test-taking skills progressively with each chapter you learn and review your understanding of key concepts in the chapter at the same time. You'll find these at the end of each chapter.

> **How to Access These Materials**
>
> To learn how to access these resources, head over to the chapter titled *Chapter 24, Accessing the Online Resources*.

To open the Chapter Review Questions for this chapter, perform the following steps:

1. Click the link – `https://packt.link/chapter12`.

    Alternatively, you can scan the following **QR code** (*Figure 12.6*):

Figure 12.6: QR code that opens Chapter Review Questions for logged-in users

2. Once you log in, you'll see a page similar to the one shown in *Figure 12.7*:

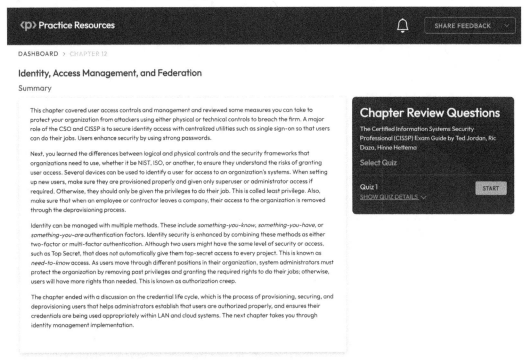

<p> Practice Resources

SHARE FEEDBACK

DASHBOARD > CHAPTER 12

Identity, Access Management, and Federation

Summary

This chapter covered user access controls and management and reviewed some measures you can take to protect your organization from attackers using either physical or technical controls to breach the firm. A major role of the CSO and CISSP is to secure identity access with centralized utilities such as single sign-on so that users can do their jobs. Users enhance security by using strong passwords.

Next, you learned the differences between logical and physical controls and the security frameworks that organizations need to use, whether it be NIST, ISO, or another, to ensure they understand the risks of granting user access. Several devices can be used to identify a user for access to an organization's systems. When setting up new users, make sure they are provisioned properly and given only superuser or administrator access if required. Otherwise, they should only be given the privileges to do their job. This is called least privilege. Also, make sure that when an employee or contractor leaves a company, their access to the organization is removed through the deprovisioning process.

Identity can be managed with multiple methods. These include *something-you-know*, *something-you-have*, or *something-you-are* authentication factors. Identity security is enhanced by combining these methods as either two-factor or multi-factor authentication. Although two users might have the same level of security or access, such as Top Secret, that does not automatically give them top-secret access to every project. This is known as *need-to-know* access. As users move through different positions in their organization, system administrators must protect the organization by removing past privileges and granting the required rights to do their jobs; otherwise, users will have more rights than needed. This is known as authorization creep.

The chapter ended with a discussion on the credential life cycle, which is the process of provisioning, securing, and deprovisioning users that helps administrators establish that users are authorized properly, and ensures their credentials are being used appropriately within LAN and cloud systems. The next chapter takes you through identity management implementation.

Chapter Review Questions

The Certified Information Systems Security Professional (CISSP) Exam Guide by Ted Jordan, Ric Daza, Hinne Hettema

Select Quiz

Quiz 1   START
SHOW QUIZ DETAILS

Figure 12.7: Chapter Review Questions for Chapter 12

3. Once ready, start the following practice drills, re-attempting the quiz multiple times.

Exam Readiness Drill

For the first three attempts, don't worry about the time limit.

ATTEMPT 1

The first time, aim for at least **40%**. Look at the answers you got wrong and read the relevant sections in the chapter again to fix your learning gaps.

ATTEMPT 2

The second time, aim for at least **60%**. Look at the answers you got wrong and read the relevant sections in the chapter again to fix any remaining learning gaps.

ATTEMPT 3

The third time, aim for at least **75%**. Once you score 75% or more, you start working on your timing.

> **Tip**
> You may take more than **three** attempts to reach 75%. That's okay. Just review the relevant sections in the chapter till you get there.

Working On Timing

Target: Your aim is to keep the score the same while trying to answer these questions as quickly as possible. Here's an example of how your next attempts should look like:

Attempt	Score	Time Taken
Attempt 5	77%	21 mins 30 seconds
Attempt 6	78%	18 mins 34 seconds
Attempt 7	76%	14 mins 44 seconds

Table 12.2: Sample timing practice drills on the online platform

> **Note**
> The time limits shown in the above table are just examples. Set your own time limits with each attempt based on the time limit of the quiz on the website.

With each new attempt, your score should stay above **75%** while your "time taken" to complete should "decrease". Repeat as many attempts as you want till you feel confident dealing with the time pressure.

13
Identity Management Implementation

Identity components are important to securing an organization's resources because you only want authorized users to access objects when they need them. For example, if attackers compromise an **identity management (IdM)** system and breach a server room that is generating millions of dollars in monthly revenue, the business will lose the critical functions of paying its staff and delivering products. It is essential to protect these critical functions by safeguarding the organization through deploying an IdM system.

The ability to connect systems has made collaboration between organizations easier, and thus more and more common. It could be sharing online resources to upskill employees or working together on short-term projects – organizations frequently need to grant and manage access to their systems for external parties. In the past, this would involve setting up new accounts and managing separate credentials, which is time-consuming and costly. To streamline these collaborations, **federated identity management (FIM)** allows one party to trust the authentication processes of another.

FIM, often implemented through **single sign-on (SSO)**, enables users to access multiple systems using a single set of credentials, regardless of which organization manages them. This reduces the friction of managing multiple identities and enhances the user experience by simplifying the login process across different platforms. It's now commonplace to use a Google or Facebook account to sign in to various unrelated services.

As organizations adopt FIM, various protocols and standards ensure the security and efficiency of identity management across platforms. This chapter looks at these protocols, exploring how they work and how they are applied in real-world scenarios to facilitate secure, cross-organizational authentication and authorization.

By the end of this chapter, you will be able to answer questions on the following:

- Implementing authentication systems

- Authentication, authorization, and accounting systems

- Access control

You will start by looking at how to implement authentication when you have third parties using your system.

Implementing Authentication Systems

There are many instances in which an organization will work with a third party to achieve a goal. It could be a training partner that shares online resources to upskill staff, online data services such as Azure or Salesforce, or it could be two companies collaborating on a short-term government contract that requires access to each other's shared resources. To make collaborations possible, each organization could set up new accounts with the correct identities and authentication, but this can also be time-consuming and costly. The solution is a system where one party can utilize or trust the authentication of another.

When you have multiple parties accessing your IT systems, such as the staff of a client accessing your learning materials, you can use FIM to allow access using SSO. In short, a single existing account will be used to access multiple systems, including yours. In FIM, multiple organizations agree to trust each other's IdM systems, allowing a user to use a single set of credentials (e.g., username and password) to access resources and services across these organizations. This is done with shared authentication processes and trust established between identity providers and service providers.

One common example is how you can now use your Facebook, Google, or Apple email to log in to multiple different platforms, such as online food delivery services, without having to create a new account.

Access control is done through three popular markup languages: **SAML**, the security assertion markup language, **OpenID Connect**, and **Open Authorization 2.0 (Oauth2.0)**. These markup languages are discussed in detail in the following sections.

SAML

Security Assertion Markup Language (SAML) is an **extensible markup language (XML)**-based framework used for SSO. A bank might offer the service of printing checks for their customers. The bank is not a printing company, but they have a relationship with the printer that prints checks for their customers. Within this relationship, when a customer (user) wants to print a check, they will be taken from their online bank account to a separate service, the printing company, to make the request.

Prior to working together, the printing company provides the bank with its **System and Organization Controls 3 (SOC3)** report, which is an audited report that assures the bank that the printer meets its minimum security standards. it can proceed to establish a secure communication framework using SAML.

In this setup, SAML will use an XML-based framework to describe the security information and exchange protocols that the bank can use to communicate with the check-printing company. SAML uses assertions, XML documents that contain information about a user's authentication, attributes, and authorization. These assertions maintain the security between an identity provider, such as the bank, and a service provider, such as the printing company. The identity provider is responsible for authenticating users and issuing SAML assertions. It maintains user credentials and handles the authentication process. The service provider relies on the SAML assertions provided by the identity provider to grant or deny access to resources or services.

The three roles of SAML are shown in *Figure 13.1*.

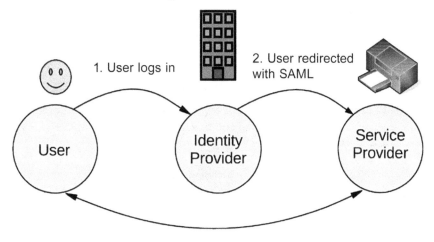

3. User logged in to service

Figure 13.1: The SAML authentication process

The preceding figure shows the three roles involved in SAML:

- **Identity Provider** – in our example, that is the bank
- **Service Provider** – in our example, that would be the check printing company
- **User** – in our example, that is the customer who needs banking and check printing services

SAML is composed of four components. First are the **assumptions** – the XML documents mentioned, which describe how the SAML request and response messages will be exchanged between the identity provider's and service provider's identity systems. Second, **bindings** describe how message exchanges are transported, such as the HTTP redirect response. Third, **protocols** describe which method to use – for example, should they use SOAP or HTTP? Finally, the **profiles** describe the rules for attribute protocols in the bindings of a session.

Open Authorization 2.0

Open Authorization 2.0 (OAuth 2.0) is similar to SAML in that it provides a framework for the customer to work with parties other than the one authenticating their identity. But, OAuth 2.0 is more secure than SAML because of the simplicity of configuring and managing it. Also, OAuth2.0 supports a more granular approach to authorization, leading to stronger security.

Like SAML, OAuth2.0 offers enhanced customer service and sales conversions. For example, a customer might be annoyed that they have to visit another website, create a new account, determine a new password, and so on, to have their checks printed. Also, the check-printing company risks losing business if the bank provides a list of printing companies for customers to call. SSO is more convenient for the customer, and better for business too.

The OAuth service uses HTTP to manage the relationship between a resource owner and a server. It differs from SAML in that there are four roles as opposed to the three in SAML, which are resource owner, resource server, client application, and authorization server, as shown in *Figure 13.2*.

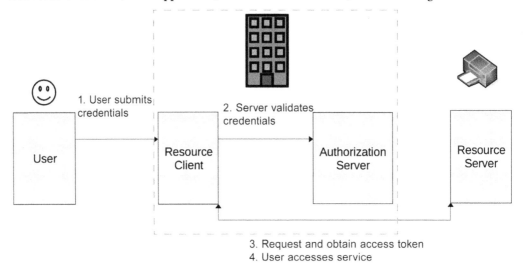

Figure 13.2: OAuth 2.0 authentication

The resource owner is the user or the customer. The resource server is the system that manages the authentication of the user. The client application makes a request for the resource owner and its authorization. The authorization server issues the tokens to the client after authenticating the resource owner.

OpenID Connect

OpenID Connect (**OIDC**) is an authentication tool used by third-party vendors such as **Google**, **Yahoo**, and **Facebook** that rides on top of OAuth 2.0 authorization. Features also include encryption and signing.

Authenticating with OIDC through an app or website is similar to SAML in that the user selects the authentication vendor (AOL, GitLab, Reddit, etc.), the request is redirected to the authentication vendor (i.e., the identity provider), the user completes the authentication process by entering the password of the authentication vendor, and then the user can use the app or website, as shown in *Figure 13.3*.

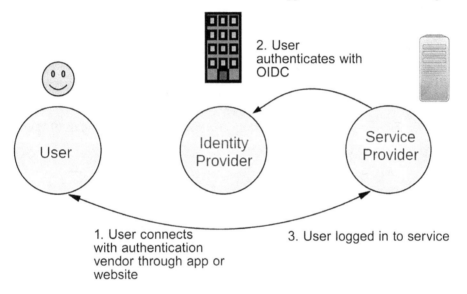

Figure 13.3: OIDC authentication

As an example of the process shown in *Figure 13.3*, imagine Chen uses the ACAL calendar app. He may be given the option to authenticate through Yahoo! via OIDC instead of creating a new username and password as part of the ACAL system. OIDC makes life simpler for the user because authentication is handled by a third party (Yahoo!, in our example). Further, the organization that requires the authentication – ACAL in this case – no longer has to develop and maintain the authentication system.

FIM is a common solution to reduce the friction of a user using multiple systems. This could be a member of an organization given access to a third party's resources, or it could be a customer accessing new services without having to create new usernames and passwords. However, in these use cases, there is still the initial step of signing in. With cloud-based systems becoming de facto, all authentication needs to be secure enough to work in remote systems, as you will see in the next section.

Authentication, Authorization, and Accounting Systems

When a user interacts with an organization's resources remotely, there needs to be a way of checking the user is legitimate, giving that user access to the correct resources, and then tracking the user's activities and resource usage. **Authentication, authorization, and accounting (AAA)** systems allow users to access a corporate network using one set of credentials, no matter where they are accessing the network from. In this section, we will look at protocols that allow all these functions in one system: RADIUS, TACACS+, and Kerberos.

RADIUS

The **Remote Authentication Dial-In Service (RADIUS)** is a protocol first published in the 1980s, when dial-up modems were popular, but it is still used today by organizations for SSO access to a company via a **virtual private network (VPN)**. The system allows the user to access the entire network with a single username and password.

The RADIUS system consists of a **supplicant** (the user), a RADIUS client, and the RADIUS server, as shown in *Figure 13.4*.

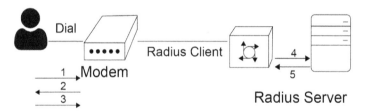

Figure 13.4: RADIUS authentication overview

When the user, or supplicant, accesses the RADIUS client, either via VPN or a wireless access point, the RADIUS client requests the user's authentication credentials via the **Extensible Authentication Protocol (EAP)**, as described in *Chapter 10, Network Architecture Security*.

Next, the RADIUS client submits an Access-Request RADIUS packet to the AAA or RADIUS server, encrypted using a shared secret over UDP port 1812. The AAA server decrypts the access-request packet using the shared secret. If the decryption fails, the server ignores the packet because the failure implies that an unethical hacker is attempting the request.

The process continues with EAP exchanging access-challenge and access-request packets between the supplicant and the RADIUS server using the RADIUS client as a pass-through. Finally, the RADIUS server sends an Access-Accept packet to complete the authentication process; otherwise, an Access-Reject packet is submitted. One limitation is that RADIUS combines authentication and authorization, limiting the control you have.

TACACS+

The **Terminal Access Controller Access-Control System Plus** (**TACACS+**), pronounced TACK-AXE-PLUS, is another protocol, and an improvement over RADIUS because TACACS+ can decentralize authentication, authorization, and accounting into three different servers to reduce the risk of a single point of failure.

Some features of TACACS+ include TCP communications over port 49 instead of UDP for reliability, and all TACACS+ packets are encrypted instead of only using a password, like with RADIUS.

Kerberos

Kerberos is an authentication protocol that utilizes SSO to allow users to authenticate once in an organization's network and gain access to the entire network without re-authenticating. Kerberos uses a system called tickets to manage authentication and authorization. Instead of repeatedly sending passwords over the network, Kerberos uses tickets to authenticate users. A ticket is a time-stamped, encrypted token that proves the identity of the user to services. How Kerberos carries out authentication and authorization is important, so the next sections will go into more detail, along with how to secure Kerberos.

Kerberos Authentication

Kerberos provides SSO network authentication and authorization for Linux, Windows, UNIX, and other operating systems. It consists of three parts: the principal or the user, which requests access; the **key distribution center** (**KDC**), which issues the tickets; and the application server, which validates the tickets for the user to access.

The process can be seen in the following diagram and is explained below.

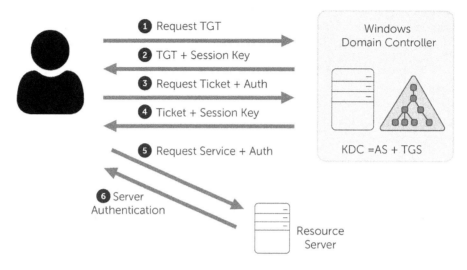

Figure 13.5: Kerberos authentication process

The principal sends a request to the KDC's authentication service for a **ticket granting ticket** (**TGT**). This TGT will later serve as proof to the application server that the user has been authenticated.

The TGT includes the encrypted date and time from the principal and uses a hash of the user's password for key encryption and decryption purposes. To prevent replay attacks, the clocks of the KDC and the principal must be synchronized within a 5-minute window; otherwise, the ticket request is denied. By default, a ticket remains valid for 10 hours before it expires.

Next, the authentication server verifies the user by decoding and matching the password hash and validates that the request has not expired. Once the user is validated to use the service, the following takes place:

1. The authentication service sends the TGT, which includes a timestamp and validity period, to the principal. The TGT is encrypted with the KDC's secret key.

2. The authentication service sends the **ticket granting service** (**TGS**) key. The TGS key is encrypted using the hash of the user's password and grants the user access to a specific privilege, such as email or intranet access.

At this point, the user does not have access to the service, but now they have the *right* to that service. Access to the service comes through Kerberos authorization, which is discussed in the next section.

Kerberos Authorization

Once the user is authenticated, the TGS session key can be decrypted. The decryption establishes that the principal and KDC know the identical shared secret, and the principal does not interfere with the TGT because it doesn't know the TGT.

After authentication, the user has access to the domain. The Kerberos authorization process is as follows:

1. To access services from within the **local area network** (**LAN**), the client requests a service ticket from the TGS. This is a token that grants access to the desired application – for example, email.

2. The client requests from the TGS a copy of the TGT and the name of the application server to be accessed with an authenticator. The authenticator is encrypted with the TGS session key and contains a time-stamped client ID.

3. To confirm that the requests are genuine, the TGS confirms that the ticket has not expired – for example, using NTP to manage synchronization of time zones – and then decrypts both messages using the secret key of the KDC for the first message, and the TGS session key for the second message.

4. Next, the TGS responds by issuing the following:

   - A **service session key** to be used by both the client and the application server and that is encrypted with the TGS session key.

   - A **service ticket** that contains details about the user, such as a **security identifier** (**SID**), IP address, session key, group, and timestamp. All are encrypted with the application server's secret key.

5. The service ticket is forwarded by the client to the application server (using a key that the latter cannot decrypt). A time-stamped authenticator is added and encrypted with the service session key.

6. The service ticket is decrypted by the application server and acquires the service session key using its secret key. This confirms that the client has sent an unaltered message. The authenticator is then decrypted with the service session key.

7. To mitigate on-path attacks (formerly known as man-in-the-middle attacks), mutual authentication can be implemented as optional extra security. The process of mutual authentication is described here:

   I. The application server responds to the client with a timestamp used in the authenticator. The timestamp is encrypted with the service session key. The timestamp is decrypted by the client and verifies that the value matches with what was sent. If there's a match, the application server is trustworthy.

   II. Finally, the server responds to the requests from the client.

To ensure consistent and reliable management of both authentication and authorization processes within a network, the Kerberos KDC has backup KDC servers to mitigate single-point-of-failure issues on the network. Similarly, the **Lightweight Directory Access Protocol** (**LDAP**) and Active Directory have multiple domain controllers running KDC services when necessary. This redundancy ensures that authentication services remain available and accurate, avoiding downtime while keeping systems secure.

Kerberos Attacks

Like any other system, Kerberos is vulnerable to attacks, and often, the aim of a malicious actor is to gain user and, preferably, administrator credentials in a network. If an attacker is able to become an administrator, they will be able to do much more damage if they are able to make changes to the permissions to access data or allow other attackers in. The three most common attacks on Kerberos are as follows:

- **Pass-the-hash attack**: Instead of attempting to gain access to the victim's password, the hacker gains access to the victim's hash. The attacker grabs the hash from the cache of the domain controller and uses the hash to authenticate as the user because the KDC validates the hash, not the password.

- **Golden ticket attack**: With a successful golden ticket attack, an attacker will have unlimited access to the company's domain, including files, devices, and other domain controllers. The attacker exploits the vulnerabilities within Kerberos' authentication protocol, using methods such as phishing or malware attacks, and bypasses the normal authentication process.

- **Kerberoasting attack**: Attempts to crack the password of services within LDAP or Active Directory are considered Kerberoasting. Kerberos' **service principal name** (**SPN**), which connects a service to a user account within LDAP or Active Directory, is connected to the attacker's spoof user account, through which the attacker acquires the encrypted password. The attacker then attempts to crack the hash via rainbow table or brute force attacks.

AAA systems such as RADIUS, TACACS+, and Kerberos make the job of administering users on your IT system more efficient than if you had to carry out every function individually. However, with more enterprises using infrastructure-as-a-service providers, there is also the possibility of outsourcing this management to a third party.

Managing the Identity and Access Provisioning Lifecycle

Identity as a service (**IDaaS**) allows organizations to outsource their identity management to cloud providers, who then take on the responsibilities of authentication, authorization, and accounting. The cloud vendor manages the login names, passwords, and authentication methods, such as **two-factor authentication** (**2FA**) or **multifactor authentication** (**MFA**).

Cloud vendors offering IDaaS ensure access and security for various business requirements – for example, access to varied utilities, federation support, SSO, user provisioning, deprovisioning, account access reviews, multi-factor authentication, and privilege escalation (users allowed to use `sudo`, the Linux administrator tool, for example). They ensure access and security with the help of the following three components:

- **Identity governance and administration**: This component provisions users to access cloud utilities and change passwords

- **Access**: This component implements authentication, authorization, and federation, including SSO

- **Intelligence**: This component implements accounting (logging) and access reporting

Popular IDaaS vendors include Azure Active Directory Federated Services, Centrify, OneLogin, and Okta.

On-Premises versus Cloud Identity Management

IDaaS offers efficiency because most of the backend administration is done by a third party. It can simplify management and reduce costs. It can also offer enhanced security if the provider has a dedicated team to ensure up-to-date information and strategies on the latest security threats, as well as cutting-edge authentication. It is also much easier to scale with a third-party provider.

However, many organizations still prefer to do identity management on-site. This could be because of specific regulatory requirements around sovereignty or even specific high-sensitivity security requirements. Top-secret facilities, for instance, might have air-gapped systems that simply shouldn't be connected to the cloud. Finally, for legacy reasons, an organization might decide it is just too complex to move IAM to the cloud.

But if an organization also wishes to use SaaS solutions such as Salesforce, hybrid solutions can be found where organizations can synchronize their on-premises LDAP or Active Directory with cloud-based IDaaS solutions, allowing for a unified identity management system that spans both on-premises and cloud environments. High-security operations could be managed on-premises, while other operations can make use of the advantages of IDaaS systems.

Whatever identity solutions are decided upon, even if the administration is done with IDaaS, organizations and security specialists should have a clear idea of how to ensure that the right people or things can access the systems and data they need – but only the ones they need. This is covered in the next section.

Access Controls

Depending on the type of organization, the needs of individuals, and even regulatory requirements, there are different systems of access control. **Mandatory access control** is strict and based on a stringent security policy. **Discretionary access control** is more flexible and gives control to the owners of resources. Risk-based controls dynamically assess risk based on variables such as user IP and resource sensitivity and role-based controls assign access permissions to users based on their roles within an organization. **Rule-based access control** and **attribute-based access control** both work in situations where access needs to be governed by a set of conditions or attributes rather than static roles or predefined classifications. The following sections discuss these in more detail.

Mandatory Access Control

Mandatory access control systems are based on clearance levels. Typically, they range from unclassified, which are systems open to everyone, to classified, which are systems that have some restrictions, through to secret and then top secret. These clearance levels will correspond with access levels of subjects, ensuring that users can get to everything they need, but also only what they need.

For example, files in a container marked top secret are inaccessible to those that have only secret or classified access. Subjects with top-secret access can access unclassified, classified, secret, and top-secret documents. This is illustrated in *Figure 13.6*.

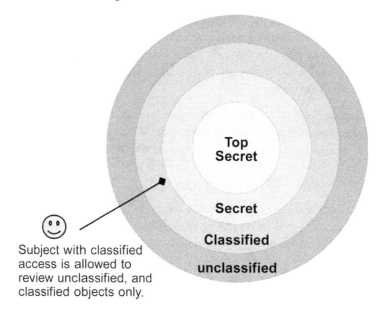

Figure 13.6: Mandatory access control

Discretionary Access Control

The **discretionary access control** (**DAC**) system model is designed for users who value availability over confidentiality and integrity by giving access controls to the individuals in an organization. DAC systems are utilized within corporate, sales, and marketing firms where quickly distributing information, such as sales quotes and agreements with customers, is prioritized, and there are lower risks associated with incorrectly allowing access to resources. Owners assign access rights to objects, rather than access being based on levels such as secret or top secret. For example, files can be set to read-write for team members of a project and to read-only for anyone else. Though there is a risk that the owner might accidentally, or even maliciously, grant write access to someone who doesn't need it, the consequences would have a low impact, especially if files are correctly version controlled.

For additional security, system administrators include features of **non-discretionary access control** (**NDAC**) by implementing malware protection and disallowing the use of USB drives. Unlike in DAC, where the **user** (**subject**) determines access to their resources, NDAC policies are set by the system, meaning the system or object determines the access outcome. Consequently, even the owner of an application cannot execute a program if the system administrator, or super-user, disallows the action.

> **Note**
> Unless a system uses DAC, it uses some form of NDAC.

Risk-Based Access Controls

Risk-based access control, also known as **risk-based authentication** (**RBA**), uses variables to determine whether further authentication needs to occur to prove the identity of the user. For example, if Luke normally authenticates from home, the **cloud service provider** (**CSP**) might only ask for his login and password. But, if the system notices that the IP address is not the same as Luke's, or that Luke's login time is abnormal, the CSP might require him to enter a code texted to his phone or answer a security question as further proof of identity.

A subset of RBA includes **just-in-time** (**JIT**) authentication, which provides temporary access to an object, such as a server, virtual machine, network, or even thumb drive. This minimizes risk by granting users access to resources only when needed and for a limited time – for example, when Marta plugs a thumb drive into a USB port in their workstation. As the default, USB ports are locked down and the system policy verifies whether she has this elevated privilege. If so, the system asks for her password to grant her access. For security purposes, she is asked to enter her password again every 15 minutes or so. The port automatically locks down until the correct password is entered. This protects the data in case the validated user walks away and another individual (potentially a threat) attempts to use the USB port.

Role-Based Access Control

Role-based access control (**RBAC**) systems group staff into different system administrator roles that determine system permissions and which resources they can access. Before RBAC was introduced, administrator rights were usually given to everyone. This approach was very insecure because if everyone had the administrator or root password, everyone had the privilege to delete all files, even if they did not own them.

With RBAC, administrator rights are apportioned according to the function of the subject. For example, a member with a **Network** role will only be able to install, remove, start, and stop networks. Similarly, the **Printer** role will only allow a member to add, remove, stop, and start printers. A **Power User** role would combine features of both the **Network** and **Printer** roles but not other roles, such as the **Hard Drive Management** role.

Rule-Based Access Control

Rule-based access control (**RuBAC**) systems allow or deny access to objects depending on a set of very specific rules. An example of this is a firewall. In a firewall, rules define whether network traffic will be allowed or denied depending on certain rules, such as whether traffic comes from a specific country, a set of IP addresses, or at a designated time.

Attribute-Based Access Control

Attribute-based access control (**ABAC**) utilizes subjects, objects, and actions like the other control systems but also adds another element to refine security options for the organization. This element could be related to environmental factors such as time, location, encryption, device, and so on. For instance, an ABAC system can allow subjects to access objects but only between 8 AM and 5 PM, or only allow access to items that are located in Dubai to address cloud data sovereignty issues.

While these systems address the technical aspects of access control, access and actions within the system need to be monitored and managed appropriately. This is where accountability comes in.

Accountability

Accountability implies that the subject (user) accessing the system is, in fact, the validated and authorized user and that they are using the system in a responsible manner. A concept included in accountability is non-repudiation. This entails that a person *cannot* deny an action or event that they carried out. This makes sure that the responsible parties are held accountable. To establish accountability, a register of user actions and events is kept in a file called a log. The log should be reviewed periodically to ensure users are working responsibly.

Summary

This chapter covered user access controls and management. It discussed how organizations can protect against attackers using either physical or technical controls to breach their systems. A major role of the chief security officer and the CISSP is to secure identity access with centralized utilities such as SSO so that users can do their jobs. Users add to security by using strong passwords.

You learned how user access to ancillary products, either within the organization or online, can be simplified if system administrators enable SAML or OAuth 2.0, which provides user identity federation. This keeps users from having to re-authenticate when working on ancillary systems or purchasing related products. These systems use a type of service provider or resource server to manage the identity on the ancillary service.

You also examined **identity as a service (IDaaS)**, a feature for managing identity federation. Through IDaaS, corporations can hire firms to ease their authentication, authorization, and accounting overhead.

Finally, the chapter discussed access control models. Government systems use **MAC**, or **mandatory access control**, to manage their subjects' interaction with some object. Most corporate organizations use DAC to manage access to objects where availability is more important than confidentiality. In the next chapter you will look at the importance of security assessment and how to design them to be effective.

Further Reading

- *SAML Security Cheat Sheet*, OWASP Cheat Sheet Series, `https://packt.link/TZBoz`, 2021

- *Evaluation of Access Control Techniques in Cloud Computing*, Devyani Patil, Nilesh Mahajan, PhD, International Journal of Computer Applications, Vol 176-No. 17, Apr 2020

Exam Readiness Drill – Chapter Review Questions

Apart from a solid understanding of key concepts, being able to think quickly under time pressure is a skill that will help you ace your certification exam. That is why working on these skills early on in your learning journey is key.

Chapter review questions are designed to improve your test-taking skills progressively with each chapter you learn and review your understanding of key concepts in the chapter at the same time. You'll find these at the end of each chapter.

> **How to Access These Materials**
>
> To learn how to access these resources, head over to the chapter titled *Chapter 24, Accessing the Online Resources*.

To open the Chapter Review Questions for this chapter, perform the following steps:

1. Click the link – `https://packt.link/chapter13`.

    Alternatively, you can scan the following **QR code** (*Figure 13.7*):

Figure 13.7: QR code that opens Chapter Review Questions for logged-in users

2. Once you log in, you'll see a page similar to the one shown in *Figure 13.8*:

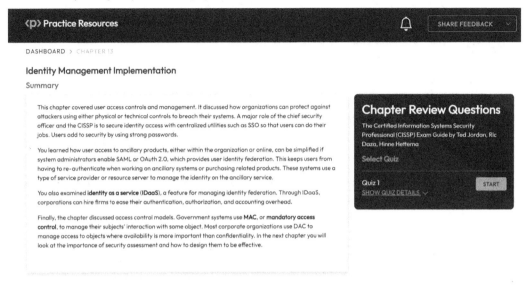

Figure 13.8: Chapter Review Questions for Chapter 13

3. Once ready, start the following practice drills, re-attempting the quiz multiple times.

Exam Readiness Drill

For the first three attempts, don't worry about the time limit.

ATTEMPT 1

The first time, aim for at least **40%**. Look at the answers you got wrong and read the relevant sections in the chapter again to fix your learning gaps.

ATTEMPT 2

The second time, aim for at least **60%**. Look at the answers you got wrong and read the relevant sections in the chapter again to fix any remaining learning gaps.

ATTEMPT 3

The third time, aim for at least **75%**. Once you score 75% or more, you start working on your timing.

> **Tip**
>
> You may take more than **three** attempts to reach 75%. That's okay. Just review the relevant sections in the chapter till you get there.

Working On Timing

Target: Your aim is to keep the score the same while trying to answer these questions as quickly as possible. Here's an example of how your next attempts should look like:

Attempt	Score	Time Taken
Attempt 5	77%	21 mins 30 seconds
Attempt 6	78%	18 mins 34 seconds
Attempt 7	76%	14 mins 44 seconds

Table 13.1: Sample timing practice drills on the online platform

> **Note**
>
> The time limits shown in the above table are just examples. Set your own time limits with each attempt based on the time limit of the quiz on the website.

With each new attempt, your score should stay above **75%** while your "time taken" to complete should "decrease". Repeat as many attempts as you want till you feel confident dealing with the time pressure.

14
Designing and Conducting Security Assessments

You can take steps to deploy security management, logical, and operational controls, but they are meaningless unless you verify and validate that the controls are working. For this reason, there are a variety of different ways of testing your architecture depending on your goals, and they will be covered in this chapter. Audits are among the most common methods for assessing the security posture of your organization, as well as that of your partners. They can be conducted internally, externally, or by a third party.

However, audits focus on the overall security landscape and are large-scale undertakings. Security managers should also consider ongoing testing, including regular code reviews, continuous monitoring of tools and operations, and conducting penetration testing to evaluate the organization's resilience to simulated real-world threats.

For these tests to be effective, you should clearly define your goals, get buy-in from key stakeholders, and select the appropriate methodology. The end of any testing involves the proper reporting of the results and deciding upon actions to mitigate identified risks.

By the end of this chapter, you will be able to answer questions on the following:

- Designing and validating assessment, test, and audit strategies
- Conducting security control testing

Let's begin by covering how to design and validate testing strategies.

Designing and Validating Assessment, Test, and Audit Strategies

When designing and developing security assessments, tests, and audits, you must consider their purpose. For example, you might want to test for vulnerabilities, assess risk management, or improve security awareness. Organizations need policies for security systems, and there are often regulatory and compliance requirements within your organization that must be followed. Testing systems is a good way to ensure these policies and regulations are being followed. Whatever your test might be for, you first need to understand the goal, so you can then work out the scope.

When planning an audit, it's important that you work with the different departments of the organization, such as operations, marketing, information systems, sales, production, manufacturing, and so on. Each department will have different priorities that need to be assessed and might also be impacted by the testing. For example, you might need to work with the human resource department if the purpose of an audit is to test security awareness. Likewise, you will have to consult with IT departments on the parameters of the test, the operations department on when and how to carry it out, and also the legal department to address any regulatory requirements.

According to the **Information Systems Audit and Control Association (ISACA)**, the cybersecurity audit process should follow these steps:

1. Set the goals of the audit.
2. Get the right business departments involved.
3. Decide on the scope of the audit.
4. Select the audit team members.
5. Plan the audit.
6. Launch the audit.
7. Document the results.
8. Communicate the results to the organization's leaders.

Audits can be conducted internally, externally, or by way of a third party. Internal audits are run by the staff of your organization. External audits are conducted by your vendors or business partners to verify that your organization meets their minimum security requirements. Third-party audits are conducted by external organizations to verify that the organization is meeting regulatory requirements. The following subsections will look into these in more detail.

Internal Audits

Internal cybersecurity audits are essential for organizations to ensure the security of their information systems and are conducted by employees of the organization. These audits involve assessing the organization's cybersecurity policies, procedures, and practices to identify potential vulnerabilities, threats, and areas for improvement.

The following steps are recommended to properly plan an internal audit:

1. **Define the scope**: The first step is to define the scope of the audit. This should include identifying the systems, networks, and applications that will be assessed as well as the information assets that need to be protected. For example, a financial institution might focus on securing customer data, while a manufacturing company may prioritize protecting intellectual property.

2. **Establish objectives**: After defining the scope, the organization should establish clear objectives for the audit. These objectives should be aligned with the organization's overall cybersecurity policies and should include complying with regulations, identifying vulnerabilities, and assessing the effectiveness of existing controls. For example, the objective could be to ensure that the online banking platform complies with relevant regulations, such as PCI-DSS for handling payment card information.

3. **Assemble the audit team**: The organization should form a dedicated audit team, consisting of individuals with expertise in cybersecurity, IT, and the organization's specific industry. This team should include a mix of internal staff and external experts, as necessary, to ensure a comprehensive audit.

4. **Develop an audit plan**: The audit team should create a detailed plan that outlines the specific tasks and activities that will be performed during the audit. This plan should include a timeline for each task, as well as the resources needed to complete them.

After the plan is completed, the next step is to conduct an internal audit.

Conducting an Internal Audit

Conducting an internal audit involves verifying that policies are being followed and evaluating systems, as discussed below:

1. **Review policies and procedures**: The audit team should begin by reviewing the organization's existing cybersecurity policies and procedures. This includes examining the organization's risk management processes, incident response plans, and employee training programs. Any gaps or inconsistencies should be noted for further investigation.

2. **Assess technical controls**: Next, the audit team should evaluate the effectiveness of the organization's technical controls, such as firewalls, intrusion detection systems, and encryption technologies. This may involve performing vulnerability scans, penetration tests, or other assessments to identify the organization's vulnerabilities.

3. **Evaluate user practices**: The audit team should assess the cybersecurity practices of employees, contractors, and other users within the organization. This may involve reviewing user access controls, password policies, and the use of personal devices within the organization.

4. **Analyze audit findings**: After completing the audit, the team should analyze the findings and identify vulnerabilities, threats, and mitigations. This information should be used to make recommendations to management (for example, the CTO) to improve the cybersecurity posture.

After conducting the internal audit, management is informed of the results and prioritizes mitigations based on assets and their risks.

Some organizations might choose to have external audits if they feel internal audits might be too biased, or that they don't have the right expertise or regulatory knowledge. They can also offer extra assurance to outside parties, and external audits might also be a requirement for partnerships and government contracts that involve security. We will look in more depth in the next subsection.

External Audits

External audits are conducted by business partners of the organization to validate that the terms of contract agreements are being met. They are typically used to provide assurance to stakeholders that the organization's cybersecurity controls are effective. These audits help build trust with customers and partners that minimum security expectations are being met.

Planning the External Audit

External audits are typically carried out by reputable firms and are requested by the organization being audited. The first step is to plan, which consists of the following stages:

-   **Define audit scope and objectives**: The organization being audited must clearly outline the scope of the audit, which may include specific systems, applications, or processes to be assessed. Additionally, the objectives of the audit, such as compliance with working agreements or identification of vulnerabilities, should be defined.

-   **Establish a timeline and budget**: A realistic timeline for the audit should be established and should take into consideration the scope, objectives, and availability of resources. The organization should also allocate an appropriate budget for the audit.

-   **Engage management**: It is essential to involve management in the planning process because budgeting is critical. This will ensure that the audit is aligned with organizational priorities and that the findings are effectively communicated and addressed.

Though these factors are ultimately decided by the organization requesting the audit, the auditor will often advise on best practices and possibilities.

Conducting an External Audit

The process of conducting an external cybersecurity audit generally involves the following stages:

1. **Pre-audit phase**: The auditor gathers information about the organization, its systems, and its processes. This may involve conducting interviews, reviewing documentation, and analyzing network architecture. This is essential for understanding the organization's context, defining the audit scope, identifying risks, planning audit activities, ensuring compliance, and building a baseline for evaluation.

2. **On-site audit**: The auditor visits the organization's premises to perform a detailed examination of its cybersecurity controls, including physical security, access controls, network security, and incident response procedures.

3. **Testing**: The auditor conducts a series of tests to identify vulnerabilities, such as penetration testing, vulnerability scanning, and social engineering simulations.

4. **Reporting**: After completing the audit, the auditor provides a detailed report for management, regulatory organizations, and other interested third parties outlining the findings, recommendations, and any identified vulnerabilities or risks.

5. **Remediation and follow-up**: The organization addresses the identified issues and implements the auditor's recommendations. A follow-up audit may be conducted to verify whether the remediation efforts have been effective.

The last type of audit to discuss is the third-party audit, in which vendors hire an auditing organization to attain a specific certification, such as ISO, OWASP, CMMI, and so on.

Third-Party Audits

Third-party audits are conducted by auditors who are not employees of the organization being audited. These audits are typically used to assess the organization's compliance with regulations or standards. Third-party and external audits can be helpful in avoiding fines or penalties for non-compliance.

These audits are typically performed by specialized firms with expertise in cybersecurity, risk management, and compliance. The primary objective of an external cybersecurity audit is to provide an *unbiased* assessment of an organization's cybersecurity posture and to identify potential vulnerabilities that could be exploited by cybercriminals. The findings of a third-party audit can be used to create a roadmap for improving security measures and reducing the likelihood of data breaches, financial losses, and reputational damage.

The goal of these audits is to provide a comprehensive understanding of the organization's security posture and ensure compliance with relevant regulations and industry standards, such as ISO/IEC 27001, PCI DSS, and GDPR.

Planning a Third-Party Cybersecurity Audit

The planning phase is crucial for the success of a third-party cybersecurity audit. It involves several key steps, including the following:

1. **Establishing objectives**: The organization must first define the scope and objectives of the audit. This includes identifying which systems, networks, and processes will be assessed and the specific regulations or standards with which the organization must comply.

2. **Selecting an auditor**: Security management will choose a reputable auditing firm with expertise in the organization's industry and relevant security standards. It is essential to verify the auditor's credentials and experience and ensure that they can maintain confidentiality and impartiality throughout the process.

3. **Preparing documentation**: The organization should gather all relevant documentation, such as network diagrams, system configurations, and security policies, and share them with the auditor. This information will help the auditor understand the organization's environment and facilitate a more efficient audit.

4. **Communicating with stakeholders**: Security management informs all relevant stakeholders, including employees, about the upcoming audit. This ensures their cooperation and enables them to prepare for any potential disruptions or changes to their daily routines.

Conducting a Third-Party Cybersecurity Audit

The execution of a third-party cybersecurity audit typically involves the following steps:

1. **On-site assessment**: The auditor visits the organization to assess its physical security, interview key personnel, and review documentation. This helps the auditor gain a deep understanding of the organization's security practices and identify any gaps.

2. **Technical assessment**: The auditor performs a series of technical tests to identify vulnerabilities in the organization's systems and networks. This may include penetration testing, vulnerability scanning, and code review.

3. **Analysis and reporting**: After completing the assessments, the auditor analyzes the findings and compiles a detailed report that outlines the identified vulnerabilities, their potential impact, and recommendations for improvement.

4. **Remediation and follow-up**: The organization must address the identified vulnerabilities by implementing the auditor's recommendations. The auditor may perform a follow-up audit to ensure that the necessary changes have been made and that the organization's security posture has improved.

Advantages of Third-Party, External, and Internal Audits

All three audit types improve your organization's security posture by identifying vulnerabilities, resulting in solutions that mitigate data breaches or other cyber incidents. Although you might run regular security checks, the goal of an audit is to be able to demonstrate an organization's security posture to third parties. This could be to customers, shareholders, potential partners, or regulatory organizations.

The three types of audits have different advantages. Internal cybersecurity audits enable organizations to better understand their risk landscape and prioritize investments in security controls and processes. Internal audits can also involve employees in the audit process. This fosters a culture of cybersecurity awareness and encourages users to take a more proactive role in protecting the organization's information assets.

External and third-party auditors bring specialized knowledge and experience, which can help organizations identify vulnerabilities that may have been overlooked by internal teams. External and third-party audits are also much more likely to be free from internal biases or conflicts of interest. Because of this, audits done by outside parties tend to be seen as more credible.

As mentioned, the main distinction between external and third-party audits is who requests them. External audits have the obvious advantage in that because they are requested by a partner or stakeholder, the auditors are less likely to cause operational issues. However, for this reason, a third-party audit, which isn't even bound by the audited organization's own requested scope, might be seen as the most credible of the three.

A cybersecurity breach can cost a firm millions of dollars in fines and damage to reputation, so if an audit uncovers a vulnerability before this happens, it should be seen as a win. However, although the gains outweigh the costs, audits can be time-consuming and expensive. To maintain security without large costly audits, security testing should be done in a systematic and regular way. This is discussed in the next section.

Conducting Security Control Testing

Security control testing is the process of evaluating the effectiveness of **security controls** in protecting an organization's information systems and data. It is conducted by assessing the implementation, operation, and effectiveness of security measures through various methods such as **vulnerability scanning**, **penetration testing**, and **security audits**. Security controls can be technical (such as intrusion detection systems), physical (such as fences), or administrative (such as security policies).

There are many different types of security control testing, each with its own strengths and weaknesses, as discussed in the following subsections.

Vulnerability Assessment

A vulnerability assessment is an evaluation of the security posture of an organization's information systems and data. It is conducted by scanning systems and applications using automated tools and analyzing the results to identify known vulnerabilities and potential security risks. It typically identifies known vulnerabilities in systems and applications, as well as potential security risks, and allows organizations to identify, quantify, and prioritize vulnerabilities in their systems, networks, and applications. The following are some key steps to follow when conducting a comprehensive vulnerability assessment:

1. **Define the scope**: The first step is to define the scope of the assessment. This includes identifying the systems, networks, and applications that will be assessed, as well as any constraints, such as time and budget limitations. It is essential to involve all relevant stakeholders in this process to ensure a complete and accurate understanding of the organization's assets and infrastructure.

2. **Develop a vulnerability assessment methodology**: Having a structured methodology for the assessment, such as MITRE ATT&CK, is critical to ensure the process is consistent, repeatable, and effective. The methodology must include the tools and techniques to be used, as well as guidelines for prioritizing vulnerabilities based on their impact and likelihood of exploitation.

3. **Collect data**: The data collection phase involves gathering information about the target systems, networks, and applications. This can include technical data, such as system configurations, patch levels, and network topology with tools such as Wireshark or Nmap, as well as non-technical data, such as policy documents and user access controls. Both automated and manual techniques can be employed for data collection, including network scanning tools and vulnerability scanners such as Nessus.

4. **Analyze and prioritize vulnerabilities**: Once the data has been collected, it is essential to analyze it to identify vulnerabilities and assess their potential impact on the organization. This involves categorizing vulnerabilities based on their severity and impact. Vulnerabilities should be prioritized according to risk so that the organization may focus on the most critical issues.

5. **Remediation and mitigation**: After identifying and prioritizing vulnerabilities, the next step is to develop and implement remediation and mitigation strategies. This can involve applying patches and updates, modifying configurations, or implementing additional security controls, such as firewalls or intrusion prevention systems. It is crucial to continuously monitor the progress of these efforts and ensure that vulnerabilities are mitigated.

6. **Reporting and continuous improvement**: Finally, the results of the vulnerability assessment should be documented and communicated to relevant stakeholders. This should include information on the vulnerabilities identified, their impacts, and the actions taken to address them. Additionally, the vulnerability assessment process should be periodically reviewed and updated to ensure that it remains effective in identifying and addressing new and emerging threats.

A vulnerability assessment is a vital tool for organizations that aim to proactively identify and address vulnerabilities. By following structured methodologies such as **MITRE ATT&CK** or Lockheed Martin's **Cyber Kill Chain** frameworks, which help to prioritize vulnerabilities based on risk, organizations can take a proactive approach to managing their cybersecurity risks and protecting their valuable assets. Nowadays, vulnerability scanning tools update the latest vulnerability databases from the **National Vulnerability Database (NVD)** and **Common Vulnerabilities and Exposures (CVE)** systems.

Penetration Testing

A **penetration test** attempts to exploit vulnerabilities in an organization's information systems and data in order to expose vulnerabilities that can be then addressed. Penetration tests simulate real-world attacks such as HTML injection and are typically conducted by security professionals.

Penetration testing, also known as ethical hacking, involves simulating cyberattacks on an organization's systems, networks, and applications to identify vulnerabilities. By conducting penetration tests, organizations uncover vulnerabilities and assess the effectiveness of their security controls to reduce the risk of real-world cyberattacks. The following are the key steps involved in conducting penetration testing:

1. **Define the scope and objectives**: Define the scope of the penetration test, including the systems, networks, and applications to be tested. This should be done in consultation with relevant stakeholders, such as system owners, human resources, legal, finance, and IT managers, to ensure a clear understanding of the objectives and constraints of the test. Additionally, it is crucial to establish the rules of engagement, including the testing methods, techniques, and tools to be used, as well as any regulatory requirements to be adhered to during the test.

2. **Reconnaissance**: The reconnaissance phase involves gathering information about the target systems, such as network topologies, IP addresses, domain names, and system configurations. This can be achieved through both **open source intelligence (OSINT)** such as online research and active (e.g., network scanning) reconnaissance methods. The goal of this phase is to develop an understanding of the target environment for the subsequent phases of the test.

3. **Vulnerability identification**: With the information gathered during reconnaissance, the next step is to identify potential vulnerabilities and map them to relevant threat actors and attack scenarios. This can involve using automated vulnerability scanners, manual testing techniques, or a combination of both. The goal is to identify vulnerabilities, which will then inform the exploitation phase of the test.

4. **Exploitation**: Then, the penetration tester attempts to exploit identified vulnerabilities to gain unauthorized access to the target systems, doing their best to cause no real harm to the organization. The goal of this phase is to demonstrate the potential impacts of the identified vulnerabilities being exploited on the organization. Exploitation techniques can include social engineering, password cracking, privilege escalation, and remote code execution.

5. **Reporting**: Next, the findings of the penetration test should be documented in a final report. This report should detail vulnerabilities, exploitation methods, and impact on the organization. The report can provide recommendations for mitigations, such as patching, configuration updates, or additional security controls. It is crucial for the organization to prioritize and address these recommendations; otherwise, unethical hackers are likely to exploit the vulnerabilities.

> **Note**
>
> Penetration testing reports contain sensitive information, such as hostnames, IP addresses, usernames, discovered passwords, and more. Restrict access to this report to authorized personnel only.

6. **Continuous improvement**: After implementing mitigations, it is important to conduct a retest to ensure that the remediation efforts have been effective. Penetration testing should be performed once or twice a year, depending on the organization's needs, or after significant changes to infrastructure to ensure ongoing security.

Penetration testing provides organizations with insights into their security posture and helps them identify vulnerabilities before they can be exploited by malicious actors. By incorporating penetration testing into their cybersecurity processes, organizations can enhance their resilience against cyber threats and protect sensitive data.

Log Reviews

A **log review** is an analysis of security logs from information systems and applications and can identify suspicious activity, such as unauthorized access attempts or data exfiltration. Log reviews are an essential component of a comprehensive cybersecurity strategy, as they provide valuable insights into the activities occurring within an organization's systems, networks, and applications.

By regularly reviewing and analyzing log data, organizations can identify suspicious behavior, detect security incidents, and ensure legal and regulatory compliance. The following are some key steps for conducting effective log reviews:

1. **Define the scope**: Determine the scope of the log review process, including the systems and applications whose logs will be reviewed, as well as the specific objectives of the review, such as detecting potential security incidents or ensuring compliance with relevant regulations. Ensure that all relevant stakeholders are involved in defining the scope and objectives for a comprehensive overview of the organization's logging requirements.

2. **Establish log management policies**: Develop clear policies and procedures for log management, including log collection, storage, and access control. This should include guidance on the types of events to be logged, the logging formats, and the frequency of log reviews.

3. **Implement log analysis tools**: Utilize log aggregation and analysis tools to automatically collect, correlate, and analyze log data from multiple sources. These tools can help analysts identify trends, patterns, and anomalies in the log data, making it easier for them to detect potential security incidents or compliance violations.

4. **Conduct regular log reviews**: Perform log reviews on a regular basis, as determined by the organization's risk profile and regulatory requirements. Log reviews should be conducted by trained analysts who are familiar with the organization's systems, networks, and applications, as well as the relevant threat landscape.

5. **Remediation and follow-up**: Upon identifying potential security incidents or compliance violations through log reviews, the IT security team can use the necessary techniques to address issues. This may include patching vulnerabilities, adjusting security configurations, or updating access controls. Document the findings and actions taken to improve accountability and facilitate future log reviews.

By implementing a structured approach to log management and analysis, organizations can enhance their ability to identify and address potential threats, ultimately strengthening their overall cybersecurity posture.

Synthetic Transactions

Synthetic transactions are machine-simulated transactions used to test the security of an organization's information systems and data and can be used to test the effectiveness of security controls, such as firewalls and intrusion detection systems. By conducting synthetic transactions, organizations can check the performance, security, and stability of certain systems. The following steps are involved in synthetic transactions:

1. **Scope**: A test begins by determining the scope of the synthetic transactions, including the systems, networks, and applications to be tested, as well as the specific objectives, such as identifying potential security vulnerabilities, performance bottlenecks, or functional defects. Relevant stakeholders should be engaged to ensure a comprehensive understanding of the organization's user requirements.

2. **Develop simulation scripts**: Next, create scripts that simulate real-user interactions, such as a customer visiting a site and clicking on a link, with the target systems, networks, and applications. These scripts should cover a variety of use cases, including common user workflows, edge cases, and potential attack scenarios. Ensure that the scripts are regularly updated to reflect changes in the organization's digital assets and the evolving threat landscape.

3. **Execute and monitor**: Next, execute the synthetic transaction scripts at regular intervals, and continuously monitor the results to identify any deviations from expected behavior. Analyze the data to detect potential security issues, such as unauthorized access attempts, data leakage, or signs of a breach.

4. **Mitigate and document**: Upon identifying potential security issues through synthetic transactions, place appropriate mitigations to address the problem. This may include patching vulnerabilities, adjusting security configurations, or updating access controls. Document the findings and actions taken to enhance accountability and facilitate future synthetic transaction exercises.

5. **Integration into cybersecurity processes**: Finally, you can integrate user simulations into the organization's existing cybersecurity processes, such as vulnerability assessments, penetration testing, and incident response. This can help you create a more comprehensive and proactive approach to detecting and addressing potential threats.

For example, a company simulates 100,000 users attempting to connect to their website instead of hiring 100,000 people to do the same to measure the system's performance. Sudden increases in traffic could cause sites to crash, costing revenue and customer retention. The development team might want to see how well a web portal holds up to prepare for a large marketing push.

By simulating real user interactions, organizations can identify potential security issues before they impact actual users, ultimately strengthening their cybersecurity posture and enhancing the overall user experience.

Code Review and Testing

Code review and testing is an evaluation of the security of an organization's software code. Code review is typically a manual check by a developer, while testing can be automated. The aim of code review and testing is to identify security vulnerabilities in code as well as potential security risks. Code testing helps organizations identify and address potential vulnerabilities and issues within their software applications.

By carefully examining source code, organizations can ensure that their applications are secure, reliable, and compliant with industry standards and best practices. The following are the key steps involved in code review and testing:

1. **Develop a formal code review process**: First, application teams need to develop a formal code review process that includes guidelines on the responsibilities of reviewers, the frequency of reviews, and the specific criteria to be assessed, such as adherence to coding standards, security best practices, and performance requirements. Ensure that all relevant stakeholders are involved in defining and implementing the code review process.

2. **Utilize static and dynamic analysis tools**: Next, utilize static analysis tools, such as SonarQube or Checkmarx, to automatically examine the source code for potential vulnerabilities, such as buffer overflows, SQL injection, or cross-site scripting. Complement this with dynamic analysis tools (such as Burp Suite) that test the application during runtime to identify potential issues, such as memory leaks, race conditions, or insecure data handling. This is covered in more detail in *Chapter 23, Secure Coding Guidelines, Third-Party Software, and Databases*.

3. **Perform manual code reviews**: In addition to automated analysis, performing manual code reviews, which involve a thorough examination of the source code by developers, can help identify complex issues that may be overlooked by automated tools, and provide valuable insights into the overall quality and maintainability of the code.

4. **Train developers in secure coding practices**: Management must train developers in secure coding practices, such as input validation, secure error handling, and proper encryption techniques. Encourage the use of well-established libraries and **application programming interfaces (APIs)** that have been extensively tested for security vulnerabilities.

5. **Prioritize and address identified vulnerabilities**: After identifying potential vulnerabilities through code reviews and testing, prioritize and address the issues based on their severity and potential impact. Retest the application to ensure that the vulnerabilities have been effectively mitigated.

6. **Promote a culture of security awareness**: Finally, encourage a culture of security awareness within the development team by providing regular training, promoting sharing of best practices, and promoting collaboration between developers and security experts.

By implementing a structured approach to code reviews, utilizing both automated and manual analysis techniques, and fostering a security-conscious development culture, organizations can significantly reduce the risk of security breaches and improve the overall quality of their software products.

Misuse Case Testing

Misuse case testing is a specialized approach to software testing and analysis that identifies potential security vulnerabilities and weaknesses in an application by examining how it could be misused or exploited by malicious actors, for example, attempting to manipulate the pricing of items in an online shop. By simulating the actions and motivations of attackers, misuse case testing allows organizations to proactively assess and improve the security of their systems and applications. It differs from penetration testing in that misuse case testing examines specific misuse scenarios to understand how targeted attacks might exploit particular weaknesses, whereas penetration testing aims to discover and demonstrate a wide range of general vulnerabilities through hands-on attack simulations. Misuse case testing consists of the following key steps:

1. **Identify potential attack scenarios and threat actors**: Begin by identifying potential attack scenarios and threat actors relevant to the application, taking into consideration the organization's industry, target audience, and technology stack. These scenarios should cover a range of attack vectors, such as unauthorized access, data theft, or denial of service.

2. **Develop misuse case models**: Misuse case models describe the actions and objectives of the threat actors, as well as the potential vulnerabilities and weaknesses in the application that could be exploited. These models should outline the techniques an attacker might use to compromise the system, such as form injecting to manipulate prices in the previous example.

3. **Execute misuse case scenarios**: Next, execute the misuse case scenarios against the application, monitoring its response to the simulated attacks. The monitoring can involve both manual and automated tools, such as vulnerability scanners or static code analyzers.

4. **Review and update regularly**: You should regularly review and update the misuse case scenarios to reflect the evolving threat landscape, and any changes to the organization's systems, applications, or business objectives.

With misuse case testing, organizations can check for known issues that might be particular to the type of component or architecture. In the previous price manipulating example, if a tester successfully changes the price, the organization could mitigate vulnerabilities with techniques such as input validation.

Test Coverage Analysis

Code testing can be done using various techniques already mentioned in this chapter, but with penetration testing, manual checks, and even misuse case testing, you are often only checking part of the functionality, and thus only part of the code is being checked at any one time. Test coverage analysis is done with tools such as JaCoCo for Java and Istanbul for JavaScript. It measures how much of an application's code, functionality, and components have been tested for potential vulnerabilities and security issues. By analyzing test coverage, organizations can ensure that their cybersecurity measures are comprehensive. The following steps are involved in a test coverage analysis:

1. **Establish clear criteria**: First, set out what constitutes adequate test coverage, considering factors such as code complexity, criticality of the application, and regulatory requirements.

   These criteria should encompass different aspects of test coverage, such as statement coverage, which measures the percentage of executable statements that have been tested; branch coverage, which assesses whether each possible branch or decision point in the code has been tested; and path coverage, which evaluates whether all possible paths through the code have been tested.

2. **Employ automated testing tools**: Tools such as static code analyzers, dynamic testing tools, and fuzz testing help identify potential vulnerabilities. These tools also provide valuable insights into test coverage, highlighting areas that may require additional testing.

3. **Strive for continuous improvement**: This can be done by refining testing strategies, methodologies, and tools. This may involve adopting new testing techniques, enhancing collaboration between development and security teams, or providing additional training and resources for testers.

Interface Testing

Modern software systems often consist of numerous interconnected components and rely on third-party services or APIs, and securing these interfaces is critical to ensuring the overall security of the application. Interface testing evaluates the security of interactions between different components, systems, or applications, using various tools and methods, including manual testing, API testers, and protocol analyzers such as Wireshark. Interface testing is carried out as follows:

1. **Design scope**: Begin by cataloging all the interfaces within the application, including internal interfaces between components, external interfaces with third-party services, and user interfaces. This should include identifying any dependencies, such as data formats, communication protocols, and authentication mechanisms.

2. **Establish clear objectives**: This could include checking that interfaces work as intended by validating input and output data, verifying error handling, or assessing the security of communication channels. These objectives should be aligned with the organization's overall cybersecurity goals.

3. **Execute test scenarios and cases**: You should carefully monitor the application's behavior and response to the simulated interactions. This can involve both manual testing techniques and automated tools. For example, using API testing frameworks such as Postman or SoapUI, you can simulate API calls such as GET and POST to verify their responses.

   Additionally, network protocol analyzers such as Wireshark can be used to inspect the data being transmitted between systems such as TCP or UDP connections, ensuring that they adhere to expected protocols. Manual testing can also play a crucial role, where testers interact with the application interfaces directly, observing and documenting any irregularities or security issues that automated tools might miss.

4. **Regular updates**: Interface testing scenarios should be checked regularly to reflect changes in the application's architecture, dependencies, or business requirements.

By systematically evaluating and securing these interfaces, organizations can reduce the risk of security breaches and enhance the overall stability and performance of their software systems.

Breach and Attack Simulations

Breach and attack simulation (BAS) tools simulate real-world cyberattacks on an organization's systems, networks, and applications. By automating the **tactics, techniques, and procedures (TTPs)** used by actual threat actors, BAS tools help organizations identify potential vulnerabilities, evaluate the effectiveness of their security controls, and improve their overall security posture.

First, develop realistic cyberattack scenarios, such as fuzz testing or phishing emails, that represent the most likely threats to the organization, considering factors such as industry, size, and technology. These scenarios cover a range of attack vectors, such as social engineering, malware, or unauthorized access.

Then, measure the effectiveness of existing security controls, such as firewalls, intrusion detection systems, or endpoint protection solutions. By identifying areas where these controls may be insufficient or misconfigured, organizations can prioritize improvements and optimize their cybersecurity investments.

Finally, regularly conducting BASs can help organizations demonstrate compliance with regulatory requirements and industry standards, such as GDPR, HIPAA, or PCI-DSS because BAS is automated testing.

BAS is an essential tool to proactively enhance cybersecurity. By simulating real-world cyberattacks, organizations can better identify potential vulnerabilities over penetration testing because testing can be done continuously because it removes the human element.

Compliance Checks

Compliance checks are an evaluation of an organization's adherence to security regulations and can identify security vulnerabilities that may not be compliant with regulations. Compliance checks typically involve a thorough review of an organization's IT infrastructure, policies, and procedures to identify areas of weakness or non-compliance.

One of the primary purposes of compliance checks is to ensure that an organization is satisfying industry-specific regulations, such as HIPAA, GDPR, or PCI-DSS. These regulations have strict requirements for data privacy, and compliance checks help organizations avoid costly penalties and reputational damage associated with non-compliance. For example, GDPR fines are in the tens of millions of euros per incident of **personally identifiable information (PII)** loss.

Another benefit of compliance checks is that they can help organizations identify areas for improvement in their cybersecurity measures. Though most organizations will have an idea about what security measures should be in place, regulatory measures can be valuable security frameworks and incentivize measures that a company might otherwise ignore.

Compliance checks are an essential component of an organization's security strategy. They ensure that an organization is adhering to industry regulations and help identify potential vulnerabilities that could be exploited by cybercriminals. By conducting regular compliance checks, organizations can reduce the risk of costly data breaches and other cyber incidents.

As you have seen, there are a number of different methods for testing security controls. The following table gives some brief strengths and weaknesses of each one:

Security Control Testing Method	Strengths	Weaknesses
Vulnerability assessment	Can identify known vulnerabilities in systems and applications	May not identify unknown vulnerabilities
Penetration testing	Can identify security vulnerabilities that can be exploited by attackers	May not be representative of real-world attacks
Log reviews	Can identify suspicious activity, such as unauthorized access attempts or data exfiltration	May not identify all suspicious activity
Synthetic transactions	Can be used to test the effectiveness of security controls	May not be representative of real-world transactions
Code review and testing	Can identify security vulnerabilities in code	May not identify all security vulnerabilities in code
Misuse case testing	Can identify potential security threats	May not identify all potential security threats
Test coverage analysis	Can identify security vulnerabilities that have not been tested	May not be accurate for complex systems
Interface testing	Can identify security vulnerabilities in interfaces	May not identify all security vulnerabilities in interfaces
Breach attack simulations	Can identify uncommon security vulnerabilities that can be exploited by attackers	Cannot simulate zero-day attacks
Compliance checks	Can identify security vulnerabilities that may not be compliant with regulations	May not identify all security vulnerabilities

Table 14.1: Comparisons of security control testing

Summary

In this chapter, we covered how effective assessment, test, and audit strategies are essential for ensuring the security and integrity of any organization's systems and data. Regular security control testing is a critical component of any security program and enables organizations to validate that security controls are working as intended and to detect any vulnerabilities that need to be addressed.

To conduct security control testing effectively, organizations must have a comprehensive understanding of their security risks and a robust testing methodology. This includes identifying the types of security controls to be tested, selecting appropriate testing techniques, and defining test objectives and success criteria. Organizations should also have a plan for addressing any issues or vulnerabilities that are identified during testing and should be prepared to adapt their security controls and procedures as needed.

Assessment, test, and audit strategies are crucial for maintaining a strong security posture, and regular vulnerability scans, penetration testing, and log analysis are essential parts of this. By implementing effective compliance strategies, organizations can ensure that their systems and data remain secure and protected from potential threats. The next chapter will look at designing and conducting security testing.

Further Reading

- *Updates - October 2022*, MITRE ATT&CK Framework Series, October 2022: `https://packt.link/76fT5`

- *Intelligence-Driven Computer Network Defense Informed by Analysis of Adversary Campaigns and Intrusion Kill Chains*, Hutchins, Cloppert, Amin, Ph. D, Lockheed Martin Corporation, 2010: `https://packt.link/TPL1t`

- *Cybersecurity Assessment Tool*, Federal Financial Institutions Examination Council, May 2017: `https://packt.link/0W2Yk`

Exam Readiness Drill – Chapter Review Questions

Apart from a solid understanding of key concepts, being able to think quickly under time pressure is a skill that will help you ace your certification exam. That is why working on these skills early on in your learning journey is key.

Chapter review questions are designed to improve your test-taking skills progressively with each chapter you learn and review your understanding of key concepts in the chapter at the same time. You'll find these at the end of each chapter.

> **How to Access These Materials**
>
> To learn how to access these resources, head over to the chapter titled *Chapter 24, Accessing the Online Resources*.

To open the Chapter Review Questions for this chapter, perform the following steps:

1. Click the link – `https://packt.link/chapter14`.

    Alternatively, you can scan the following **QR code** (*Figure 14.1*):

Figure 14.1: QR code that opens Chapter Review Questions for logged-in users

2. Once you log in, you'll see a page similar to the one shown in *Figure 14.2*:

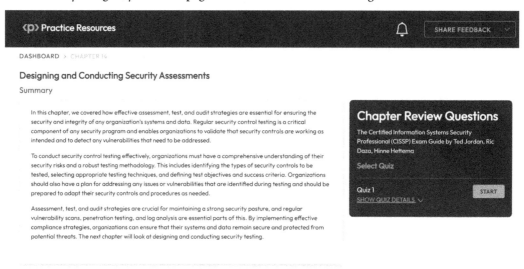

Figure 14.2: Chapter Review Questions for Chapter 14

3. Once ready, start the following practice drills, re-attempting the quiz multiple times.

Exam Readiness Drill

For the first three attempts, don't worry about the time limit.

ATTEMPT 1

The first time, aim for at least **40%**. Look at the answers you got wrong and read the relevant sections in the chapter again to fix your learning gaps.

ATTEMPT 2

The second time, aim for at least **60%**. Look at the answers you got wrong and read the relevant sections in the chapter again to fix any remaining learning gaps.

ATTEMPT 3

The third time, aim for at least **75%**. Once you score 75% or more, you start working on your timing.

> **Tip**
> You may take more than **three** attempts to reach 75%. That's okay. Just review the relevant sections in the chapter till you get there.

Working On Timing

Target: Your aim is to keep the score the same while trying to answer these questions as quickly as possible. Here's an example of how your next attempts should look like:

Attempt	Score	Time Taken
Attempt 5	77%	21 mins 30 seconds
Attempt 6	78%	18 mins 34 seconds
Attempt 7	76%	14 mins 44 seconds

Table 14.2: Sample timing practice drills on the online platform

> **Note**
> The time limits shown in the above table are just examples. Set your own time limits with each attempt based on the time limit of the quiz on the website.

With each new attempt, your score should stay above **75%** while your "time taken" to complete should "decrease". Repeat as many attempts as you want till you feel confident dealing with the time pressure.

15
Designing and Conducting Security Testing

In the previous chapter, you gained an understanding of different security control testing models, such as vulnerability assessments and penetration testing. In this chapter, you will explore security audits in depth – specifically, how to conduct audits and analyze the output.

An important part of security assessment and testing is collecting technical and administrative data to make sure that our systems are secure. With regular reviews and testing of systems and processes, gaps in knowledge and practices can be found and remediated. The data collecting process during such reviews includes checking that identity management and access control are being implemented correctly, reviewing training and awareness programs for scope and effectiveness, and ensuring that backup data is properly collected and stored in the event of a disaster.

This chapter covers disaster recovery, disaster recovery planning, and business continuity. You will examine aspects of disaster recovery planning, such as management involvement and verification of systems backups. You will also review the differences between disaster recovery and business continuity.

By the end of this chapter, you will be able to answer questions on:

- Collecting security process data
- Analyzing test output and generating reports
- Conducting or facilitating security audits

Collecting Security Process Data

To maintain an effective security posture, organizations need to **collect and analyze data** in their technical and administrative security processes. This data can be used to identify areas where security controls are weak or missing, allowing them to make informed decisions about how to improve security.

There are a number of different sources of data that can be used to collect information about security processes. These include the following:

- Security logs

- Audit reports

- Interviews with security staff

Once data has been collected, it needs to be analyzed to identify areas where security controls are weak or missing. This analysis can be done manually or using automated tools. These tools include **security information and event management** (**SIEM**) systems such as Splunk, vulnerability detection tools such as Nessus, **endpoint detection and response** (**EDR**) solutions such as Carbon Black, and **data loss prevention** (**DLP**) solutions such as Digital Guardian. This will be discussed in more detail in *Chapter 17, Security Operations*.

Once vulnerabilities have been identified, organizations must take steps to improve security. This can involve implementing new security controls, strengthening existing controls, or providing additional training to security staff.

The next section covers security analysis, from reviewing accounts and backup verifications to planning for disaster recovery and business continuity. The section will look at all these factors, starting with account management.

Management Review and Approval

When you carry out security testing, an organization's management should be involved to make sure the activities are conducted comprehensively and effectively. The management will not only help to ensure that the right tests are being carried out but also ensure that all parties, such as other departments and employees, are fully engaged and cooperating in the process.

When carrying out a testing program, management should review and approve the following:

- The scope of the assessment or test

- The methodology to be used

- The qualifications of the individuals conducting the assessment or test

- The budget for the assessment or test

- The schedule for the assessment or test

Management should also review and approve the results of the assessment or test. This will help to ensure that an organization takes appropriate action to address any security vulnerabilities identified.

There are several benefits of management review and approval for security assessments and testing, including increased confidence in results, as they will better track accuracy and reliability, and improved compliance in meeting industry regulations and standards.

By following the guidelines of management, organizations can ensure that these security assessment activities are conducted comprehensively and effectively.

Account Management

Regular **security reviews** of account management practices are essential to ensure compliance with organizational policies and industry regulations. Such reviews require assessments of **account creation**, **modification**, **termination**, and **access controls**. By examining account provisioning and de-provisioning procedures, organizations can detect issues such as unauthorized access, privilege escalation, and data leakage risks.

A critical aspect of **account management** reviews includes evaluating the principle of **least privilege**, which dictates that users should only be granted the minimum necessary permissions to perform their job functions. For example, you might be given the privilege to access sales data because you work in the sales department. Then, you get transferred to marketing and are provided access to focus group data. If your previous rights have not been removed, this is known as **account sprawl**. Account management reviews can detect such issues and resolve them. In the case of this example, your access to sales data is removed. By detecting and resolving account sprawl, firms reduce the attack surface and limit damage if an account is compromised.

Reviews should also assess **password policy effectiveness**, including **complexity requirements**, **expiration cycles**, and **password reuse** restrictions. Strong **password hygiene** is critical to preventing unauthorized access. Regular account reviews verify account ownership and validation of active user accounts. Removing inactive and orphaned accounts mitigates the risk of unauthorized access.

Account monitoring is a vital component of an account management security program. By continuously analyzing account activity, companies can identify threats such as unauthorized login attempts, data exfiltration, or lateral movement.

To ensure the effectiveness of account management security, organizations should establish a culture of security awareness among staff. Training programs should emphasize the importance of protecting account credentials, recognizing phishing attacks, and reporting suspicious activities.

Regular security awareness campaigns can reinforce good account management practices and empower employees to be the first line of defense against cyber threats. To do so effectively requires a way to measure performance and risk. How this is measured is covered in the next section.

Key Performance and Risk Indicators

While planning a security review, it is useful to work out some metrics that can accurately quantify key features of a system, such as availability or number of threats detected. **Key performance indicators (KPIs)** and **key risk indicators (KRIs)** can be used to clearly communicate an organization's security posture, as well as set goals for any further action. KPIs are used to evaluate the effectiveness of an organization's cybersecurity controls, while KRIs are used to identify potential risks and vulnerabilities that may pose a threat to the organization's security.

KPIs are quantitative measures typically used to track the effectiveness of security controls, such as firewalls, intrusion detection systems, and antivirus software. Examples of cybersecurity KPIs include the number of cyberattacks prevented successfully, the average time to detect and respond to a cyberattack, and the percentage of vulnerabilities identified and remediated within a given timeframe.

Conversely, KRIs are qualitative measures that help to identify emerging threats and potential gaps in an organization's security posture. Examples of cybersecurity KRIs include the number of security incidents detected, the level of employee awareness regarding security, and the effectiveness of security awareness training programs. KPIs and KRIs should be set in line with the organization's overall goals, as well as specific risks and regulatory requirements.

Backup Verification Data

Backup verification data is a critical component of any data backup and recovery plan. It is the data that is used to verify that the backup is complete and accurate and can be restored successfully. Backup verification data can include a variety of items, such as the following:

- The date and time of the backup
- The name of the backup file
- The size of the backup file
- A checksum or hash value of the backup file
- A list of the files that were included in the backup

> **Note**
> Backup verification data should be stored in a secure location, separate from the backup itself. This will help you ensure that the data is not lost or damaged in the event of a disaster.

There are several ways to verify that a backup is complete and accurate. One common method is to use a **checksum** or **hash value**. A checksum is a mathematical value calculated from the contents of a file. If the file is changed, the checksum will also change. This makes checksums and hash values useful tools for verifying the integrity of data files.

Another common method for verifying backups is to restore them to a test environment. This will allow you to verify that the data can be restored successfully and that it is in the correct format. For best security, employ separation of duties such that the person verifying the backups is different than the person making the backup.

Training and Awareness

Assessing the effectiveness of training and awareness is critical in ensuring that employees have the knowledge to protect corporate assets. Important metrics include behavior change and incident reduction. For example, gamification such as quizzes or **phishing campaigns** can evaluate knowledge retention and behavior change.

In a phishing campaign, the security department creates a simulated phishing attempt to test how employees respond to it. The goal is to see which employees fall for the fake phishing email by clicking on a malicious link or providing sensitive information. The results of this test can help determine whether further education and training on phishing awareness is needed. If a significant number of employees fall for the simulated phishing attempt, it suggests that more training or awareness efforts are necessary to strengthen an organization's security posture.

Ultimately, a compelling indicator of success is the reduction of security incidents by analyzing trends. A reduction in security incidents can mean a successful awareness program. Security incidents can help identify where training needs to be improved.

Qualitative feedback, including surveys and focus groups, can provide insights into training relevance and uncover areas of improvement. Observing staff carrying out daily tasks, especially if they routinely handle sensitive information, can help reveal knowledge gaps.

Both quantitative and quantitative data help organizations determine the impact and results of their security training and awareness programs.

Disaster Recovery and Business Continuity

Business continuity (BC) and **disaster recovery** (DR) planning help organizations understand how much data they can afford to lose, how quickly they need to be able to recover from an outage, and how much downtime they can tolerate. Both concepts are closely related and aim to ensure the resilience and stability of an organization in the face of potential disruptions. However, they address different aspects of the disaster response process.

BC focuses on maintaining the essential functions of an organization during and after a disruptive event, such as a natural disaster, cyberattack, or major equipment failure. BC planning includes identifying critical processes, identifying **single points of failure** (SPOFs), evaluating potential risks, and developing strategies to minimize the impact of disruptions on an organization.

DR is an integral part of BC. It is a subset of BC that specifically deals with the recovery of IT systems and infrastructure following a disruptive event. DR planning involves creating detailed procedures to restore hardware, software, and data as quickly and efficiently as possible, ensuring the resumption of critical operations.

The key metrics used in BC/DR planning that help organizations plan for and measure their ability to recover from disruptive events are **recovery point objective (RPO)**, **recovery time objective (RTO)**, **work recovery time (WRT)**, and **maximum tolerable downtime (MTD)**. These are described in detail as follows:

- RPO represents the maximum acceptable amount of data loss an organization can tolerate in the event of a disruption in units of *time*, referring to the age of the latest backup or data replication that can be used for recovery. A lower RPO indicates a lower tolerance for data loss, hence the need for more frequent backups. Sensitive and valuable data such as patient records or payments might have short RPOs, measured in minutes. Other data, such as an inventory, might have RPOs of days. Although it might seem obvious to have all data with shorter RPOs, backups require resources, and the shorter the RPO, the higher the expense.

- RTO is the target time to restore critical systems and data after a disruption before data and system verification. If an organization can meet its RTO, it is better prepared to quickly recover from a disruptive event.

- WRT starts after the RTO ends and is the time it takes for a business process to return to normal functioning levels. It considers the time needed to catch up on any backlog of work that may have accumulated during the disruption, returning systems to their original integrity to ensure that malware is removed and the latest data is recovered. Where WRT ends, normal operations begin.

- MTD is the maximum amount of time that an organization can tolerate a disruption before it leads to business failure and includes factors such as staff availability, third-party support, and recovery site readiness. For example, an online bookshop might be able to go down for a few hours or even a couple of days, but after a while, it will lose customers to competitors. MTD could be two days. MTD is greater than the sum of RTO and WRT and must never be exceeded.

The preceding metrics are essential for developing effective business continuity and disaster recovery plans. They help organizations set recovery objectives, evaluate their preparedness, and prioritize efforts to minimize the impact of disruptions on their operations.

Make sure to understand the differences between these metrics, and choose appropriate values for each of them in order to protect a business from the consequences of a disaster. The key differences between the metrics are shown in *Table 15.1*.

Metric	Definition
RPO	The maximum amount of data that can be lost in a disaster before it becomes unacceptable to the business in units of time
RTO	The maximum amount of time that a business can be down before conducting validation of data, verification of critical systems, and evaluation of processes
WRT	The amount of time that it takes to validate data, critical systems, and processes, and for employees to return to their normal work routines after a disaster
MTD	The maximum amount of downtime that a business can experience before it suffers irreparable harm

Table 15.1: Comparisons of the key metrics for business continuity and disaster recovery

In order to determine the appropriate values for RPO, RTO, MTD, and WRT, it is important to conduct a **business impact analysis** (**BIA**), as discussed in *Chapter 3*. A BIA is a process that identifies the critical business functions and the impact that a disruption would have on a business. The results of a BIA can be used to determine the appropriate values for RPO, RTO, MTD, and WRT. This is done by considering not only the impact of downtime of different functions or units but also the cost of keeping it to a minimum.

For instance, if a business unit incurs a cost of $1,000 per day when it is non-operational, but maintaining the resources required to keep the RTO under 24 hours costs only $100 per day, it may not be cost-effective to invest in those resources. (In practice, calculations are more complex, considering risk factors and probabilities as well.) Understanding these key metrics enables businesses to make informed decisions about disaster planning.

Analyzing Test Output and Generating a Report

Test outputs are the results generated after conducting security audits on a computer system or application (audits were discussed in more detail in *Chapter 14, Designing and Conducting Security Assessments*). These tests can range from penetration testing and vulnerability scanning to code analysis and intrusion detection. Analyzing the results helps security professionals identify vulnerabilities, assess their severity, and prioritize remediation efforts. This section explores the importance of these steps and delves into remediation, exception handling, and ethical disclosure within the context of cybersecurity audits.

Remediation

Remediation is the process of addressing identified vulnerabilities to minimize the risk of exploitation. To effectively remediate vulnerabilities, security professionals should follow these steps:

1. Prioritize vulnerabilities based on their severity, potential impact, and ease of exploitation. This will help you to allocate resources efficiently and address the most critical risks first.

2. Regularly update software and apply security patches to fix known vulnerabilities. Often, vendors provide patches to address security issues. If no patch is available, consider workarounds such as network segmentation or enhanced monitoring.

3. Review and adjust system configurations to limit potential attack vectors. Ensure that security features are enabled and unnecessary services are disabled.

After placing mitigations, implement continuous monitoring to detect any changes in a system or attempts to exploit vulnerabilities. Finally, validate systems after remediation to confirm that vulnerabilities have been addressed and no new issues have been introduced.

Exception Handling

Exception handling refers to the process of addressing unexpected events or irregular findings that occur during a security audit. The process involves developing a systematic approach to deal with situations that deviate from standard audit procedures or expected results.

Common exceptions include unexpected failures of systems or networks, account access restrictions, and vulnerabilities found outside of the initial audit scope. The steps for handling audit exceptions are as follows:

1. Identify and log exceptions as they occur.

2. Analyze exceptions and determine the impact on the audit objectives.

3. Develop and execute a plan to address discovered exceptions, such as adjusting the audit schedule, acquiring additional resources, or creating a workaround.

4. Document an exception and how it was addressed. State how this might impact the audit findings.

5. Communicate results and implications to stakeholders.

By effectively managing exceptions, security auditors maintain the integrity of the audit process, provide important insights on security posture, and aid in complying with security standards such as ISO, GDPR, or HIPAA.

Ethical Disclosure

Ethical disclosure (also known as responsible disclosure) is the process of informing affected parties and the public about security vulnerabilities in a timely and responsible manner. This practice helps protect users from potential harm while giving vendors the opportunity to address the issues. For example, a security flaw in a database that allows unauthorized access to records would need to be reported quickly so that the vendor can start remediation efforts. However, this report should be done confidentially to stop other malicious actors from exploiting the flaw.

Security auditors must limit the distribution of vulnerability information to those who need to know, such as affected vendors or parties responsible for remediation. Also, they must report high-impact vulnerabilities as soon as possible to facilitate prompt remediation.

If a security auditor detects violations during an audit, they face an ethical and legal responsibility to address the situation appropriately. The auditor must verify the findings before taking any action to avoid overreacting to a situation. For instance, in the case of suspicious money transfer records, they need to first verify that those records were not created by a bug before reporting suspected fraud. Auditors should report their findings to their management. This allows an organization to take appropriate action, such as launching an internal investigation or contacting law enforcement. Of course, clear felonies such as finding media depicting the abuse of minors must be immediately reported to law enforcement by the auditor.

Ultimately, the auditor's actions in response to detected criminal activity should be guided by ethical principles, professional standards, and legal requirements to protect an organization and its stakeholders.

As mentioned in *Chapter 14*, audits can be conducted internally, externally, or by a third party. Internal audits are run by an organization's staff, external audits are run by a vendor or business partners, and third-party audits are conducted by outside organizations to ensure that a firm meets regulations.

Summary

This chapter covered the process of collecting security data during a security assessment. Planning your assessment and getting the right data is key, as well as ensuring that your organization follows security processes and that they're verified and validated. This planning should be done systematically with the full involvement of management. The important parts of this assessment include ensuring that the right people have the right accounts with the right permissions, as well as checking for any accounts that are orphaned or not in use. Also important is checking that training is effective, with gamification such as phishing attacks, as well as ensuring that backups are done properly. Key performance and risk indicators provide a way to ensure that we're properly following our audit plan.

KPIs are also important in disaster recovery and business continuity, giving data on an organization's and department's tolerance for downtime and data loss, as well as targets for recovery. We looked at the differences between RPO, RTO, WRT, and MTD and why these are critical metrics if there is a major incident, such as a flood or earthquake.

Finally, the chapter discussed post-audit procedures, including remediation, and the differences between internal, external, and third-party audits. You saw what to do when exceptions were raised in the process of auditing. We emphasized the importance of addressing high-impact vulnerabilities and ensuring that audit recommendations are implemented. The next chapter, Planning for Security Operations, will cover what needs to be done to prepare for security incidents, and subsequent investigations.

Further Reading

- The following are some good references for disaster recovery, running a cybersecurity audit, and understanding the differences between primary and third-party audits:

- NIST Special Publication 800-34 Rev. 1, *Contingency Planning Guide for Federal Information Systems* by National Institute of Standards and Technology: `https://packt.link/Tt0zY`.

- NIST Special Publication 800-115, *Technical Guide to Information Security Testing and Assessment* by National Institute of Standards and Technology: `https://packt.link/fYtEn`, 2021.

- *IS Standards, Guidelines, and Procedures for Auditing and Control Professionals*, ISACA: `https://packt.link/BmUkD`, January 15, 2009.

Exam Readiness Drill – Chapter Review Questions

Apart from a solid understanding of key concepts, being able to think quickly under time pressure is a skill that will help you ace your certification exam. That is why working on these skills early on in your learning journey is key.

Chapter review questions are designed to improve your test-taking skills progressively with each chapter you learn and review your understanding of key concepts in the chapter at the same time. You'll find these at the end of each chapter.

> **How to Access These Materials**
>
> To learn how to access these resources, head over to the chapter titled *Chapter 24, Accessing the Online Resources*.

To open the Chapter Review Questions for this chapter, perform the following steps:

1. Click the link – `https://packt.link/chapter15`.

   Alternatively, you can scan the following **QR code** (*Figure 15.1*):

Figure 15.1: QR code that opens Chapter Review Questions for logged-in users

2. Once you log in, you'll see a page similar to the one shown in *Figure 15.2*:

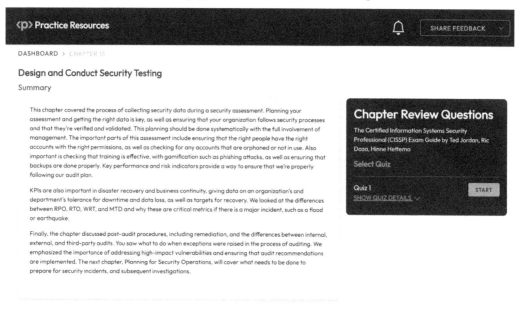

Figure 15.2: Chapter Review Questions for Chapter 15

3. Once ready, start the following practice drills, re-attempting the quiz multiple times.

Exam Readiness Drill

For the first three attempts, don't worry about the time limit.

ATTEMPT 1

The first time, aim for at least **40%**. Look at the answers you got wrong and read the relevant sections in the chapter again to fix your learning gaps.

ATTEMPT 2

The second time, aim for at least **60%**. Look at the answers you got wrong and read the relevant sections in the chapter again to fix any remaining learning gaps.

ATTEMPT 3

The third time, aim for at least **75%**. Once you score 75% or more, you start working on your timing.

> **Tip**
> You may take more than **three** attempts to reach 75%. That's okay. Just review the relevant sections in the chapter till you get there.

Working On Timing

Target: Your aim is to keep the score the same while trying to answer these questions as quickly as possible. Here's an example of how your next attempts should look like:

Attempt	Score	Time Taken
Attempt 5	77%	21 mins 30 seconds
Attempt 6	78%	18 mins 34 seconds
Attempt 7	76%	14 mins 44 seconds

Table 15.2: Sample timing practice drills on the online platform

> **Note**
> The time limits shown in the above table are just examples. Set your own time limits with each attempt based on the time limit of the quiz on the website.

With each new attempt, your score should stay above **75%** while your "time taken" to complete should "decrease". Repeat as many attempts as you want till you feel confident dealing with the time pressure.

16
Planning for Security Operations

As cyber incidents are inevitable, operational security teams should ensure they are prepared to handle them so that response can be swift and efficient. Incident handling involves a prescribed cycle, investigation management, which will be covered in detail in *Chapter 17*. This chapter will discuss the planning that needs to be done prior to an incident and will focus on the specific operational practices of the incident response cycle that organizations must have in place before an incident even occurs.

A key element of pre-incident activities that organizations need to undertake is preparation for detecting and responding to incidents. This entails ensuring that events are visible and traceable and that the right people, processes, and technologies are in place, tested, and ready for deployment when an incident occurs. The key element of incident planning is operational resilience; the organization should be able to survive any incident, with the impact of the incident as minimal as possible.

This chapter will focus on incident planning and related policy activities and cover the following:

- Understanding and complying with investigations
- Conducting logging and monitoring activities
- Performing configuration management
- Applying foundational security operations concepts
- Applying resource protection

You will start by looking at your and your organization's role in cyber investigations.

Understanding and Complying with Investigations

A cyber investigation can be complex and lengthy. For those reasons, it is important that cybersecurity leadership in an organization has the requisite knowledge of local legislation and law enforcement obligations, as well as the security team's responsibilities, roles, evidential standards, and strategies needed to collect evidential information on a wide range of artifacts.

It should go without saying that these elements need to be in place before any incident occurs since creating them during the incident will likely result in errors and carry significant risks; incidents may not be handled as well as they should be, and important steps may get missed.

A security program should operate from a clear set of responsibilities, guidelines, and understanding of its own legal obligations and should develop a robust set of practices that are predictable, repeatable, and well practiced so that they can be executed in the heat of the moment during an incident.

The following sections will outline some principles and policy considerations that each team must consider when creating an incident response capability.

Evidence Collection and Handling

Digital evidence is fragile and easy to destroy or alter to the point where it becomes unusable in further investigations.

Additionally, it is important to recognize that the vast majority of cyber incidents investigated by a team will never become a case in a court of law. Nevertheless, there are good reasons to maintain robust evidential standards during investigations. Examples of such standards are published by the Scientific Working Group on Digital Evidence (https://www.swgde.org/). Among these reasons are the following:

- Adherence to robust evidential standards will help an operational security team maintain a high level of documentation and repeatability of their actions during an investigation.

- In the early stages of an investigation, it is often not known whether a case will end up in court, and although the vast majority do not, for the few that do, robust evidential standards will be required.

- Even cases that do not make it to court become part of a process of threat landscape assessment done by the cyber team in the organization. A robust incident handling process will improve the quality of threat landscape assessment and the depth to which an organization understands its attackers (which is a prerequisite for handling successor attacks from the same and similar groups).

Operational security teams should always be aware of new legislation, for instance, on privacy, that will affect how they do things during an incident. Other recent changes in legal approaches focus on whether an organization is part of the critical infrastructure of a country or region, such as an energy provider, road infrastructure, or healthcare systems. Legislation regulates various aspects of incident response such as reporting requirements and broader government involvement. In general, it is not possible to outline what the legal requirements around incident handling currently are in each region of the world, and even less what they will be in the future. However, it is vital that operational security teams inform themselves and consider their obligations during incident planning.

With a potential for legal involvement, it should not come as a big surprise that some principles from policing have made it into digital investigations so that IT security incidents are now treated as digital crime scenes. The most well-known of these is **Locard's exchange principle**.

Locard's Exchange Principle

In criminal investigations, evidential scenes are handled according to Locard's exchange principle. Locard's exchange principle states that every time someone enters or exits a crime scene, an exchange takes place. Such exchanges then leave traces. The perpetrator of a crime will take something to the crime scene and leave with something from it. Both can be used as forensic evidence.

In the physical world, this exchange is often something tangible—a fingerprint, DNA, a scent, threads of clothing, or similar. In the digital world, the *exchange* is often the exchange or alteration of information on evidence. The key point is that Locard's exchange principle involves anyone coming into contact with evidence—whether they be a criminal, an investigator, or a report writer.

In digital investigations, this means any interaction with a piece of digital evidence—such as the simple act of shutting something down or even pulling the power cord on a device—will have the consequence that the state of the evidence changes. A regular shutdown is a good example. During the regular shutdown of a device, files are closed, handles are destroyed, the contents of the RAM disappear, entries are added to the log files on disk, registry keys may be overwritten, and files may be deleted.

Locard's exchange principle involves anyone coming into contact with evidence. This is related to the concept of a chain of custody, as discussed next.

Chain of Custody

A corollary of Locard's exchange principle is that all digital investigations (especially those where devices are involved) must have a documented **chain of custody**. A chain of custody details how a device was handled after the investigation began. Found in a place by an incident responder, taken in, put under lock and key, and handed to a law enforcement official a day later would be one example.

Ephemeral Evidence

Another aspect of digital evidence is that it is **ephemeral**. This implies that it can be easily destroyed. Examples include information held in RAM or swap files and temporary registry keys. Now that many cyberattacks involve *living-off-the-land* techniques or employ fileless malware, the contents of RAM are important pieces of evidence that can no longer be ignored in investigations.

> **Note**
>
> Living off the land involves the use of native system administrative tooling by an attacker during an incident. Examples are PowerShell in Windows systems, with the use of `netcat`, `net use`, `net group`, or `dsquery`, to mention some examples. Living off the land has several advantages for attackers. There is no specific tooling to bring into the victim environment, reducing the risk of detection by detection and response software, and often, the use of such native tooling is not logged, ensuring that fewer traces are left.

Policies and Guidelines

In the collection and handling of evidence, you also need to be aware of the legal requirements of the jurisdiction in which you operate. It is best to discuss this with someone from your jurisdiction who has prior experience with those requirements, and it is also a good idea to obtain legal advice on what the obligations of the local team are in each jurisdiction.

The following is a brief guideline of the relevant laws present in most jurisdictions that impact cyber investigations:

- **Privacy laws**: These come into play during the investigation of the behaviors of a specific person or group of persons.

- **Criminal legislation**: This outlines what is considered a crime and what should be notified to law enforcement.

- **Evidential standards**: These may differ between jurisdictions but are crucial to all investigations.

- **International treaties**: These regulate extradition.

- **Notification framework**: Different types of cyber incidents may be notifiable to different agencies, depending on the jurisdiction. Be aware of what needs to be notified and the notification process, such as which agency to report to.

This list is by no means complete and will depend on the nature of the organization that the team functions in as well as the jurisdiction.

Reporting and Documentation

If you wait for an incident to happen before working out what to report and document, you run the risk of missing crucial steps and details, such as regulatory requirements, and not meeting the needs of key stakeholders. So, the **reporting and documentation of cyber incidents** is something to think about in advance, with reference to regulatory bodies and management. A report on a cyber incident should meet the following requirements:

- It should provide an accurate statement of events, actions taken, and the result of those actions
- It is often a good idea to provide a timeline of the event that can be used to measure the effectiveness of the response as a service, with details such as time to detect, time to respond, and time to resolve
- It should contain some narrative that will allow the incident to be communicated to business stakeholders if required
- Any notifications should be documented

It is often also a good idea to store away the artifacts related to an incident should a more in-depth investigation become necessary in the future. Another reason such storage is a good idea is that some attacker groups create a string of incidents, and having the data on past incidents handy helps correlate and determine whether an individual attack is part of a bigger picture or not. Such *bigger pictures* are almost always of strategic importance to both the team itself and the organization.

A reporting system should ensure that reports can be searched and incidents for a given time period can be mapped out.

Investigative Techniques

When conducting or assisting with investigations, you will likely have to assist in gathering evidence from or allowing access to devices and storage, and any mobile, embedded, **internet of things (IoT)**, or **industrial control system (ICS)** infrastructure on your network. This sort of investigation is sometimes referred to as eDiscovery and can involve servers, desktops, laptops, mobile devices, and embedded systems. Each of these has specific technical procedures for how this is done. We do not cover the technical details here but limit ourselves to a general overview of approaches.

For many types of devices, the collection procedures can be complex and involve software and hardware that is not commonly used during *business-as-usual* IT operations. It therefore makes sense to understand the principles of some of the common procedures.

Disk Forensics

Disk forensics is the process of analyzing hard disks to find evidence of malware intrusion or **advanced persistent threat** (**APT**) implants. Disk forensics analyzes entire hard disks and can find deleted files, reconstruct (some of) the data in these files, and provide a timeline of write and delete activities on files and areas of the drive. Hard disk forensics is still one of the mainstays of law enforcement tools such as Encase.

Generally, hard disk forensics involves making a full copy of each block of the target disk in a read-only disk enclosure to prevent the disk evidence from being contaminated. For traditional platter disks, a specialized hardware device was used that would mount the disk in a read-only mode so that an accurate copy could be made.

By contrast, most hard drives these days are **solid-state drives** (**SSDs**), which are more ephemeral than the once ubiquitous platter drives. The controller in an SSD performs significantly more complex tasks than those of a controller in a platter drive. An SSD controller allocates files across multiple pages in its flash memory and also performs read/writes of partial pages by reading a page, modifying it, storing it on a different page, and then queueing the original page for deletion. More advanced and expensive SSDs have automatic garbage collection (i.e., the deletion of unused file space) that minimizes forensic traces.

A significant disadvantage of hard disk forensics is that it is very time-consuming as the size of modern-day hard disks has grown. As a result, hard disk forensics is likely of little help with the analysis phase of resolving an incident since, in general, you need tools that can get to answers quicker than hard disk forensics can to deal with an incident sooner.

Memory Forensics

Not all evidence of a malicious intrusion is kept on disk. Memory forensics is necessary when incidents involve *living-off-the-land* techniques or fileless malware.

Memory forensics can find evidence of running processes and the command lines with which they were started, any techniques of process injection, and malicious drivers running on the system. An advantage of memory forensics is that the amount of data that needs to be analyzed is significantly smaller than what needs to be investigated in disk forensics.

Compared with disk forensics, memory forensics is much less automated and requires a highly skilled investigator. However, when incidents need to be resolved quickly and accurately, such as if you have a critical breach that needs investigating to get a system up and running again, it is well worth investing in.

ICS, IoT, and Mobile Devices

In addition to servers, laptops, and desktops, many organizations also have additional devices on their network, such as ICS and building management devices, IoT devices, and mobile devices, which may need to be collected and analyzed in case of an incident. Security leaders should understand what devices they have and have a plan for dealing with them should an incident involving these devices occur.

Extended Detection and Response

The old antivirus software has moved on to now include **endpoint detection and response** (**EDR**) capabilities, as well as **extended detection and response** (**XDR**) capabilities. In general, the current generation of anti-malware tooling stores a large digital record of the actions taken on a workstation, sites visited, files opened and closed, and so on. Sorting these records is an important addition to pre-existing forensic techniques when it comes to security data. In these data lakes, analysts can use a query language (and, increasingly, artificial intelligence) to query the fleet of endpoints for the occurrence of certain events or the presence of certain running software.

EDR and XDR software is an extension of traditional antivirus software, which was largely based on checking files and executions against a signature database of known bad files and software. EDR and XDR software still retain this layer of signature-based detection but, in addition, integrate more deeply into the endpoint system to monitor network configurations, changes in the registry or configuration files, devices inserted, or data uploaded to external locations, to mention just a few. Many EDR systems can also take a copy of executable files and execute them in a sandbox.

In addition, XDR also includes telemetry from network devices to gain a deeper insight into the network traffic generated by potential malware.

For an analyst, the main distinguishing feature of EDR and XDR systems is that they can keep a record of causal change of events on an endpoint, which helps the analyst determine whether an intrusion has happened and what actions to undertake next. Since EDR and XDR systems also contain a query language associated with the data lake, analysts also need skills in developing basic and advanced queries to support practices such as incident response and threat hunting.

Digital Forensics Tactics, Techniques, and Procedures

Similar to how **tactics, techniques, and procedures** (**TTPs**) are ascribed to attackers, the actions of an incident responder can also be categorized by a set of defined TTPs. **Tactics** are high-level objectives, such as detection and containment, and correspond somewhat to the stages of incident management discussed in *Chapter 17, Security Operations*. **Techniques** are the specific tools and operations, such as log management or containment, that help achieve those objectives. **Procedures** are usually well-documented, technology-dependent steps that are followed to get a quality result, such as using antivirus software or SIM management.

> **TTPs and MITRE ATT&CK**
>
> TTPs are enumerated in the Mitre ATT&CK framework, which you should be familiar with. You can find the framework here: `https://attack.mitre.org/`. An example of a tactic is privilege escalation, where an attacker aims to gain higher-level permissions than the ones they currently have. Under tactics, we have techniques, which refer to the specific methods that attackers use to achieve this gaining of permissions. Examples of techniques are access token manipulation or token forgery. Procedures are the specific steps, tools, and methods attackers will follow to execute the technique.

In cyber investigations, it is important that teams follow procedures precisely. It is easy, in the heat of the moment, to overlook or skip necessary steps.

Artifacts

Cyber investigations involve many different artifacts (essentially, data or equipment to be analyzed), and each type has its own collection techniques. Some of the artifacts that teams need to consider are as follows:

- **Endpoints**: These can be either desktops or laptops, running Windows, macOS, or Linux. Each of these operating systems has its own disk formats, kernel, and memory structures.

- **Mobile devices**: These often run a version of either Android or iOS.

- **Servers and virtualization environments**: These run a **hypervisor** operating system with a number of guest operating systems. At the hypervisor layer, the memory of the guest operating system is often mapped to disk.

- **Storage devices**: These often use varieties of **Redundant Array of Independent Disk** (**RAID**) to efficiently lay out data across multiple disks with varying levels of redundancy. Some knowledge of the internal workings of these devices, especially controllers, may be required in investigations

- **Operational Technology** (**OT**) **and automation**: This involves its own devices, such as programmable logic controllers, which run a proprietary operating system.

The planning for handling cyber incidents must consider whether these devices will need to be investigated, how they will be investigated (especially where the specialized technical knowledge to do so will come from), and what tools will be required to reliably collect forensic data from these devices. These tools must be acquired and collected before any incident. Care must also be taken that these tools are regularly tested and updated during the usual operations of the team.

Jump Bags

It is often a good idea to collect all the tools and devices needed for incident response in a jump bag that can be taken to the site of the incident. A **jump bag** is a package (usually a bag) of everything needed for incident response, including hardware and software tooling, cheat sheets, communication equipment, and any additional storage devices, such as USB and hard disks, all of which have been tested and can just be taken to the site of an incident (the *jump* aspect of the jump bag). The jump bag concept ensures that all tools are available to the responder at the time they are needed. There are no standard best practices for the creation of a jump bag since the tools needed will depend on the environment and technology that needs to be investigated. It is important that the tools taken in the jump bag are regularly inspected and tested to make sure they are ready to be deployed when needed. There is nothing worse than arriving at the site of an incident and finding half the tools missing and the other half non-functional.

For industrial and operational technology environments, team members often need safety training and specialized clothing and equipment to access the site. In these cases, the jump bag may be large and have a variety of items such as safety equipment, a cheat sheet of processes and procedures, management software for industrial devices, as well as all the equipment needed for IT environments.

For cloud environments, one can also consider a digital jump bag, which is a collection of investigation tools and scripts that have been tested and maintained and are ready to go in case of an incident. This can be kept on encrypted USB drivers, secure cloud locations, or any other storage medium. Multiple copies can be kept, with each member of the security team having their own USB jump bag, for example, to allow quick response.

It is also important to ensure that all the training and access requirements are met for all incident responders that you would expect to go onsite in case of an incident, so as not to delay response or endanger people and assets.

Before a security team can respond to an incident, it must detect it. Detection is only possible with a robust logging and monitoring approach, which we will discuss next.

Conducting Logging and Monitoring Activities

Logging and monitoring are key to ensuring that an incident gets detected as soon as it occurs. Many organizations implement logging and monitoring but do not do so consistently because they have not developed and implemented a log management and monitoring strategy. Such a strategy outlines what logs need to be collected, where they will be stored, how long they will be kept, and what happens afterward.

Log retention and storage is often a crucial component of compliance frameworks. The problem is that compliance frameworks tend to focus on historical incident data and, consequently, sometimes have an outdated view of what constitutes an incident.

With new types of attacks rapidly evolving, the needs of the security team do not always line up with what a strict compliance regime requires. Specifically, compliance requirements tend to lag behind the needs of incident response. A log management strategy needs to carefully distinguish which logs are kept for compliance reasons, as well as for how long, and which ones serve the immediate needs of incident response.

An example of this dichotomy may be the following. Many compliance frameworks require organizations to keep a record of failed authentication attempts, with the idea that a common attack is password stuffing (trying many different passwords on an already known account). While password stuffing attacks still occur, by far the most common approach to account compromise these days is credential phishing, and the real risk of successful phishing is not a *failed* authentication attempt, but a *successful* one.

From a compliance perspective, organizations still need to record and store failed authentication attempts. However, from an incident response perspective, it is much more interesting to consider the *context* of *successful* authentication attempts: did the attempt come from an unusual place at an unusual time? Was it part of a "batch" of both successful and unsuccessful authentication attempts coming from that same place, which would indicate that the phisher is "testing" a number of responses to the phishing email? In this attack context, just the record of failed authentication attempts is not very interesting. Although password stuffing attacks happen continuously, they do not usually result in successful intrusions, and a list of *failed* authentication attempts will not enumerate the *successful* attempts that result from successful phishing and that *do* lead to an intrusion.

It is this underlying dichotomy that is now sometimes expressed by stating that log management is a compliance problem, whereas incident response is a graph—that is, an event and context problem. The discussion on SIEM will dive into the details of this.

Monitoring and alerting done well will detect intrusions, but we can do more by looking specifically at some of the changes that are made on an endpoint when intrusions occur. This is the role of intrusion detection and prevention systems.

Intrusion Detection and Prevention

The aim of logging and monitoring is to get visibility on intrusions and provide alerts when incidents take place. In addition to logs, we can also monitor for changes that are being made on endpoints by attackers and set up systems that may disallow such changes. This is the approach taken by intrusion detection and prevention.

Three broad approaches to intrusion detection and prevention will be discussed in the following subsections.

Host-Based Approach

Host-based intrusion and detection focuses on collecting logs from the host, as well as watching for the appearance of unusual files or the modification of key system files, such as those in the /bin directory on Unix-based systems or the system32 directory in Windows. It also monitors the loading of drivers, registry modifications, or autostarts.

Host-based intrusion detection systems can monitor systems for the aforementioned changes or, in the case of host based intrusion prevention systems, can take more active steps to prevent such changes from being made. An example of the latter is the SELinux system, which provides additional fine-grained permissions for running executables on the host, preventing these executables from making some specific types of changes to the operating system or applications on the host.

Network-Based Approach

Network-based detection is done through a **Network Intrusion Detection System** (**NIDS**), which monitors packets on the network and compares them to a (known) set of signatures for malware or bot-type behavior. Network intrusion detection complements host-based intrusion detection, especially in cases where malware involves living-off-the-land techniques.

NIDSs mostly contain a collector with one or more network cards dedicated to monitoring the passing traffic. There are several layers on which network intrusion detection can be performed:

- **Application layer**: Some intrusion detection systems (especially those used for monitoring industrial control systems) have a certain amount of understanding of application layer traffic and will apply this knowledge to pick up anomalies. An example of intrusion prevention at the application layer is a **web application firewall** (**WAF**).

- **DNS**: At the DNS layer, one can monitor for resolutions to known malware sites and take action (i.e., alert or block) accordingly. Monitoring DNS traffic can be a useful method of monitoring what sites your users connect to.

- **Packets**: The standard intrusion detection system monitors packets and compares them to a database of known bad behavior, such as bot traffic, malware scanning, or abuse of insecure protocols (such as passwords being sent in the clear). An open source system that monitors network traffic and alerts on any known bad or suspected traffic is Suricata (`https://suricata.io/`).

- **Netflow**: The last level at which one may do network intrusion detection is at the level of network flows—a record of sessions to individual addresses on the network with some higher-level statistics on that session, such as the encryption used, the amount of data transferred, and the network protocols used. Netflow data can be very useful for gaining a contextual understanding of traffic patterns that are normal, and then mining that for any outliers that may need further investigation.

Which of these methods you should use depends on the individual situation: packet data is voluminous and hard to store for any length of time, whereas flow data is relatively small and can be kept for a long time.

XDR-Based Approach

It's important to separately discuss methods based on the concept of detection and response (XDR). As previously mentioned, XDR methods integrate various sources of data to provide a comprehensive view of security events. Specifically, XDR combines event and log data and telemetry data. The goal of XDR is to not only detect and log security events but also to provide context around these events. This means understanding how different data points (e.g., logs, network traffic, and device telemetry) interrelate, which helps in more accurately identifying and responding to potential threats.

XDR methods collect individual events, turn them into alerts, and then present these alerts to the analyst with a significant amount of context, such as the following:

- **Causality**: This addresses the cause of the event, recorded users at the time of the event, and the chain of events that was set in motion

- **Consequence**: This addresses what the impact of the event was—for example, whether a network connection opened, a file was transferred, a new was account created, or a service was started or stopped

- **Reputational context**: This addresses what the reputation of the IP addresses contacted in a subsequent network session was, or whether that file that was just downloaded related to known malware

XDR methods are becoming widely adopted and are being increasingly associated with the ability to consume and produce threat intelligence, as well as with the ability to perform targeted hunting for the presence of threats. XDR solutions keep a large amount of historical data in a data lake (a large pool of relatively unstructured data) that can be mined for the presence of threats.

Most XDR solutions are accompanied by a cloud-based console since the amount of data collected is too large to store on-site, and so the use of XDR can bring in considerations of data sovereignty and access. Additionally, systems that use XDR typically need to be connected to the internet, which is a consideration for industrial control systems.

Continuous Monitoring

The ubiquity and unpredictability of cyber incidents mandate that event monitoring be continuous and not rely upon collection at incident time. Therefore, teams need to ensure that solutions are producing logs and storing them somewhere centrally and as close as possible to the time they are being generated on the system, especially for key systems in the business.

Continuous monitoring also has a human component. Teams need to decide whether they want their infrastructure and applications monitored around the clock (which has staffing implications) or whether business hours monitoring will suffice to meet their needs. It is also possible to distinguish between systems being monitored around the clock versus (less important) systems that are only being monitored during business hours.

All these decisions have an impact on how an organization designs and implements its **Security Incident and Event Monitoring (SIEM)** system and how it staffs its security operations.

For staffing, many organizations choose a hybrid model in which a **managed security services provider (MSSP)** is responsible for round-the-clock monitoring augmented with a local team that holds more contextual knowledge about the business. The local team can then be on call for the MSSP.

Egress Monitoring

Egress monitoring refers to monitoring the traffic and information that *exits* the organization. With all the focus on attackers coming from outside, it is easy to forget that egress monitoring is one of the more effective measures an organization can take to strengthen its security posture.

Egress monitoring can indicate the following:

- The presence of malware inside an organization's network
- When and where internal users are responding to malicious emails by monitoring whether their DNS traffic matches that of links received in malicious emails
- Large data uploads to locations outside the organization
- The presence of compromised hosts or web shells inside an organization's network

Egress monitoring is generally more indicative of security problems inside an organization's network than ingress monitoring; the latter monitors a large number of automated attacks that have a low chance of being successful but may still end up consuming significant amounts of team time and resources.

Log Management

Systems for continuous monitoring and egress monitoring generate a large amount of data. In many organizations, logs need to be kept for compliance reasons within a time frame set by the legislation or policies that the organization needs to comply with.

All this sets the following requirements for a log management system:

- A consistent logging format such as **Open Cybersecurity Schema Framework (OCSF)** (https://packt.link/45XDc)
- Large amounts of disk space available to keep logs
- The distinction between fast (or "hot") storage for recent logs and slow(er) ("warm" or "cold") storage for historical logs, with logs automatically rolling over into slow storage after a certain amount of time has expired
- A NoSQL architecture to ensure logs can be searched for a variety of fields and under conditions that may not have prevailed when the log was collected and stored
- An index architecture that matches the needs of the business and the security team

Examples of modern systems that meet these requirements are Splunk (`https://packt.link/0QTCN`) and Elastic (`https://packt.link/XfIqp`)

SIEM

Once logs and network data are collected, they end up in a SIEM system, which is a software solution such as Splunk or Qradar. SIEM systems are the mainstay of many security teams, and they are not likely to disappear in the near future, although they will change considerably.

The Role of SIEM Systems

SIEM systems play a crucial role in guiding security teams and organizations by helping them effectively manage and respond to security threats. These systems enable the real-time collection of security event logs and telemetry, which are essential for detecting threats and ensuring compliance. By analyzing this telemetry both in real time and over time, SIEM systems can identify attacks and other significant activities. Furthermore, they assist in investigating incidents to assess their potential severity and the impact they may have on the business.

SIEM systems consist of compliance requirements for log collection and storage and incident detection and response needs. Note that these different use cases impose a significant design dichotomy on SIEM solutions, which makes them both expensive and difficult to operate and maintain.

A SIEM system needs to satisfy the following distinctive use cases: compliance, incidence response, and historical analysis. For compliance, the focus is on log collection and retention for long periods of time; this requires large amounts of cheap storage. For incident response, the focus is on the delivery of timely and accurate incident data within a very short time frame with as much incident context as possible to determine severity and impact. Historical analysis requires historical data to hunt for past intrusions.

SIEM typically ingests and normalizes log data (i.e., puts it in the same time zone and categorizes and classifies the data), and then applies a set of correlation rules in an attempt to detect incidents by using the log data from either the same or different devices (cross-device correlation) to add additional context to the event so that it can be identified and investigated.

Operating a SIEM System

The operation of a SIEM system, at a minimum, requires a number of activities from the analysts:

- Looking at the dashboard for the occurrence of events that need investigating (or monitoring an alert stream if the SIEM system is integrated with other systems such as Slack)
- Investigating and resolving alerts and incidents
- Editing and tuning the rules that perform correlations to increase the fidelity of the system, especially to address alert fatigue

- Monitoring feed health to ensure that all the data feeds being used by the SIEM system are up and sending data

- Editing and tuning the normalization rules to ensure that the SIEM system is capable of correctly parsing and categorizing the events

Additional tasks may consist of tuning and monitoring threat intelligence feeds into the SIEM system, setting up and configuring additional data feeds, managing the databases and search engines in the system, and understanding the details of the business context of alerts to correctly determine what the response should be. The latter cannot be outsourced to an MSSP.

Many SIEM systems also have a case management system that allows security teams to open and close cases and store playbooks.

As you have learned so far, the operation of a SIEM system requires a number of disparate tasks ranging from monitoring to alert and incident resolution through to the development of rules (detection engineering) and, in some cases, even covering tasks usually associated with threat intelligence. Since SIEM is usually seen as part of security, these disparate jobs usually fall to the security team alone, which is not an easy task.

Depending on the size of the team and the organization, it may help to have a security engineering division inside the security team that is primarily responsible for optimizing and tuning the SIEM.

The (Likely) Future of SIEM

Notably, many of the tasks traditionally performed by a SIEM system are now also performed by XDR-type monitoring solutions, and so the job of detecting and contextualizing, and an increasing number of incident handling strategies no longer involve SIEM systems. A likely future of the SIEM is a separation into a log storage and retention facility with hunt capability, whereas the incidents requiring immediate attention could be handled by an XDR-type solution.

Threat Intelligence

Threat intelligence refers to the information about current threats, usually in a format that allows teams to take action on the information provided.

Threat intelligence is a wide topic that is often misunderstood. Intelligence deals with scenarios and likelihoods, not certainties. That means that intelligence works on the strategic level, not on the operational level. Organizations cannot take intelligence data and turn it, for instance, into an automated block list. But you can take threat intelligence and turn it into a threat hunt scenario, where you systemically mine the data from your security tooling for matches to intelligence, and then further investigate these cases.

Threat intelligence concerns information about threats and threat actors and provides sufficient understanding to mitigate the threat. Cyber threat intelligence limits the scope of that intelligence to the cyber domain. This section will briefly discuss the types of information that make up threat intelligence, the process by which threat intelligence is created, threat hunting, the ATT&CK framework to categorize aspects of intelligence, and finally, how to integrate threat intelligence into a security program.

We'll consider some aspects of threat intelligence in more detail.

Types of Threat Intelligence: the Pyramid of Pain

Threat intelligence can come in many flavors, from a simple list of file hashes or bad IP addresses to comprehensive inside information about the operations of a known threat group and anything in between. The most evocative characterization of threat intelligence is David Bianco's Pyramid of Pain, depicted in *Figure 16.1*:

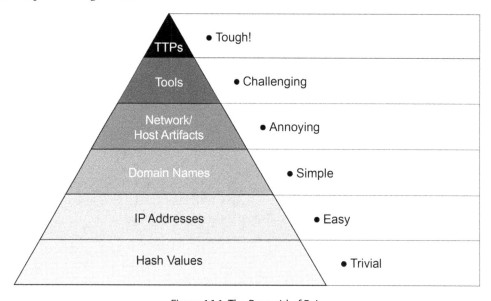

Figure 16.1: The Pyramid of Pain

> **Note**
> This was first introduced in a 2013 blog post by David Bianco: `https://packt.link/I1Fd8`.

At the bottom of the Pyramid of Pain are indicators that are easy to observe and measure but that also change quite frequently—things such as hash values, domain names, IP addresses, and so on. Toward the top of the pyramid lie the things that translate more into the business model of an attacker and that are thus difficult to change.

The point is that for the low-level indicators (often referred to as **indicators of compromise** or **IOCs**), it is easy to write blocklists or detections for firewalls and endpoint protection software. In contrast, the high-level indicators, such as the specifics of attack tools or the TTPs, require some work on the part of the team to detect, since a lot of context (sometimes also called **enrichment**) needs to be added to a collection of low-level indicators to get this correct. Essentially, what a security team attempts with the design and implementation of defenses for a high-level indicator is to choke off the attacker's business process with either a detection or a block.

Consider the following example. Compromised accounts resulting from phishing runs need to be tested before they are sold. Can you write a specific detection that picks up such testing? To get that right, it is necessary to first spend some time on the system to work out what a testing run looks like on the attacker's side. Then, you need to make sure that the correct data to detect it is being collected. After that, you can write and refine a detection. This sounds like a lot of work (and in practice, it is), but it does pay off. Attackers focused on the sale of compromised accounts will find it hard to bypass this step in their process or to engineer around it with only limited knowledge of your environment.

At the highest level of intelligence lies strategic intelligence, which focuses on the threat landscape in, say, your industry vertical (such as education, government services, or manufacturing) and how that threat landscape changes over time. This information can shape your decisions about the future of the security program, the types of tools you invest in, the size of the team, and the budget for the program.

Intelligence Cycle

Intelligence is a somewhat particular product. It is what you know or suspect about an adversary who takes great pains to hide that information from you. Specifically, intelligence isn't knowledge. In intelligence agencies, the gathering, analysis, and dissemination of intelligence is often viewed in terms of the intelligence cycle. The intelligence cycle is depicted in *Figure 16.2*, and more details can be found at `https://packt.link/qABxW`.

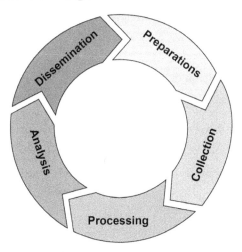

Figure 16.2: The intelligence cycle

The intelligence cycle has several phases, which are briefly summarized in the context of cyber threat intelligence as follows:

- **Preparation** or **Direction**: In this phase, an organization decides on the need for a threat intelligence program, and, most importantly, the type of threat intelligence they want to collect. The direction phase narrows down the field of potential questions that need to be answered through the intelligence process.

- **Collection**: In this phase, the organization acquires or collects the raw data that it needs for intelligence assessment.

- **Processing**: In this phase, the organization decides what data to use in what manner, and how different data sources will be combined into a single view.

- **Analysis**: In this phase, the data is mined for answers to the questions posed in the direction phase.

- **Dissemination**: This is the phase in which the new knowledge is applied to the environment. In cyber threat intelligence, this can take the form of **tasking** your equipment to either detect or block the threat, writing detections and blocks for higher levels in the Pyramid of Pain.

- Sometimes, a separate step is added for **Reporting**. This entails reports on the activities of the program and may feed back into the direction phase.

According to the threat intelligence cycle, intelligence is a specific product that is the end result of direction (i.e., narrowing down the questions), collection, processing, and analysis. That is, intelligence is not merely facts; an intelligence program lays out possible scenarios with assigned trustworthiness and probabilities.

Often, in the wake of an especially high-profile attack, organizations rush to implement blocks on a small number of known IOCs. While it may make sense in some cases to do this, in the vast majority of cases, organizations are better off spending their time working out whether the TTPs of such attacks would work in their environment, and what would be needed to detect or stop such attacks.

Threat Hunting

One of the ways in which organizations can operationalize threat intelligence is through a threat-hunting program. Threat hunting is the process of searching through (often) historical data for patterns of attacks that might only have become known after that data was collected. It is usually recommended to ensure that a robust threat-hunting capability exists before an organization embarks on a threat intelligence program so that the output of the threat intelligence can find some use immediately.

Usually, hunts start with a **hypothesis**—something that could be true in the environment, such as the presence of a specific threat actor. The hunt then involves systematically combing through existing data to either prove or disprove the hypothesis.

Hunts focused beyond a search for specific IOCs often involve the development of analytic queries or statistical analysis on the security data in order to get an answer.

ATT&CK Framework

Many of the known and prevalent attacks are now collected in the MITRE ATT&CK framework. Modern security tooling is also increasingly characterizing its observations in terms of ATT&CK.

> **Note**
>
> The ATT&CK framework can be found on the MITRE website: `https://packt.link/2vncA`.
>
> The ATT&CK framework categorizes attack data in terms of TTPs.

Tactics correspond to the objective that an attacker wants to achieve and often lead to a step toward that objective. Tactics that are recognized and categorized in the ATT&CK matrix are elevation of privilege, establishing persistence, and actions on objectives, to name a few.

Techniques are the specific (platform-dependent) methods employed to achieve the tactical objective. Techniques can be as diverse as phishing, process injection, and data exfiltration, and one or more techniques will usually suffice to achieve the tactical objective. Procedures are low-level specific implementations of techniques. ATT&CK also recognizes sub-techniques as subsets of techniques.

The ATT&CK framework, briefly mentioned previously, has developed into the core of **threat-informed defense**, such that organizations can rate themselves and their security tooling against their desired ability to handle specific types of cyberattacks. Under each technique, the ATT&CK framework lists the data sources and data components that must be collected to analyze and detect any use of that technique on the network of the target organization. This also provides defenders with a chance to develop a threat-driven collection strategy for network logs and system data.

The implementation and execution of a threat intelligence program can be a delicate undertaking as CISOs need to consider some aspects of governance and management in their program. We'll briefly outline these next.

Strategy and Threat Intelligence

The consumption of threat intelligence carries some risks for how CISOs operate their cybersecurity program and there is a risk of wanting too much too soon or having over-reliance on threat intelligence as a source of information. The ability to consume threat intelligence and take action on the information provided is not an excuse to neglect the delivery of a robust security posture or the development of robust system maintenance and administration procedures. Threat intelligence is not a stopgap measure to make up for deficiencies in team practices and maturity. Rather, it is the opposite as it will mostly be the mature teams with robust practices that will be able to profit most from a threat intelligence program.

It can be beneficial to start by developing and implementing a threat-hunting program before launching a threat intelligence program. Threat hunting involves a process similar to the intelligence cycle but simplifies it by skipping the often-complex data collection phase and focusing on generating and testing hypotheses instead. By building a strong threat-hunting capability first, you lay a solid foundation for developing a more advanced threat intelligence program in the future.

The last aspect of security operations we will consider is **User Entity Behavior Analytics (UEBA)**.

UEBA

UEBA focuses on the behavioral patterns of users in any organization. UEBA programs look for outliers in known behavior patterns, such as logons outside normal working hours (for that user), unusual services or servers that are being accessed, large data transfers, or the use of unusual software.

UEBA platforms usually monitor the traffic in your organization for a set period of time, and then learn what is normal for your organization. After the learning period is over, the UEBA platform will start alerting you to unusual behaviors. When in operation, users might feel UEBA platforms are more intrusive because they monitor employee behavior. They also lead to a higher proportion of false positives.

We will now move on to configuration management, which is part of the activities a team undertakes to prevent incidents from occurring.

Performing Configuration Management

Configuration management (CM) is making sure that devices on the network are configured according to the standards set by the organization. As an example, if the organization requires every laptop to run XDR, then making sure that XDR agents are installed and operating correctly is part of CM. Similarly, hardening systems is part of CM.

A significant amount of work is required on all your systems before you can be ready for incident response. For example, you may need to deploy endpoint protection and monitoring tools, do some baselining to understand what is normal on a given system, and think about automating some of the responses. CM is systematically organizing and controlling changes to system configurations to ensure consistency, stability, and compliance across all IT systems. It includes provisioning, backup, endpoint protection, vulnerability and configuration management, logging and monitoring, and baselining. These are discussed in the following sections.

Provisioning

When a system is provisioned, you need to think about how to ensure that the data on that system remains confidential and available, and in some cases, how unauthorized data modifications may be detected. To that end, organizations typically deploy agents on a system that are responsible for local monitoring and reporting to a central server.

Agents are typically deployed for the following:

- Backup
- Endpoint protection
- Vulnerability and configuration management
- Logging and monitoring

Each of these points is discussed in detail in the following subsections.

Backup

Backup agents are deployed to ensure that a system is backed up and can be restored in case of an incident. Backup agents on Windows now often interact with the **Volume Shadow Copy Service (VSS)**, which creates snapshots of the system to be backed up. Similarly, many XDR agents have a (limited) capability to roll back bad activity, such as mass file encryption by ransomware. Backup agents can then restore the system to a certain predefined state.

With the increase in virtualization across the server fleet, it is becoming more common to run backups at the virtualization host layer rather than rely on individual backup agents on systems.

Endpoint Protection

Endpoint protection agents are what were traditionally called **antivirus**. It is now more common to refer to these agents as endpoint protection. Endpoint protection agents usually monitor the filesystem for the presence of malware but can also monitor network communications, manage the host-based firewall (in some cases), inspect running processes, and perform a certain amount of system hardening.

Endpoint protection agents collect a large amount of telemetry that is made available to the security team to assist with incident response. Many endpoint protection agents allow systems to be queried with a specifically designed query language that will inform the security team of the actions taken on that system. For this reason, teams must ensure that they have good coverage of endpoint protection agents across their fleet.

> **Note**
> An example of a *causal chain* from an endpoint protection agent can be found at https://packt.link/qXdnX.

Vulnerability and Configuration Management

Vulnerability and configuration management agents keep a record of the installed software on a device and report this to a central console. Some agents can also scan for the presence of vulnerable software or configuration errors on a device that might negatively affect its security or allow reporting against existing frameworks such as the CIS hardening guides, which can be found at `https://packt.link/pVYJh`.

Logging and Monitoring

Logging and monitoring agents ensure that the logs on the device are parsed and events are forwarded to a centralized logging server. Usually, the configuration of the agent includes the events to be collected and forwarded as well as details on how logs need to be forwarded to the central server.

Baselining

Monitoring the performance and events on a device allows the security or systems management team to create a baseline of what is normal for that device. In security operations, this baseline can then be used to determine whether there are any outliers that need to be further investigated.

For managing system configurations, a baseline is necessary to understand which settings are likely to interfere with normal user behavior. As an example, an organization may decide that they will only use TLS 1.3 and disable support for older versions. A baseline helps the organization to understand whether such a step would stop business activity, for instance, by blocking users from sites needed for the business.

Automation

Automation is an important aspect of configuration management. Agents and settings are usually deployed via automated tools to ensure good compliance and wide coverage (in addition to saving a significant amount of work). Automation also helps with making rapid changes when needed, for instance, in the case of an incident.

There is a wide range of configuration and automation tools available to assist with this, but even lacking a specific tool, many settings can be configured in Windows domains with Group Policy Objects, which will ensure a consistent configuration across a Windows domain.

In the last section of this chapter, we will focus on the application of some foundational security concepts.

Applying Foundational Security Operations Concepts

The behaviors of security teams are governed by policies, norms, and best practices. For security teams, there are a number of best practices that are specific to security operations and are focused on keeping the security operations themselves secure (these practices are sometimes referred to as **OpSec**). The aim of operational security practices is to keep the details of how a team operates out of the hands of attackers.

This section will discuss the following:

- Need to know and least privilege
- **Separation of duties (SoD)** and responsibilities
- Privileged account management
- Job rotation
- **Service-level agreements (SLAs)**

In small teams, it can be hard to meet the requirements of SoD and job rotation since there are generally not many team members to divide the duties with. However, that difficulty does not mean that teams should not consider what the requirements are and come up with the most efficient solutions for their situation.

Need to Know and Least Privilege

Need to know and least privilege are important operational concepts that govern the daily operations inside security teams. Not everyone needs to know everything about everyone, and this applies even more when a security team may be working on sensitive investigations, involving persons or private data.

Some key determinants of **need to know** are whether someone needs to know something in order to take a certain action (for example, blocking a website) or whether someone needs to know about an investigation as a matter of policy or legal obligation.

Most organizations employ access controls to ensure that only people who need to know information can access that information. Access control lists are technical artifacts that describe whether a certain group of users has access to information and the type of access that they have (e.g., read, write, delete, or full control).

Least privilege refers to the principle that the scope of action of the security team is restricted to only what the team needs to do its job and nothing more. For example, because the security team needs to monitor a large number of systems, it may be considered viable to just give team members administrative access to firewalls, certificate authorization systems, and every server and desktop in an organization. However, for many scenarios, all the security team needs is read access. As a general principle, security operations should request the level of permission that is required to do the job and nothing more, not even for a *just-in-case* scenario.

The **Traffic Light Protocol** (**TLP**) (`https://packt.link/jdMMG`) is a widely used framework that governs the exchange of data between security teams. The current 2.0 version of the framework contains a few additions that give guidance to managed security service providers about what can be shared with their clients.

The TLP has four levels (as detailed in the preceding page reference):

- **TLP:Red**: This cannot be shared and is intended for the recipient of the information only.

- **TLP:Amber** and **TLP:Amber+STRICT**: This sharing level is for information to be shared within an organization. Amber can also be shared with clients of an organization (e.g., where that organization is an MSSP; Amber+STRICT means it cannot be shared with clients and is intended for the organization only.

- **TLP:Green**: This represents limited disclosure where recipients can share the information within their community. The community in this case may include other organizations in the same area of business (e.g., tertiary education) and thus includes competitors. Sometimes, security teams have a working relationship with security teams from other organizations who can also be included in TLP:Green sharing.

- **TLP:Clear**: This has no limits on disclosure and can be freely shared, although it is customary to mark the information with TLP:Clear during dissemination.

Under the conditions of the TLP, the originator of the information is responsible for its classification and the communication of that classification to the recipient.

> **Note**
>
> The need-to-know model is extended to the exchange of threat intelligence between teams or between teams and threat intelligence providers in the **Information Exchange Protocol** (**IEP**), a copy of which is on the First website: `https://packt.link/vEOeL`. This sort of information sharing is increasingly important with the emergence of threat intelligence platforms where threat intelligence is stored and can be looked up both by human analysts but also security tooling.
>
> The IEP specifies not only the classification level of the information (in terms of the TLP) but also extends the TLP to include statements on what action a recipient may take on the information, whether it can be shared, and whether affected party notifications are permitted by the originator of the information.

Separation of Duties and Responsibilities

It is a security risk if someone can initiate an action, approve it, and also execute it. To prevent this scenario, organizations use SoD in which, as much as possible, the initiation of actions, approvals, and execution fall to different people in the organization. Accordingly, auditing is usually structured as its own separate activity.

For example, consider certificate management. Certificate management entails requesting certificates, approving them, issuing the certificate, and deploying it. Auditing certificate management entails reviewing the certificates that have been issued under the domain of the organization. If the lines between requesting, approving, issuing, deployment, and audit are blurred, or, worse, if these tasks all fall to the same person, an attacker could use compromised credentials to issue certificates on behalf of the organization and be able to build fake sites with correct certificates.

As you will learn in the next section, we can implement SoD between persons as well as between accounts. This brings in privileged account management.

> **Note**
>
> More information on this and SoD can be found in *Chapter 19*.

Privileged Account Management

Privileged account management is a key issue that organizations must address prior to audits, since auditors and cybersecurity insurers need to focus on this area for audit findings and decide whether they will provide cover, respectively.

Privileged accounts are accounts that are used to administer systems, cloud environments, or applications. For that reason, privileged accounts have elevated privileges on a system. Privileged accounts can usually change the configurations of systems, create new users, set up new API keys, and so on. Therefore, privileged accounts are a key area where teams need to provide protection and a key target for attackers. In fact, a significant number of techniques in the ATT&CK framework are dedicated to elevating privileges from normal users to privileged users.

Privileged users can be of various types including regular admin accounts and service accounts, domain administrators, users who can make configuration changes on applications, developers, and application administrators.

However, before you can dive into the details of privileged accounts, it is important that you have an understanding of attack graphs.

Attack Graphs

A graph is a structure consisting of nodes and links between nodes (also called vertices). Graph structures can be used to map interactions between things, along with their types. We encountered an example of a graph in the causal graph from EDR previously, which links "nodes" (events) with actions undertaken by that node, such as *a process opened a registry key*.

Attack graphs are a useful and popular way in which defenders structure the approach an attacker might take on their network. The reason for this is that these graphs allow you to think in terms of specific attack paths and how they might appear on your own turf. Once you have a view of these attack paths, you can start thinking constructively about how to plug the holes.

This development is relatively new. Initially, defenders used to think in terms of lists, where the lists were (mostly) derived from audit findings based on various best practice frameworks and the presence of various controls. The problem with such lists is that defenders need to consider the entirety of their environment, while attackers only have to find a single vulnerability that can be exploited. This will always be true, regardless of how you think of your environment. However, thinking in terms of graphs makes it easier for defenders to focus their efforts in areas where it makes the most difference.

Figure 16.3 shows an example of an Active Directory graph for a fictitious Active Directory created with Bloodhound:

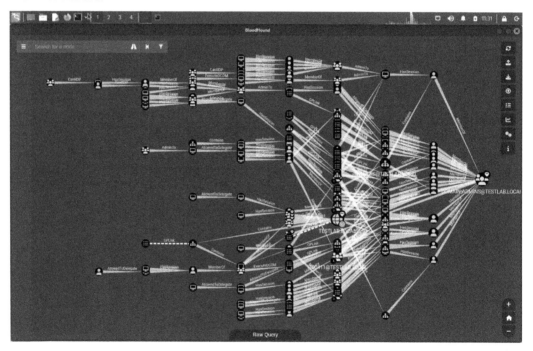

Figure 16.3: Example of an Active Directory graph created with Bloodhound

The preceding graph outlines all the possible attack paths from various points in the graph to a target, which, in this case, is the domain admin group.

Thinking in graphs in this context involves focusing on the existing security dependencies and where they occur, as well as how an attacker might traverse the graph along its edges to hop from one compromised system to another. In general, the length and the cost of an attack path will determine how easy that path is to traverse. Each system has a *lowest-cost* path that is the easiest to traverse for an attacker. The job of defenders is then to increase both the length and the costs of the existing attack paths in an organization.

The point is that when compared to thinking in lists, thinking in graphs allows defenders to focus on raising the cost of compromise to attackers, which simultaneously focuses on real improvements in the overall security posture of the organization.

Thinking in graphs is especially important once you consider the issue with privileged accounts. Privileged accounts are favored by attackers as single-stop items of interest that, once compromised, offer access to many areas of the network. The problem for defenders is that there are various types of privileged accounts within a network, each with different levels of access and different roles, making administration difficult and time-consuming.

The next section will discuss admin (or general administrative) accounts, service accounts, domain administrators, and developers, as well as an ICS corner case: the engineering workstation. This is primarily to illustrate the many flavors of privileged accounts and how it is easy to overlook areas of the business where privileged accounts might occur.

Admin Accounts

In line with the SoD discussed earlier in this chapter, it is useful to create a separate set of admin accounts for the administration of systems so that access to these accounts can be monitored. When it comes to privileged account management, it is usually a good idea to deploy a specialized password store. The password store can also be used to store other business secrets such as API keys, SSH keys, and private encryption keys. Access to the password store can be monitored, along with who accesses what secret.

A special type of administrative account that is often overlooked is an endpoint management account. Endpoint management accounts are used by the service desk to maintain endpoint systems, often remotely. A compromise of these credentials will allow attackers to install or disable software on many endpoints at once or compromise the endpoint used by other system administrators to then compromise more administrative accounts or administrative accounts of a different type, especially domain administrators.

Service Accounts

Service accounts often have elevated privileges in systems so they can do their work. Moreover, in many organizations, service accounts are over-permissioned; that is, they have been given too many permissions to get the service to work. This is partially the fault of system administrators, but it was also not uncommon in the past to have vendors ask for service accounts that had too many permissions.

Since service accounts usually do not reset their passwords and have a long lifetime, they are key accounts to consider when looking at privileged accounts on a network.

Service Principal Names (**SPNs**) accounts are a special form of service account in Microsoft Entra ID that are vulnerable to **Kerberoasting** attacks. In a Kerberoasting attack, an attacker obtains the hash of the victim's account and then performs an offline crack of the hash to recover the password. The hash of an SPN can be obtained by any user of the domain by abusing the Kerberos protocol through an offline attack called Kerberoasting. Moreover, SPN accounts have often elevated privileges and can be attractive targets. Kerberoasting is covered in *Chapter 13, Identity Management Implementation*.

Domain Administrators

Domain administrators are one of the highest permissioned groups in Windows Active Directory. Domain administrators can create users, change permissions on users, add and remove domain controllers, change configuration settings on the Active Directory, and, in general, perform any administrative task that is necessary to keep a Windows domain healthy and functioning.

For this reason, domain administrator accounts are a desirable target for attackers. The known modus operandi of many of today's ransomware groups is to elevate privileges to domain administrator after the initial compromise. They can then proceed to encrypt files, destroy servers, and even remove the domain to deny access to their victims.

Developers

Developers usually have a large amount of direct access to code repositories, databases, filesystems, and **user acceptance testing** (**UAT**) systems, which may also contain valuable business data that should not be leaked. In that sense, developer accounts and development workstations should be considered privileged access accounts.

Engineering Workstations

Privileged accounts can also occur in places where we do not really expect to find them. A special type of privileged access in industrial and industrial or OT environments is the **engineering workstation**, which is used to program the **programmable logic controllers** (**PLCs**) in the environment. Usually, engineering PCs run the programming software for the PLC, and there may be a need to access this workstation remotely. Moreover, the software running on the engineering workstations used to program the PLCs is often left in a running state by the engineers maintaining the installation, meaning that these machines are very valuable to attackers.

Job Rotation

Job rotation is another tool that may help increase the robustness of the security operations in an organization. In a job rotation scheme, different team members rotate between different aspects of security operations. It has the following two benefits:

- As, over time, most members of the security team go on to perform different roles, it helps with creating a broader skill base in the team that is difficult to get without job rotation. Core competencies are familiarity with systems that are shared, creating resilience in the organization. If one team member is absent during an incident, there are trained members of the team that can step in.

- It helps detect any fraudulent activities. If people spend more time on a particular role, they might be committing fraud, which will be revealed once the next person takes up the role.

Service-Level Agreements

The security team provides a service to the organization, and this service can be measured with SLAs, which are written formalized agreements that outline the expected level of service. They lay out agreed metrics, often measured in the time it takes to carry out a key function.

Perhaps the most visible service that the security team provides is incident response. Moreover, the incident response service is also uniquely time-critical. Most organizations demand and need a fast response from their security teams once they are compromised.

SLAs for an incident response service can be measured with the following set of parameters:

- **Time to detect**: The time it takes before an intrusion is detected. This measure is retrospective in the sense that it requires an incident to be analyzed first before a reliable measure is available.

- **Time to respond**: The time taken to respond to the incident after detection.

- **Time to contain**: The time taken to contain the incident.

- **Time to resolve**: The time taken to resolve the incident.

A security team and its host organization can draw a firm agreement in place that outlines what sort of detection, response, containment, and resolution times are acceptable. An SLA usually takes the form of a certain percentage of incidents resolved within a certain amount of time, such as *90% of endpoint virus incidents contained within 30 minutes and resolved within 24 hours*. The acceptable times are a matter of agreement between the business and the team and will depend on the nature of the business and the cost of outages.

In the final section of this chapter, we will consider resource protection approaches for security teams.

Applying Resource Protection

Both forensic and backup data are sensitive and must be protected against leaks, destruction, or modification. This section will discuss some resource protection approaches such as media management, media protection techniques, and "break-glass" (or emergency access) scenarios.

Media Management

Media with forensic data is subject to a chain of custody and must be kept in a secure physical location when not needed.

Backups are usually stored in a secure location offsite so that they can be accessed even when the data center is unavailable. Cloud backups often use replication in the cloud to ensure that a copy is kept in several locations.

Media Protection Techniques

A number of techniques can be used to protect media when not in active use:

- **Offline backup**: This approach focuses on ensuring that the backups cannot be modified through a compromise of the network by keeping the backup system offline or on an airgapped network. The backups may still be stored in the same location.

- **Offsite backup**: This is a form of physical protection of media where media is stored in a different physical location, which ensures that attackers do not easily gain physical access to backup media. Offsite backup also helps the organization with access to its data should the main data center become unavailable.

- **Encryption**: This ensures that anyone accessing the media will need to know the decryption key in order to access the data.

- **Duplication**: This can also be used to ensure that the corruption of a single copy does not make the entire data unavailable.

Some backup devices also implement the idea of immutable filesystems, in which data can be added to a filesystem but not deleted or modified once it is written. This is intended to protect against ransomware attacks, which often aim to overwrite the backups in an effort to get victims to pay the ransom.

Break-Glass Scenarios

Organizations also need to think about *break-glass scenarios*—last-resort scenarios in which, for instance, an authentication system has become unavailable. To this end, sometimes, organizations configure (and regularly test) break-glass accounts, which can be stored offsite in a data capsule that is only opened under extreme circumstances.

Summary

This chapter focused on all the preparation necessary to resolve incidents. As you must have understood by now, there is a lot to think about when it comes to incident resolution. Some of the policies required for security teams were discussed, as well as the corresponding foundational security concepts. This chapter also discussed privileged account management, approaches to logging and monitoring, endpoint protection, and device forensics.

These areas are continually evolving as new technologies and approaches are introduced into the technology landscape, and managers of security programs need to consider the impact of new technologies in light of the requirements also discussed in this chapter.

In the next chapter, we will focus on security operations itself—that is, how security teams put all the things we have discussed in this chapter into practice.

Further Reading

Security operations is a wide topic, and a potential reading list can get very long. A good consolidated resource is the MITRE guide *11 Strategies of a World Class Cybersecurity Operations Center*, which discusses the details of running security operations extensively. A copy in PDF format is available at `https://packt.link/G5r7u`.

Exam Readiness Drill – Chapter Review Questions

Apart from a solid understanding of key concepts, being able to think quickly under time pressure is a skill that will help you ace your certification exam. That is why working on these skills early on in your learning journey is key.

Chapter review questions are designed to improve your test-taking skills progressively with each chapter you learn and review your understanding of key concepts in the chapter at the same time. You'll find these at the end of each chapter.

How to Access These Materials

To learn how to access these resources, head over to the chapter titled *Chapter 24, Accessing the Online Resources*.

To open the Chapter Review Questions for this chapter, perform the following steps:

1. Click the link – `https://packt.link/chapter16`.

    Alternatively, you can scan the following **QR code** (*Figure 16.4*):

Figure 16.4: QR code that opens Chapter Review Questions for logged-in users

2. Once you log in, you'll see a page similar to the one shown in *Figure 16.5*:

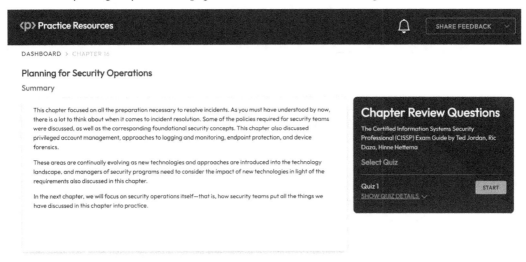

Figure 16.5: Chapter Review Questions for Chapter 16

3. Once ready, start the following practice drills, re-attempting the quiz multiple times.

Exam Readiness Drill

For the first three attempts, don't worry about the time limit.

ATTEMPT 1

The first time, aim for at least **40%**. Look at the answers you got wrong and read the relevant sections in the chapter again to fix your learning gaps.

ATTEMPT 2

The second time, aim for at least **60%**. Look at the answers you got wrong and read the relevant sections in the chapter again to fix any remaining learning gaps.

ATTEMPT 3

The third time, aim for at least **75%**. Once you score 75% or more, you start working on your timing.

Tip

You may take more than **three** attempts to reach 75%. That's okay. Just review the relevant sections in the chapter till you get there.

Working On Timing

Target: Your aim is to keep the score the same while trying to answer these questions as quickly as possible. Here's an example of how your next attempts should look like:

Attempt	Score	Time Taken
Attempt 5	77%	21 mins 30 seconds
Attempt 6	78%	18 mins 34 seconds
Attempt 7	76%	14 mins 44 seconds

Table 16.1: Sample timing practice drills on the online platform

Note

The time limits shown in the above table are just examples. Set your own time limits with each attempt based on the time limit of the quiz on the website.

With each new attempt, your score should stay above **75%** while your "time taken" to complete should "decrease". Repeat as many attempts as you want till you feel confident dealing with the time pressure.

Security Operations

This chapter will focus on the actual mechanics of security operations, how to manage cybersecurity incidents, and the steps involved in incident resolution.

The chapter will start with key aspects of managing security incidents and take you through the incident response cycle, focusing on detection, response, mitigation, eradication, and recovery. You will also examine how to conduct a "no blame" lessons learned review and some points on reporting and communication.

You will then learn how to operate and maintain common security defenses and review the practicalities of vulnerability management and change management approaches. While the previous chapter focused on the policy and planning side of things, this chapter focuses more on the practical implementation of all the techniques in day-to-day operations.

By the end of this chapter, you will be able to answer questions on the following:

- Conducting incident management
- Operating and maintaining detective and preventive measures
- Implementing and supporting patch and vulnerability management
- Understanding and participating in change management processes

This chapter will start by looking at conducting incident management

Conducting Incident Management

The management of an incident involves a number of distinct stages that, in combination, form the incident response cycle. The stages are as follows:

- **Detection**
- **Response**
- **Mitigation** or **Containment**

- **Eradication**

- **Recovery** or **Remediation**

- **Lessons learned**

- **Reporting**

It would be a mistake to read these stages as a linear progression from one to the other. Sometimes, it is necessary to go backward to go forward. A good example of this is that, during mitigation or containment of an intrusion, you may discover that the attackers have managed to reach a different area in your environment, or the intrusion is larger and more serious than originally anticipated. This will generally kick off another response and mitigation step in a different area, although these occurrences will still be deemed part of the same cyber incident.

Although the incident response process is highly dynamic and can be adapted depending on the situation, it is useful to discuss it in terms of these stages. The next sections will go into more detail about what each stage looks like and some of the considerations that come into play.

Detection

The first stage is where an incident is detected. Detection is usually the result of the efforts discussed in *Chapter 16, Planning for Security Operations*, to conduct logging and monitoring activities. The detection of an incident is usually the result of one or more alerts from the SIEM or a baseline anomaly, a third party contacting us about a breach, or a user reporting suspicious behavior on their computer.

It may be useful to briefly recap the ways in which an incident can be detected and relate that to the likely stage to which the detected intrusion has progressed. In the case of a user report, it is likely that the intrusion is in an advanced stage, and the attackers have moved on to establish some amount of control over our systems or have started executing their actions on objectives (e.g., data exfiltration, cryptolocking, or similar). The user may also report things such as ransom notices appearing on their screen.

In the case of a SIEM alert or anomaly detection, more analytic work is required to determine the stage of the intrusion. In the detection phase, many people somewhat freely use the terms event, alert, and incident and use them interchangeably. Nevertheless, the terms differ in their meaning, and it is good to understand the differences.

An **event** is something that happens on a network or endpoint, for instance, when a user logs in, opens an application, or a service starts or closes down. In general, an event has a corresponding log entry against it.

An **alert** is a (usually automatically generated) notification that the occurrence of some events has exceeded a set of predetermined policy settings, such as a number of failed logon events. An alert can therefore be composed of more than one event and is usually a series of events in a predetermined sequence. Alerts generally prompt a security team to do a brief investigation.

An **incident** can be the result of multiple alerts, and usually also has some sort of consequence that is visible, such as malware installed on an endpoint. An incident usually begins with alerts, followed by a brief investigation, and an observation of some consequence. At this point, an incident is usually **declared**.

In most organizations, incidents are triaged, and the response is based on the initial assessment of the seriousness of the incident. A **triage** process involves a small group of people reviewing the initial detection data, determining the scope and potential impact of an incident, and then classifying the incident on a scale usually ranging from minor to critical.

The seriousness of the incident is usually determined on the basis of the criticality of the affected infrastructure, the potential financial and/or reputational impact on the organization, the potential for privacy breaches, and the impact on business continuity—to mention just a few. The following figure provides an overview of the possible impacts of security incidents and ranks them according to their seriousness to the business. In general, each organization will have its own triage matrix because each organization is different. The **CISO** will usually be part of its creation. A matrix will help management make decisions about resource allocation in relation to the risk and seriousness of IT events.

The following table is an example of triage focusing on the duration of an outage, its financial impact, media impact, and other impacts.

Level	Low	Medium	High
Outage duration	5 minutes or less	5 minutes to 2 hours	2 hours or more
Financial impact	$1,000 or less	$1,000-$10,000	$10,000 or more
Media impact	No social media or media reports	Social media reports only	Newspaper articles, requests for interviews
Other impact		Potential data loss	Potential regulatory breaches

Table 17.1: Possible impacts of an outage, and their seriousness

Response

This section focuses on the initial response after an incident is detected. In most cases, right after detection, very little is known about how the incident occurred and the amount of damage that it caused. In most cases, it is not known who the attacker is or what they want (a ransomware attack running to its very conclusion may be an exception), or how deeply they have penetrated your systems.

In this sense, responding to an incident is usually very similar to providing first aid in the case of an accident. As in first aid, a team responds quickly on-site and follows a standard set of procedures to stabilize the situation as much as possible. It sometimes involves running out to the site where the incident occurred and looking at how bad it is. In case of a cyber incident response, that running may also be done digitally, for example, by a security engineer gaining remote access, starting log collection from a centralized logging system, or looking at the events on an endpoint as collected by an **XDR system**.

The response involves taking some quick and immediate measures to provide digital first aid. Possible actions are to isolate endpoints, turn systems off where necessary, remove network connections, and inform the analysis team of any findings and actions taken so far. Special care must be taken in these procedures to preserve forensic evidence where possible.

As was discussed in the previous chapter, it is a good idea to prepare a jump bag that contains both the physical and digital tooling needed to respond to incidents. Especially in industrial environments, teams also need to consider any training and certifications that may be required for personnel to enter dangerous environments or prearrange with the on-site teams to have team members escorted during the initial response.

Incident responders may also collect systems to take back to the security office for further analysis. This **collection** may be physical (as in removing the machine) or virtual, as in capturing memory while the machine is powered on, taking photos and screenshots of anything that may be related to the incident, and ensuring that digital evidence is secured. If the incident involves evidence that may become part of a criminal investigation or legal discovery process, teams also need to consider the requirements of evidence handling—especially the chain of custody.

Generally, incident response **plans** contain the list of people who can act as "first responders" in the case of an incident, as well as a set of procedures that must be followed in the case of an incident.

Mitigation or Containment

Mitigation or containment is the next step in handling an incident. In incident mitigation, the initial response actions—turning off hosts, for instance—are extended to include all of the initially affected systems and further investigation of other systems that might have been affected. The main aim of mitigation is to stop an incident from spreading further. The focus is on using what little knowledge is available to stop an attacker from moving beyond their initial foothold and to contain the incident to a known area. Mitigation is therefore also often called containment; it ensures that the "blast radius" of an incident is as small as possible. It may also be possible to mitigate some of the damage done, for instance, by implementing workarounds for some of the affected systems so that the business can continue.

A number of considerations make up part of a robust mitigation response. These considerations will be discussed next.

Analysis

After the response and collection steps, the incident data and collected artifacts need to be analyzed to figure out what happened, how it happened, and what to do next. The main aim of analysis at this stage is to work out how far the incident has spread, what systems and artifacts are caught up in it, and what is required to eradicate this particular intrusion on the network. In the analysis step, you have to pay particular attention to malware, lateral movement, and living off the land techniques and the traces they may leave. During analysis, you may also decide you need to contain other areas than those initially affected because your analysis indicates they were also targeted by an attacker, or an intrusion may have spread there.

Malware

Any **malware** identified as part of the intrusion should be analyzed with a view to determining its functionality. While a full report on the incident requires the identification and analysis of any malware used in full detail, at this stage, the incident management team must perform a quick analysis of any malware to determine what its functionality is, what it was likely being used for, and how, and then determine the next steps.

Triage of malware is nowadays mostly performed in a sandbox such as the kind available online, bundled with modern XDR tooling. In a sandbox environment, the malware is executed in an isolated environment while its actions on the system are monitored in real time by tooling made available in the sandbox. Such tooling may, for instance, monitor registry or system file changes, file creation and deletion on disk, network connections started or stopped, internet domains contacted, processes started or stopped, and can even extend to capturing memory images of a running system for memory forensics.

Sandbox execution is also sometimes called dynamic analysis. Dynamic analysis is the opposite of static analysis, in which the file-based format of the malware is used to determine clues as to what its functionality may be. Static analysis looks at file hashes, strings in the malware, imports and exports (libraries loaded during the execution of the malware that may indicate what the malware was supposed to do), packers and sections in the PE file, as well as any hard-coded IP addresses or DNS entries.

These indicators can then be used in conjunction with your knowledge of your own environment and give you pointers to where to look next for more evidence of the intrusion. In this way, teams can cycle through a repeated cycle of mitigation, analysis, and further responses if they find that other areas of the business are also affected. Later on, this chapter will discuss sandboxing and what we can learn from it in greater detail.

Lateral Movement

An important step to consider in mitigation is the potential **lateral movement** of a threat from the affected systems to other systems. Limiting lateral movement is one of the main steps that teams can take to stop the further spread of malware or affected systems. You need to think about lateral movement steps quite carefully since they include more than just malware spreading from one machine to the next via the network (although this is certainly included among potential lateral movements). Among lateral movement steps, you should also consider the following:

- Account compromises where one compromised account may give access to other accounts, perhaps with a higher level of privilege

- Remote code execution lateral movement paths, using PowerShell or other scripting languages

- Installations of further malware on affected machines

To perform the lateral movement analysis correctly, teams need to understand the potential attack paths through their own systems, especially those that lead to systems of high importance.

Mapping out such attack paths is an important part of defensive security operations, and there are now a number of tools available to teams that allow them to create feasible maps when they are planning mitigation steps. The following two tools in particular stand out:

- The first tool is the open source tool **BloodHound**, which allows teams to map attack paths through configurations of Active Directory. Since Active Directory is ubiquitous, these attack paths are feasible in many organizations.

- The second tool is **attack simulation**, which allows defenders to simulate realistic cyberattacks on their network and identify the detections and preventions they may have in place for lateral movement.

Another important function of containment is to ensure that all backdoors created by an adversary are discovered so that they can be closed during an eradication step.

Living Off the Land

Many modern attacks do not use bespoke malware to perform intrusions but instead rely on **living off the land** techniques to move laterally. Living off the land involves using tools already available on the system, such as **scripting engines**, **PowerShell**, and system administration tools to drive the intrusion further into the network. Usually, such system tools are not flagged by anti-malware solutions as malware, and the attackers hope to stay under the radar in this way. Living off the land also makes things simpler for an attacker because there are no new tools to install or configure.

Detection and analysis of living off the land is key to successful containment because it will lead to any new areas where attackers may have installed tooling or left a footprint. But living off the land is not usually picked up by malware detectors looking for suspicious executables. Detection of living off the land involves working with **behavioral** detections, such as looking at the various commands that are run on systems (for instance, obfuscated or encoded commands *may* indicate malicious intent), looking for unusual network connections, or analyzing logs on systems to determine if certain activities (such as enumeration of users, drives, endpoints) are occurring that are considered abnormal.

Eradication

Eradication is the complete removal of an adversary from the network and follows the containment of the adversary. When eradicating an adversary, the key measure of success is that adversaries are completely removed and have no avenue back into the environment. To this end, consider the following as a guide to things to consider.

Complete Removal of Malware

All malware installed or used by the adversary should be completely removed. A lot of malware is designed in such a way that it is easy to miss critical components, and some malware can reinstall itself and reappear on the network if a single component is left behind. From this perspective, with a malware-infected machine, a complete rebuild of the system should always be an option that is considered.

Disabling or Removing Backdoor Accounts

During the lateral movement stage of an intrusion, attackers may have created or accessed backdoor accounts in the system and may be able to use these accounts to regain access. Hence, accounts reactivated, created, and accessed during the known time of the intrusion, especially if they are system or administrative accounts, deserve a further look. It is also advised that teams remain cautious, so they don't overlook administrative accounts in key systems in the business, the database servers, and networking equipment.

Accounts that are identified as backdoors created by intruders should be closed. In **Active Directory**, a Microsoft product that governs Windows domains and that is used in most enterprises, there are many aspects beyond account creation and deletion that may indicate trouble, such as accounts with membership of groups with elevated privileges, changes to group policy objects, and ACL permission changes.

Disabling or Removing Persistence Mechanisms

Beyond backdoor accounts, there are various other persistence mechanisms that can be used by attackers that should be considered and disabled. Attackers deploy persistence mechanisms to ensure they can get back into your systems in the future, sometimes even after a team has eradicated the original intrusion.

> **Note**
>
> A list of persistence mechanisms is given under the ATT&CK "persistence" tactic (`https://packt.link/bti15`), where each technique of this tactic has a recommended detection.

Where there is evidence that some of these persistence mechanisms have been used, teams need to carefully look for the evidence of those mechanisms and ensure they are disabled so that attackers cannot use them to get back into the system in the future.

Blocking Paths for Data Exfiltration or Remote Control

Attackers usually have a set of objectives, whether that is the exfiltration of data, remote control, encryption of the infrastructure, or something else. If objectives can be reliably guessed through hints left in malware or through threat research based on initial findings, it is sometimes a good idea to block these pathways to ensure that attackers meet greater difficulty in achieving their objectives. In this way, consider the kill chain with the attackers' end in mind.

Closing the Barn Door

In cases where the original infection was exploiting a vulnerability, ensure that this vulnerability is addressed, either by patching, disabling services, or shielding through firewalls or proxies before putting the system back online.

Recovery

Recovery is sometimes also called remediation. During this phase, we reinstall systems and gradually bring an environment online as it is deemed clean and safe. As we've already discussed, a part of recovery is ensuring attackers don't get back in by patching the vulnerability originally exploited or making sure that the original exploit path is no longer viable.

Reporting

At the end of the incident, you prepare a report. The report states what happened, what the impact on the business was, and how the incident was resolved. It can then make a number of recommendations for a blameless review of the incident.

An incident report does not usually contain an attribution, that is, a specific naming of the attack group that is deemed responsible for the attack. Attribution is a complex business, which depends mostly on circumstantial evidence (there are usually few, if any, direct pointers to the specific **people** behind a cyberattack). Attributions in most cases (and certainly in most businesses) are of limited value because intruders are often in difficult-to-apprehend jurisdictions.

The report may also outline the attack path in terms of the ATT&CK TTPs that have been used during the attack and their relative success. Moreover, the report can contain an enumeration of our own detections outlining which ones worked and which ones may need improvement.

It is important that the report includes a timeline that is as accurate as possible because this will allow the security team to measure its effectiveness. This is discussed in the next section.

Lessons Learned

The **lessons learned** section of incident response must take the form of a blameless review, in which all the facts pertaining to the incident are tabled and solutions to any problems are discussed. It is important that the review is blameless: make the "lessons learned" section about apportioning blame and people will not speak up, information will be incomplete—and even worse, likely incorrect—and the "solutions" that you come up with are not likely to be solutions to the real problems you face. The review is not the time to discuss an individual's or group's performance.

The aim of the blameless review is twofold:

- To table and discuss any improvements that could prevent future incidents or lead to a speedier or more complete recovery afterward
- To measure the effectiveness of the detection and response function

Environment Improvements

As part of the blameless review, teams can make recommendations for improvements in the environment, the processes, or even their own response that will allow better prevention of incidents, limitation of the blast radius, or improvements in the incident response playbook itself.

These improvements are recommendations, and it is up to the security leadership to determine which ones will be implemented, the timeframe, and the form of the implementation. In some cases, this might involve simple fixes that can be implemented almost immediately, but in other cases, a multi-year multi-stakeholder project might be required.

It is important that teams do not try to solutioneer recommendations, that is, engage in design discussions or debate the merits of specific technologies or vendors in depth during the review. If there is a variety of opinions at this stage, note down all the ideas and move on. The time to debate the specifics of the improvements will come later, when there is a committed project or budget to implement the recommendation.

Measuring Effectiveness

It has already been discussed that the report should include a timeline of events. Ideally, after the incident, the team collects the following data from the timeline:

- **Time to detect**: This is the time that elapsed between the first intrusion, if one was found in the analysis phase of the incident response process, and the time a detection was generated. For example, it may take 24 hours of receiving malicious emails before your mail gateway detects them and starts blocking them. If you have an incident as a result of a user opening, for instance, a malicious attachment, then in the analysis phase, you will be able to determine that you have been receiving these malicious emails for the last 24 hours before you were alerted to their presence.

- **Time to engage**: This is the time elapsed between detection and the acknowledgment of the detection by the security team.

- **Time to respond**: This is the time between the first detection and the moment the response actions start. The response may involve isolating an endpoint, doing a forensic collection, and other actions, as discussed in the *response* section earlier in the chapter.

- **Time to contain**: This is the time it took between the declaration of the incident, the determination of its extent, and the stabilization of its blast radius. Once the intrusion no longer spreads, the intrusion is contained.

- **Time to recover**: This is the time between the declaration of the incident and the recovery of the system to its normal operating state. Recovery in this measure also includes the eviction of the attacker.

- **Dwell time of the attacker**: This is the time an attacker remains undetected in the network. Dwell time is the time between intrusion and detection.

There is a certain arbitrariness in how these metrics are defined. The choice of how the different metrics are measured can be open to interpretation, and different teams may use slightly different definitions for each of these metrics. That does not matter much as long as you use the same metrics to measure each incident so that it is possible to compare different types of incidents or the evolution of a single incident type over time.

Figure 17.1 shows the various stages of an incident response process, from intrusion to recovery. The timeline begins with an intrusion event, followed by detection, then engagement response, containment, and finally recovery. The "dwell time" refers to the period between intrusion and detection.

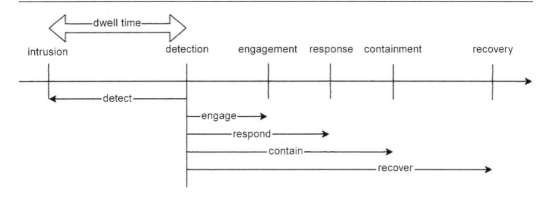

Figure 17.1: Time to detect, time to engage, time to respond,
time to contain, time to recover, and dwell time

Operating and Maintaining Detective and Preventive Measures

In order to successfully detect intrusions in your network, you need to install and operate the infrastructure necessary to prevent incidents from happening and have the capability to detect incidents when they do occur.

This section will discuss some detection and prevention technologies, focusing on **endpoint security**, **network security**, **whitelisting** (or **allowlisting**), **blacklisting** (or **denylisting**), **sandboxing**, **honeypots** and **honeynets**, the role of **machine learning**, and the **zero trust concept**. This list is not complete, and there is a very large number of new security vendors with new tooling appearing every year. However, this list contains most of the tooling that is important.

Endpoint Security and Anti-Malware

The role of **endpoint security** is to ensure that **malware** and any malicious activity has a short lifespan on your endpoints, such as end user devices, servers, and mobile devices. Some of the conceptual and design considerations of endpoint security and anti-malware were discussed in *Chapter 16, Planning for Security Operations*.

Originally, endpoint security was implemented through antivirus products, which, in their simplest form, maintained a set of "signatures" for executables and compared anything written to disk or opened on the endpoint to these signatures. Execution could be blocked, files could be deleted, and processes could be killed when something was deemed to be malicious. The more modern versions of antivirus products extended this concept to network traffic and running processes (i.e., memory), the latter to combat fileless malware—that is, malware that would be resident in memory only.

In response to the slowly growing capabilities of antivirus software, attackers started increasingly using the "living off the land" technique—in other words, using already existing system tools such as scripting languages, PowerShell, and system tooling to malicious ends. Regular antivirus software typically classified these executables as benign. Current antivirus products are capable of monitoring and reporting on the malicious use of anti-malware products.

As a result of these new products and the responses from attackers to these new products' capabilities, the manner in which teams operate these products has changed considerably. The so-called "next-generation" anti-malware products require significantly more from analysts and teams than just monitoring consoles for matches to signatures, although this is still part of what teams do.

The key operational questions focus on coverage and detection as well as capabilities, such as the capability of a team to dynamically query endpoints, perform hunts with one-off queries, establish process causality, and gather and analyze endpoint forensic data such as memory and executables.

Agent coverage

A key question for operational teams is whether their entire environment is covered and has endpoint agents running. An endpoint lacking an agent is not protected by the "signature layer" or the behavioral capabilities of the endpoint protection software but can also not be queried, be part of threat hunts, or have forensic data gathered.

A good way to measure endpoint coverage is to perform a network scan with fingerprinting, which gathers detailed information on all devices. From there, you can determine what percentage of endpoints do not have the endpoint protection software installed (provided the endpoint supports the installation).

The coverage question touches upon the question of **asset management**, which is one of the key areas of most cybersecurity frameworks, such as the NIST Cybersecurity Framework (`https://packt.link/oikVK`). Asset management is one aspect of cybersecurity that is easy to implement in theory (which involves looking at the assets a company currently has) but can in practice derail a cybersecurity program that gets stuck in this phase. The fact is that most organizations do not know what they have or how important it is for the business, due to a combination of reasons such as lacking documentation, poor visibility, and poor understanding of the components that are critical to the business. Formal asset management programs run a high risk of being vulnerable to these obstacles.

A practical approach teams can take to determine which assets they have is **comparisons**. Comparisons, which is the process of analyzing and contrasting different datasets or records from various systems to identify and understand the assets present in an organization's network. Most security and infrastructure software manages internal databases of the assets that fall under the aegis of these systems. Some examples are the number of virtual guests managed on a VM farm, the networks visible in router configurations and firewalls, and the results of network-based scans. Teams can ingest all these databases and combine them to gain an overall view of what is on the network and also determine the coverage of the security tooling.

Causality

Modern enterprise endpoint detection and response software can draw causal chains of what happened on an endpoint—which user started a process, what the process did subsequently, and what changes to data or the system configuration were caused as a result of that. The investigation of such causality chains is a key component of a modern defense capability that has to combat not only advanced malware and malware obfuscation techniques but also living-off-the-land techniques that exploit customarily available system software. The following diagram shows an example of a causality chain.

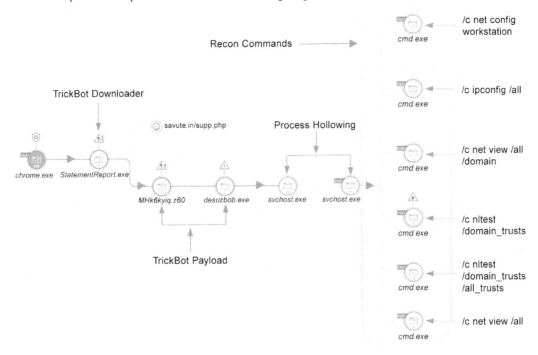

Figure 17.2: Example of a causality chain associated with TrickBot malware

The preceding diagram is based on a diagram created by Paloalto Cortex, but XDR products in general are capable of analyzing attacks in this way. This diagram depicts the initial download, the payload, process hollowing, and commands executed by the attacker. You can find more details at https://packt.link/6wRqu.

Causal chains of attack also play a big role in two other key capabilities that teams must possess with regard to security operations: the ability to "arm" defense software with one-off capabilities and the ability to systematically hunt for the presence of adversaries in their environment. Since causal chains of attack provide a "map" of attacker activities, they help teams understand where to deploy such specific detections, or what sort of behavior to hunt for. In cases where the organization has a functioning threat intelligence program, it is necessary to have both a one-off signature (or "tasking") capability as well as a functioning threat-hunting function.

One-off signatures

Organizations with a robust security operations capability will become aware of threats in their environment that may not have been detected by endpoint security software. In these cases, it is useful to be able to write "one-off" signatures that enable the endpoint detection and response software to detect the threat and prevent it from spreading further.

A generalized language in which one-off signatures can be expressed is *YARA*. YARA can, for instance, search for files that match certain hashes, files that have certain file headers (such as PE files or executables), or files that contain certain string values, as in the following example from *Security Intelligence* (`https://packt.link/hX2ZE`).

```
rule TestExample
{
    strings:
        $test_string = "foobar"

    condition:
        $text_string

}
```

Hunting and Endpoint Queries

Modern detection and response software is usually also capable of querying endpoints with a query language. A good example of such a query language is **osquery** or the **VQL** query language that comes with the Velociraptor package.

Such queries allow investigators to surface endpoints or servers that are running specific software, set particular registry entries, have files with a certain signature on their disk, and a number of other characteristics. In this way, teams can translate information from earlier in their investigation, or information they have researched on the web, into a specific hunt for further intrusions.

During incident investigations, teams usually take the characteristics of a compromised host and turn them into specific queries that can be run across the entire environment to determine if the intrusion has spread wider than initially thought.

Endpoint Forensics

In the case of a serious intrusion, it is sometimes necessary to perform endpoint or server forensics. Traditional forensics focuses on the hard disk of the device, and forensics software has the capability to inspect historical changes to the disk, such as deleted files and altered files, and investigate what software has been run on the device.

The traditional process for disk-based forensics is that the drive is forensically imaged (i.e., the investigator performs their analysis on a copy of the data rather than the data itself), and then forensics software is used to perform the analysis. On large hard drives, the analysis may take a long time—days or even weeks. Hence, full disk forensics is of limited use during the resolution of an incident.

With the growth in the size of common disk drives, disk-based forensics is becoming harder to perform and it is also not capable of inspecting volatile data stores such as computer memory. Hence, memory forensics is becoming more common. In opposition to disk forensics, memory forensics is usually carried out on a much smaller amount of data and is therefore faster. Moreover, memory forensics focuses on what is actually running on a machine instead of the history of its file system and can thereby provide direct insights into the actual state of the machine during an intrusion.

This is not to say that disk forensics is inferior to memory forensics; the two approaches are complementary, and both are required for the complete forensic investigation of an endpoint.

Forensics is sometimes also performed with a set of "forensic" endpoint queries, using osquery or another detection and response query language.

Network Security and Firewalls

Firewalls are primarily a preventive measure and can range from simple packet filters to sophisticated devices performing attack analysis and prevention. The nature of a firewall depends, to a large degree, on the OSI layer of the network at which it is operating.

Packet filters, the simplest form of firewall, operate on the network layer of the OSI model and "filter" packets based on origin, destination, and ports involved in the communication. They are neither capable of inspecting the traffic that traverses the firewall nor of comparing the traffic to a set of "known bad" signatures. The only decision a packet filter can make is to block or allow.

It should not come as a surprise that this simple model allows a number of improvements, which lead to various varieties of firewalls.

One improvement is to give the firewall some ability to inspect the traffic that traverses it, for instance with a signature set of "known bad" traffic. This includes malware traffic and botnet command and control traffic. The firewall can then make a decision based on the outcome of traffic analysis. This concept can be taken further to also include traffic that looks like it is aimed at exploiting a known vulnerability, which enables the firewall to perform host intrusion prevention.

One can also consider variations in which firewalls are capable of inspecting traffic at higher layers in the OSI stack, especially application traffic. Firewalls that can, for instance, inspect web traffic can also have a set of signatures that focus on the exploitation of known web application vulnerabilities, such as SQL injection. This family of firewalls is also known as **Web Application Firewalls (WAFs)**.

WAFs typically also terminate and reconnect TLS connections so they are capable of inspecting encrypted web traffic. WAFs can then be used to manage the organization's connections to the outside world, for instance by not allowing connections with outdated TLS protocols or disabling certain insecure cipher suites.

The secure operation of firewalls needs to consider the following frequently overlooked aspects:

- Who has administrative access to the firewall and can make changes in the rules?

- How often are the rules reviewed? Such a review may focus on why the rules are required. Can they be matched to a change management process and a business request that was approved?

- Are stale or old rules no longer required removed and what is the removal frequency?

- How are administrators authenticating to the firewall and where are the credentials stored?

- Where are the logs of the firewall sent?

- Are the logs and events being blocked on the firewall ever inspected?

All of this may also drive a review of the internet "footprint" of an organization, including the domains that the organization owns, the systems these domains point to, the open ports on those systems, the certificate infrastructure, ciphers accepted on secure websites owned by the organization, and even known organizational credentials that in the past have leaked to the cyber underground.

A final aspect of network security that teams should consider is protection against **distributed denial of service** or **DDoS**. A DDoS attack is a type of attack where an attacker attempts to temporarily remove network connectivity by flooding the network with spurious traffic in the hope of overloading firewalls or servers. Firewalls may assist with DDoS prevention if they are configured to drop traffic after a (usually short) detection span of spurious traffic, but in general, the best DDoS protection is when an organization works with its network provider or an outside service to ensure that this spurious traffic never reaches the network.

Intrusion Detection and Prevention Systems

Intrusion Detection Systems (**IDSs**) and **Intrusion Prevention Systems** (**IPSs**) are used to detect and prevent intrusions on both the network and the individual host, be that a server or an endpoint.

A **network-based intrusion detection system** (**NIDS**) *listens* to the traffic on the network and may use **Switched Port Analyzer** (**SPAN**) capabilities on networking gear and compares it to a list of *known* bad traffic to determine whether an exploit or intrusion is in progress. Because NIDSs can work with a copy of the traffic, they can be placed out of band—out of the network traffic stream—so that performance issues on the system do not affect the production network. NIDSs are capable of detecting botnet **command and control** (**C2, CnC,** or **C&C**) traffic, passwords being used in the clear, that is without any encryption, and other security risks.

Network intrusion prevention systems do not just listen to a copy of traffic; they are also capable of blocking traffic and so must be placed in line or in the network traffic stream itself. For this reason, firewalls are a good platform to perform a prevention function and many firewalls are capable of blocking packets if they determine that the packet is part of an intrusion attempt or intrusion in progress.

Something similar happens on the endpoints where host-based intrusion prevention is usually performed as part of the detection, response, and prevention software, as opposed to only detection. This is usually the normal mode in which such software is operated. Host-based intrusion detection systems can still be used and can alert, for instance, when important system files, registry settings, or autostarts are modified. Modern detection and response software usually includes these capabilities alongside its prevention capabilities, and whether to configure these capabilities in "detect" or "prevent" mode is part of the considerations security teams have to make when this software is deployed.

Whitelisting and Blacklisting

An **allowlist** or a **whitelist** is a list of things that are allowed to run, whereas a denylist or a blacklist is a list of software that is not allowed to run. Usually, in environments with high sensitivity, allowlisting is preferred. In this case, an organization determines in advance what software is allowed to run in that environment, and sometimes even locks down a specific version of that software.

An example where such enforcement might be in place is in industrial control systems, which often have low levels of change after they are initially deployed and tend to remain static over time. Furthermore, with the risk typically associated with such systems, organizations may be very restrictive in what is allowed to run in them.

Allowlisting gives an organization a lot of control over what is allowed to run in an environment, without having to monitor any extra applications that might be introduced because they are automatically blocked. However, at the same time, an allowlist is hard to maintain in environments that are subject to frequent change due to required updates to allowlists as new applications are required.

Denylisting or **blacklisting** is often used in less sensitive environments, where, for instance, an anti-malware program may maintain a list of software that is considered dangerous and prohibited from executing. The advantage of denylisting is that it is relatively easy to maintain in a dynamic environment as IT only has to decide which applications to deny while others can be installed without barriers. However, a static denylist provides only limited protection as the attacker can and will bypass such lists with relative ease in the case of more advanced attacks.

Another application of denylists is domain blocking, where organizations may block access to internet domains that contain illegal, harmful, or sometimes just distracting content.

For an IT system, you know that there are programs that are essential, such as a trusted web browser, or your CMS, and there are programs that are dangerous, such as viruses or torrent clients. In between, there is a gray zone with software and tools that are neither explicitly needed nor (known to be) dangerous. With an allow-listing approach, these things will be blocked, whereas with a denylisting approach, they will be allowed. This is an important consideration in risk management.

For example, a news agency with a design team might have a designer who needs to download a specific tool to upscale a photo for an urgent story. If the agency uses an allowlist approach, this tool would be blocked from running until an IT administrator added it to the allowlist, potentially causing unreasonable delays. In contrast, a denylist approach would permit the tool to run immediately, assuming it was legitimate software.

On the other hand, a social media administrator might download free software to help schedule marketing posts, not knowing it is buggy and insecure. If the agency uses a denylist, the software might still run if the denylist hasn't been updated to include it, potentially allowing attackers to gain access to social media credentials. However, an allowlist would prevent this software from running by default, as only pre-approved applications would be allowed to execute.

Third-party Security Services

Managed security service providers (**MSSPs**) are external organizations that provide, for instance, security operations center services to organizations that do not wish to maintain their own capability. For many organizations, working with an MSSP is a time- and cost-efficient way to acquire security skills without having to go through the effort of developing and building such services yourself. A disadvantage of an MSSP service is that they will as a rule never know your business as well as you do, and from that perspective, the services offered may be generic.

A hybrid model is often a good solution. In a hybrid model, you use an MSSP for things like "round the clock" monitoring of your security tooling and initial incident triage and also have a team based in your organization that deals with the specific incident context for incidents that MSSPs cannot deal with. MSSPs can also provide robust technical knowledge in areas such as incident response, incident analysis, and threat intelligence that is hard to source for many organizations on their own.

From a security governance perspective, it is important to critically evaluate the service contracts in place with an MSSP and gauge them against the needs of the business to ensure that the business has the coverage that it needs.

Sandboxing

A **sandbox** is an environment in which you can safely detonate malware to get a better insight into its inner workings. In malware analysis, a sandbox allows you to perform the **dynamic analysis** of a sample. Sandboxing is a useful technique in the mitigation stage of an intrusion when it is important to work out the functionality of the malware deployed into your environment. This was briefly mentioned in the *Conducting Incident Management* section (especially the sub-sections on *Response* and *Analysis*).

However, sandboxes have wider applications too. Sandboxes are now often part and parcel of both mail filtering systems and endpoint security systems, where they operate more or less seamlessly and without explicit interaction with the security team or incident responders. Emails with attachments or unknown executables are uploaded to the sandbox and executed (or "detonated") while the changes they are attempting to make to the system are observed.

Sandboxes provide an (often virtual) environment in which a sample is executed or a document is opened and all actions, such as file creation, registry writes, network traffic generated, processes started, and logs written to the event log are recorded for the analyst to examine. Many sandboxes will also provide the static details of executables that are uploaded to it, such as hashes and executable details such as imports and exports.

Public Sandboxes

There are sandboxes publicly available to which a team can upload their file, execute it in a controlled environment, and then get a report on the result.

Zeltser.com has a list of **publicly available sandboxes**. (`https://packt.link/rC1wK`)

Figure 17.3 shows **ANY.RUN**, which is a **public sandbox** and describes itself as a **cloud-based malware analysis service**. You can find it at `https://any.run`.

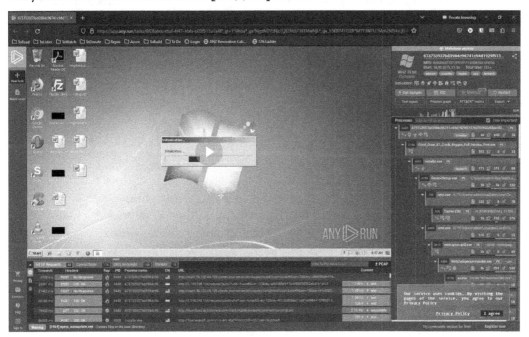

Figure 17.3: Output of ANY.RUN, a publicly available sandbox

The figure shows the software generating a report. You can find the report at `https://packt.link/iFypG`.

Sandboxes may not always be foolproof. Sometimes malware attempts sandbox evasion by using timers. Usually, sandboxes have to return a verdict within a small amount of time (say, a minute or less). Some malware will attempt to evade the sandbox by having a longer wait time till execution so that the sandbox does not have anything to observe. Another way in which malware authors attempt to evade detection and dynamic analysis is by determining whether the malware is running in a virtualized environment, assuming that normal endpoint execution is usually not carried out on a virtual machine.

Honeypots and Honeynets

Honeypots are devices that pretend to be something of interest to an attacker. Honeypots can pretend to be a file server with documentation that might be attractive to an attacker, such as a valuable database server. Honeypots are then closely monitored for any access attempts to files or the generation of network traffic. Because honeypots hold no real business data and are not used in the normal course of doing business, any access to them is of some interest.

A variation of the honeypot is the **honeynet**, which is a simulated network that similarly does not hold any business data.

Some endpoint security products now create **honeyfiles** on endpoints, which are hidden from the normal user on the endpoints, and which should never be touched by the operating system or applications during the course of normal operation. Any attempted access to these files is then alerted and investigated. Another name for these is honeytokens.

To combat spam and credential phishing, some teams have honey accounts. When a phishing attack occurs, the security team gives the credentials of the honey accounts—normally an email and a password. Any subsequent access attempts on these accounts are alerted on. This can help the security team to detect and combat credential theft by detecting when accounts are "tested" by the attackers as the honey account is likely to be part of the test run, and teams can then use the contextual data of the logon attempt to determine any other such tests.

Machine Learning and Artificial Intelligence (AI)-Based Tools

Machine learning and AI tools for security operations come in two types:

- Tools focused on the use of AI to investigate and characterize **end-user behavior analytics** (**EUBA**)
- Tools focused on improving the quality of the alert stream by enriching and performing initial analysis on events and alerts so that the human analyst gets a higher quality of alerts to work on

To overcome the alert fatigue and burnout associated with the work of operational security, many organizations seek to adopt AI and machine learning to improve the quality of alerts for their team. In this case, the focus of the tool should be on narrowing down the alert funnel, that is, filtering out unimportant alerts.

All of these measures can be deployed in an environment that already exists. But we can also consider how a new environment would need to be designed to make it as secure as possible. This leads us to the zero-trust concept.

The Zero Trust Concept

In traditional network security architecture, the network is **zoned**—divided into security zones—and separated by a firewall or any other type of policy enforcement point such as a proxy server or web application firewall. The zone of the network then has an implied level of trust based on its perimeter. For instance, in traditional networking architectures, a "client" zone is where the endpoints reside, and a security monitoring solution will take the trust level of that zone into account to determine whether an observed event constitutes a security incident or not.

In **zero-trust architecture**, no such implied level of trust in network segments exists. Zero-trust architectures do not assume any trust level based on where a device is in the network. Instead, a zero-trust architecture assumes that a device can prove its security status at the point of access or execution. The focus of a zero-trust architecture is on users, assets, and resources.

It is easy to see why **zero-trust** architecture has become highly preferable over time to the traditional perimeter-based network architecture.

Zero-trust concepts are particularly important once an environment starts incorporating a large amount of cloud infrastructure in the form of **software as a service (SaaS)**, **platform as a service (PaaS)**, and **infrastructure as a service (IaaS)**. In these cases, the sort of workloads that traditionally ran in a data center are now run in the cloud and typically accessed and managed through a web browser.

We usually access the cloud via the internet, which is an untrusted network. From this perspective, any trust that we place in network zones as a root of trust is misplaced. What matters in cloud computing is the trustworthiness of the endpoint and the cloud, as well as the security of communications between them.

With IT architectures moving from on-premises to the cloud, remote work, and an increasing amount of work being done from mobile devices, security based on network perimeters makes less and less sense. In the modern working environment, with a large proportion of remote work, many devices connect to the network from outside locations and cannot be trusted by virtue of whether they are "inside" or "outside" the network.

In a zero-trust architecture, devices have to prove their security posture before they gain access and, subsequently, all network communications are secured via end-to-end encryption. Moreover, policy definition and enforcement points are unified, making it easier to define and maintain a security posture that is compatible with the risk appetite of the organization.

Figure 17.4 shows the logical components of a zero-trust architecture, showing the inputs, control plane, and data plane.

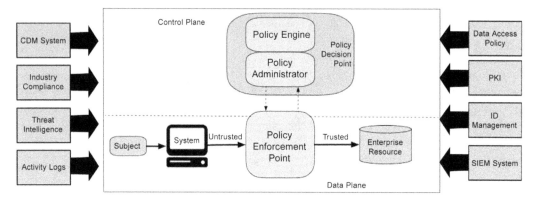

Figure 17.4: Logical components of a zero-trust architecture based on an NIST whitepaper

> **Note**
> You can find the NIST whitepaper at `https://packt.link/1A9qh`.

Policies are a key working component of a zero-trust architecture. Each execution, access, and authentication must be able to prove to a security enforcement point that it complies with the policy.

In the next section, we will be focusing on another important aspect of security operations, the management of vulnerabilities.

Implementing and Supporting Patch and Vulnerability Management

Vulnerability management is the process of identifying, assessing, and addressing vulnerabilities within a defined environment. Vulnerabilities typically result in a risk to an organization, and if not managed, the risk may result in the compromise of information and operational assets. Consequently, such a compromise may lead to financial losses, reputational damage, or legal ramifications.

The vulnerability management framework consists of five phases, of which a number are performed by the security function:

- Identifying the vulnerability
- Assessing and prioritizing the risk
- Stakeholder engagement

- Remediation
- Monitoring and reporting

These are discussed in detail in the sub-sections that follow.

Identifying the Vulnerability

Vulnerabilities are identified through a number of security processes, which may include (but are not necessarily limited to) vulnerability scanning, external notifications, and implementation of industry knowledge.

Vulnerability scanning involves using a specific tool that inspects open network ports, software versions on hosts, and, in some cases, configuration files to unearth places where a vulnerability may be present. Vulnerability scanners usually generate a large amount of data, and, especially when organizations have just initiated a vulnerability management program, generate a large list of things to fix.

Other sources of vulnerability data can be information about high-risk vulnerabilities from news or trust groups. They may share information on *zero-days* or other vulnerabilities that can be so new that the capability to scan for them has not been added to commercial scanners yet.

Since the number of discovered vulnerabilities can be large and their context complex, security teams need to conduct an assessment phase where they determine which vulnerabilities need fixing immediately and which are lower priority.

Assessing and Prioritizing the Risk

Once a vulnerability is identified, the security team must assess the risk posed by the vulnerability. This is done by following the risk management process, which takes into account risks to business objectives, existing controls, control gaps, and the impact and likelihood of the vulnerability being exploited or used in the exploitation of adjacent systems. *Figure 17.5* illustrates the different factors that play a role in transforming a vulnerability into a risk.

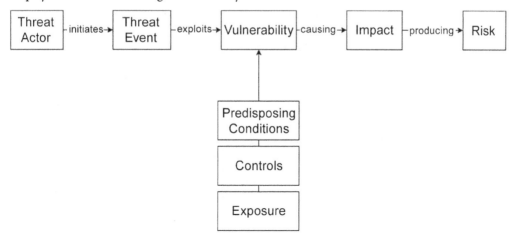

Figure 17.5 From vulnerability to risk

Not every vulnerability is created equal. The seriousness of a vulnerability, without the context that translates it into risk, is usually captured in the **CVSS** score (the **common vulnerability scoring system**), currently on version 4. The CVSS considers a number of factors to compute a score, such as the attack vector, the attack complexity, the prior privileges required, whether an attack can be launched over the network, and what the impact of an attack is on the CIA properties of information. It considers aspects such as whether the exploit leads to a compromise of confidentiality, integrity, and availability, or a combination of all three. Based on these factors, which are inherent in the technical and exploit details for each vulnerability, the CVSS formula computes a vulnerability score on a scale of 1 to 10, where 10 is critical.

The CVSS score is a useful starting point but does not consider several factors that may be dependent on the context in which the organization has deployed software. For example, a vulnerability may be exploitable via the network, but is this network path actually available at all in the organization, is it only available to a limited number of people, or is it not available at all? Has the organization deployed the vulnerable components or configuration?

Another factor in determining what risk is associated with a vulnerability is whether it is currently exploited by malicious actors, or "exploited in the wild." CISA makes available a useful list of vulnerabilities that are currently exploited in the wild by known attacker groups or malware. This provides another useful list indicating which vulnerabilities to prioritize.

Other factors, such as whether an exploit code is available or cases in which threat events based on this vulnerability are trending upwards can also be factors in the assessment of the seriousness of a vulnerability.

This sort of assessment at scale (i.e., for a large number of vulnerabilities) is an emerging discipline of vulnerability prioritization. In most organizations, this work should be automated, since a manual assessment of these factors for anything but the one or two top vulnerabilities is usually beyond the capabilities of any team and, moreover, is an inefficient use of time.

Most commercially available scanners now include a usually proprietary assessment that takes all these factors into account and generates a modified score based on the actual threat that a vulnerability poses in the current threat landscape. Some commercial scanners also include an assessment of whether that vulnerability could be reached by an attacker in a real-world attack.

Stakeholder Engagement

After this triage or assessment step, the security team presents the findings of the risk assessment to the owners of the change management process and platforms teams with the aim of collaborating on a remediation plan. Once the remediation plan is agreed to, it may be carried out under **business as usual** (BAU) processes or as a project. For most common vulnerabilities, this plan is a pre-determined patch process, but individual remediation plans may be required for high-priority, high-impact vulnerabilities for which patches need to be deployed urgently.

In the case of a zero-day, an exploited vulnerability where no vendor patch is currently available, the plan may also include utilizing a temporary workaround (which in turn may include disabling a service) until the vendor makes a patch available.

In some environments, applying patches is the responsibility of the security team, but it is usually better to delegate this responsibility to a server, endpoint, or platform management team to ensure a consistent integration between day-to-day system maintenance and configuration and keeping systems up to date (e.g., patched). The responsibility for applying patches may vary between organizations.

Remediation

Remedying vulnerabilities may involve a number of things, usually involving patching, workarounds, or temporarily disabling components.

For some vulnerabilities, a vendor patch is not available at the time the details of the exploit become public. This *becoming public* may also include *proof of concept* code that implements a successful exploit that can be weaponized and exploited by attackers. These types of vulnerabilities are usually called *0 days* or *zero-days*, though as a term it is somewhat overused in the industry. A zero-day is a known vulnerability for which a patch is not (yet) available.

Patching involves applying updates, usually supplied by the software or system vendor, to the system so that the vulnerable components of the software (e.g., libraries or executables) are replaced by a version that is not vulnerable. Patching operating system components usually requires a reboot of the system, which is why patching is usually done under a **change control** process, since it involves both a small outage of the system as well as the risk that the new software components may have an impact on the functionality or performance of the system.

Many organizations have a regime where patches are applied to a small subset of systems first, then to the test and development systems, and finally to the production systems. In today's threat landscape, this can be a dangerous approach that leaves production systems vulnerable for longer than acceptable. Ultimately, this comes down to how the business evaluates the risk of the patch versus the risk of the vulnerability, and where this fits with the other risk priorities of the business. Generally, it is not advisable to have a large number of open vulnerabilities in business systems since this increases the number of feasible attack paths and decreases the cost to the attacker of many of these paths (e.g., it makes the business easier to attack).

In cases where patches are not available, or where patches are applied late, a **workaround** can be deployed that renders the attack path to the vulnerability inoperable for an attacker. A workaround may involve closing a firewall port, disabling a vulnerable component if it is not business-critical, and deploying additional monitoring to determine whether an attack utilizing the vulnerability is in progress (this may be a workable approach in cases where a vulnerability requires significant privileges before it becomes exploitable and where other mitigating controls are in place).

Monitoring and Reporting

Monitoring and reporting involve checking whether the vulnerability is still present and usually scanning again to see if the vulnerability reappears.

Sometimes vulnerabilities have a half-life, especially in cases of a fleet of endpoints. Endpoints that were not present on the network during the initial scan may be on the network during the control scan and may not have the fix applied. In other scenarios, the fix may require a reboot in environments where users are usually just putting their computers to sleep when they are not working, and hence endpoints are seldom rebooted.

Especially for endpoints, fixing high-risk vulnerabilities can take a significant amount of (elapsed) time as new vulnerable endpoints appear on the network and need the fix applied.

Monitoring and reporting should also include some metrics to measure the effectiveness of the vulnerability management program. A monitoring and reporting function should aim to answer questions such as the following:

- What is the average lifetime of high-risk or critical vulnerabilities in your environment?

- On any given day, which percentage of your fleet has "open" (i.e., non-remediated) vulnerabilities?

- How many of your currently open vulnerabilities are listed in the CISA list of currently exploited vulnerabilities?

Security teams also commonly take part in the change management processes, which is discussed next.

Change Management Processes

Security teams usually participate in **change management** processes as reviewers and approvers. Change management is the process by which organizations make changes in their IT environment, such as the introduction of new software or platforms, the retirement of old software or platforms, and anything else that might constitute a change.

Change management is one of the disciplines in the well-known **Information Technology Infrastructure Library (ITIL)** and many organizations follow this approach. ITIL does not contain a specific security section; security teams need to determine their role in change management themselves. How this is done depends on the nature of the organization, its culture, and how it has organized changes in its IT environment.

Hence what follows is only some generic pointers. Security functions usually interact with change management at two levels: security architecture and change management.

Architecture

Security architecture was discussed in *Chapter 8, Architecture Vulnerabilities and Cryptography*. The role of managing changes in the security architecture is taken by the security architect, who evaluates changes from the viewpoint of the alterations in the security posture that they introduce. An important part of this review process is to assess how a new change introduces weaknesses in the configuration and potentially makes the organization more vulnerable to attack. Threat modeling is one of the tools that helps with this assessment.

In the change management process, architects also usually confirm that any introduced changes follow the agreed architecture patterns and do not significantly deviate from what was supposed to be built.

Another area that needs to be addressed at the architecture stage is how any change introduced at the operational level changes the logging and monitoring requirements. Architects need to talk to security operations to better understand this area.

Many organizations do not have a defined logging and monitoring strategy that defines which logs are monitored, how they are monitored, and where they are stored. For standard systems such as operating systems and common application and platform software, many standards (such as the CIS standards) will define what should be logged and monitored. For enterprise applications—especially ones that are developed in-house or are based on heavily configured versions of ERP platforms—the security architect should define these parameters together with the operational security team and make sure that they are implemented during change management.

Change Management

Change management is the approach taken by an organization to implement changes in its environment. "Changes" in this context usually refer to the introduction of new software or new software components, important changes in the overall configuration of devices and services, the introduction of new hardware, or the implementation of operational technology (cameras, automation, robotics) in the environment.

Most organizations have a change management process owned by a person responsible for change management.

A security architect's participation in change management usually involves acting as an approver and technical expert for security. As an approver, the security operations team reviews proposed changes and approves (or declines) them. Security operations should work with change management to determine which changes will be preapproved, and which ones will need further review. Alongside the security architect, security operational teams will generally be members of the **change advisory board** (**CAB**), which approves the most complex changes in the organization.

Summary

In this chapter, you explored the nitty-gritty of security operations and what an actual implementation looks like. The chapter discussed the management of cybersecurity incidents and some of the settings for operating and maintaining an operational security practice. How this works in practice depends, in many ways, on the nature of the organization, its culture, and how it generally manages its IT environment when it comes to monitoring, incidents, and changes.

Security teams need to work with the rest of the organization to ensure that their practices align with what is already established practice in an organization but also ensure that in doing so they can and will maintain an acceptable security baseline. That is a difficult undertaking.

Despite all the best efforts of the security and IT team, disasters, in the form of system compromise or natural disasters, sometimes do occur. That is the topic of the next chapter.

Further Reading

- *Zero trust*: https://packt.link/y6oi2
- *Mitre: 11 Strategies of a World-Class Cybersecurity Operations Center*: https://packt.link/1XLJl
- *CVSS*: https://packt.link/dgNJL
- *CISA list of exploited vulnerabilities*: https://packt.link/vEWII

Exam Readiness Drill – Chapter Review Questions

Apart from a solid understanding of key concepts, being able to think quickly under time pressure is a skill that will help you ace your certification exam. That is why working on these skills early on in your learning journey is key.

Chapter review questions are designed to improve your test-taking skills progressively with each chapter you learn and review your understanding of key concepts in the chapter at the same time. You'll find these at the end of each chapter.

> **How to Access These Materials**
>
> To learn how to access these resources, head over to the chapter titled *Chapter 24, Accessing the Online Resources*.

To open the Chapter Review Questions for this chapter, perform the following steps:

1. Click the link – https://packt.link/chapter17.

    Alternatively, you can scan the following **QR code** (*Figure 17.6*):

Figure 17.6: QR code that opens Chapter Review Questions for logged-in users

2.   Once you log in, you'll see a page similar to the one shown in *Figure 17.7*:

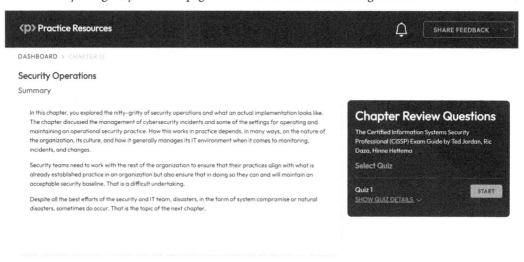

Figure 17.7: Chapter Review Questions for Chapter 17

3.   Once ready, start the following practice drills, re-attempting the quiz multiple times.

Exam Readiness Drill

For the first three attempts, don't worry about the time limit.

ATTEMPT 1

The first time, aim for at least **40%**. Look at the answers you got wrong and read the relevant sections in the chapter again to fix your learning gaps.

ATTEMPT 2

The second time, aim for at least **60%**. Look at the answers you got wrong and read the relevant sections in the chapter again to fix any remaining learning gaps.

ATTEMPT 3

The third time, aim for at least **75%**. Once you score 75% or more, you start working on your timing.

> **Tip**
> You may take more than **three** attempts to reach 75%. That's okay. Just review the relevant sections in the chapter till you get there.

Working On Timing

Target: Your aim is to keep the score the same while trying to answer these questions as quickly as possible. Here's an example of how your next attempts should look like:

Attempt	Score	Time Taken
Attempt 5	77%	21 mins 30 seconds
Attempt 6	78%	18 mins 34 seconds
Attempt 7	76%	14 mins 44 seconds

Table 17.2: Sample timing practice drills on the online platform

> **Note**
> The time limits shown in the above table are just examples. Set your own time limits with each attempt based on the time limit of the quiz on the website.

With each new attempt, your score should stay above **75%** while your "time taken" to complete should "decrease". Repeat as many attempts as you want till you feel confident dealing with the time pressure.

18
Disaster Recovery

The previous chapters primarily focused on the security operations, confidentiality, and integrity of the infrastructure. This chapter on disaster recovery will focus on the availability component of the CIA triad. Specifically, this chapter focuses on the resilience and reliability of the IT infrastructure. This is especially important for any on-premises infrastructure. Data in the cloud is managed by the cloud provider, although organizations must consider the amount of resilience that they purchase with their storage and compute infrastructure and whether that amount meets their needs, which is usually decided during the design phase of a cloud infrastructure.

All information systems are built with potential disasters in mind, and resilience is about not only being able to protect against attacks, accidents, and natural disasters but also how prepared a system is to recover from such incidents with minimal data loss and downtime. Disaster recovery, therefore, is a wide topic, where many technical options are available, although these usually carry a price tag. In practice, choosing the right disaster recovery option is dependent on what the business needs, how much they want to spend on it, and whether the chosen option can be delivered, maintained, and regularly tested by the business. This chapter discusses some of the topics that organizations need to consider when planning disaster recovery scenarios.

By the end of this chapter, you will be able to answer questions on the following:

- Implementing disaster recovery strategies
- Implementing disaster recovery processes
- Testing disaster recovery plans
- Simulating disasters

Planes are designed to be resilient and can withstand the failure of single components. They do this with redundancy, for instance, having two pilots in case one is unable to operate. But planes also occasionally have accidents, and in those cases, disaster recovery becomes important, with emergency services dispatched quickly. In this chapter, you will review both disaster recovery and resilience strategies. As the case of the airplane shows, these processes are very different. The next section will start with disaster recovery.

Implementing Disaster Recovery Strategies

Availability as a feature refers to IT services remaining available to the business in the event of an attack or disaster. It is an important part of computer security. To ensure availability, security teams are often involved in the planning and development of a disaster recovery strategy.

An important aspect to consider is whether potential data loss results from an attack or a natural event. In the case of a natural event such as flooding or power outages, resilience should be a key element of your availability strategy.

In the case of a cyber attack, you need to consider the intent of the intruder as an additional factor in the potential damage. Disaster recovery planning is especially becoming more important with the prevalence of ransomware attacks, where attackers encrypt the onsite data so that it becomes unusable to the business. They then ask for a ransom to supply the decryption key. Instead of paying this ransom, organizations may regain access to encrypted data through a recovery from backup. To make this recovery strategy work, it is important to ensure that the backups themselves remain protected from the ransomware attacker. However, this is not always assured with backups to disk rather than tape. Ransomware attackers often also attack backup systems to make recovery from backup impossible.

Recovery strategies can thus become complex and need to cover a lot of potential failure scenarios as well as malicious intent on the part of an attacker. In all discussions, planners need to put the needs of the business first. That means considering key questions such as how long the business can function without this system, how quickly this data needs to be available after an outage, and what the cost per hour is of this system being unavailable. The rest of the sections will focus on some specific considerations that planners need to take into account when developing a recovery strategy.

Backup Storage Strategies

Backup storage strategies focus on where and how the business will store dated copies (backups) of its business systems and data.

When considering backup storage strategies, organizations need to consider a number of fundamental questions, such as what data needs to be backed up, the size of that data, and the frequency with which it is backed up. For example, it may be acceptable to back up a file server storing mostly word-processing documents only once a day, while a transactional database server might have a much more frequent schedule for which its data needs to be available.

The underlying question, in business terms, is whether an organization can afford to lose a certain amount of work from each employee or lose a certain number of transactions. In many industries, organizations have a very low or no tolerance for data loss and the backup storage strategy needs to consider this. The consequence of a low tolerance for data loss is that data always needs to be recoverable. Another question to consider is how quickly data must be restored in the event of a disaster or attack.

These questions are usually captured in two parameters that organizations need to define on their various data stores:

- **Recovery Time Objective (RTO)** measures how quickly an organization's data (and the systems and services necessary to access it) must be available after a disaster. In general, some systems or data in an organization need to be available quickly while other data can be restored in due course. For example, the details of stock prices in an exchange need to be restored as quickly as possible – where possible, within minutes – because any discrepancy between a recorded price and the real price can have massive financial impacts. However, the same exchange's internal staff intranet is not as essential, and recovery of that data could wait for days if need be.

- **Recovery Point Objective (RPO)** is the point in time that must be restored. This measures how much data an organization can afford to lose if the data needs to be restored from a backup. What is acceptable here depends on the data and the use of that data. As an example, an airline may decide that they can afford to lose 1 minute's worth of bookings, so the RPO for transactional data from the booking website is that 1 minute of data loss is acceptable. For a system doing bank transactions, that may go down to seconds or less. The RPO is likely to be much larger than this for a file server with memos on it since the worst that can happen is that people have to create the memos again.

Organizations can also consider whether it is a functionality that needs to be available or data, and this can sometimes lead to interesting approaches in the disaster recovery strategy.

Consider a business's email services. During normal operations, emails are coming in and sent out by the business, some of which are loan applications. When email services are unavailable, the business is not able to send and receive emails, including loan applications. The RTO specifies how quickly email services must be restored. The RPO specifies the amount of loan applications that can be missed because email services are not available. A business might specify that they must have an email back within 3 hours but are not willing to lose more than 1 hour of missed loan applications because they could be picked up by a competitor. That means that these applications must have another means of arriving in the business, or they must be queued somewhere until the email server is ready to receive them.

Such considerations can impact the chosen strategy, so it is important to work through the potential impact of an outage first and then work out a solution that applies.

Lastly, organizations should also consider the legal requirements they have for data retention and data loss. Especially in the case of financial transactions, medical data, or other highly sensitive data, organizations should consider what their legal obligations are when designing a data recovery strategy. For example, in a medical scenario, organizations usually need to retain their full history with a patient and cannot afford to lose any data, since this may put lives at risk. Similarly, tax obligations may require a trading organization to keep financial records for a set number of years. Government organizations often have to deal with members of the public requesting official information, which also leads to specific requirements for data storage and recovery.

Recovery Site Strategies

Backup strategies focus on the **recovery of data**. What happens when organizations need to restore their systems and services in the event of a disaster? In that case, recovery sites may be needed and a site recovery strategy is required to work in tandem with the data recovery strategy.

As with backups, businesses must consider a number of business-related questions when deciding on an approach to recovery sites. In 2024, it is easy to set up a recovery site in the cloud, but in the past, many businesses had multiple data center locations to have access to a recovery site in case of a disaster. The next sections will briefly discuss some of the technology used in the design of recovery site strategies.

Setting Up Secondary Data Centers

A **recovery site** or **secondary data center** can be a **hot**, **cold**, or **warm** standby. Deciding between the different options is a matter of cost versus business needs versus the capabilities of the organization.

Each type of data center and its possibilities are discussed here:

- **Hot standby**: A hot standby site is a fully functioning data center that, in most cases, will have continuous replication of critical business data in case of a disaster so that it can take over from the main data center at very short notice (often in minutes).

- **Cold standby**: A cold standby data center has the necessary equipment to act as a secondary data center, which may be turned off and may need no continuous data replication. In the case of an emergency, the cold standby data center needs to be started, and data loaded from a tape or backup.

- **Warm standby**: A warm standby data center would be somewhere in the middle of these two extremes, with only a small subset of equipment powered up during normal operation. The plan is then to power up the entire data center if needed.

Of course, with cloud technologies such as **Infrastructure as a Service** (**IaaS**), it is also possible to provision a secondary data center in the cloud.

Replication

Replication is a strategy where data is replicated between different geographic locations to ensure that a recovery site has as much of the data and services as is available on the main site. If a disaster occurs, with a replication strategy, you can "fail over" to the replicated site and find (most of) the data needed by the business there.

Data replication between data centers is of key importance in scenarios where the business has decided to invest in a hot standby data center. When planning data replication, the following need to be considered:

- **Network performance**: Replication usually relies on the existence of a network connection between the primary and secondary data centers. Can the network between the two sites cope with the volume of data to be replicated?

- **Latency**: Is the network fast enough to guarantee that data will arrive at the recovery site within the parameters set by the business? In determining this, consider that, during a disaster, some data may be lost in transit.

- **Network resilience**: Is the replication network sufficiently separated from the main network so a disaster does not affect both?

- **Cost**: Running replication networks is costly. Is the cost worth it to the business?

For example, for database transactions, you need to consider how a copy of these transactions will be delivered to a disaster recovery site so that the disaster recovery location can take over in case it is needed. The options available to do that are transactional replication (replication at the database level, with the database server responsible for the replication), disk block replication (replicating altered blocks on disk to the disaster recovery site), which is the responsibility of the operating system, or copying backup files, which would be a process performed by a script or a human. One would not use all of these three options at the same time. The best option is the one that matches your business needs to have services restored in case of a disaster, also considering how many transactions could potentially be lost and what the business is willing to spend on disaster recovery.

It is precisely this interplay between costs, processes, and technical options that makes disaster recovery such a wide topic.

Virtualization

Virtualization uses a software application (often called a *hypervisor*) to run multiple *virtual* servers, each with its own operating system and settings on a single hardware device. The technology is mostly used for server farms. The hypervisor is also called the *host* and the virtual operating system the *guest*.

Virtualization allows significant technical simplifications in the design of disaster recovery sites. Instead of failing over to entire servers with all their specific idiosyncrasies and particulars (registry settings, for example), you can focus on failing over to the virtualized infrastructure at the host layer—that is, failover to the virtualization hosts, the virtualization software, and the storage as needed. The virtual machines—or the *guests*—can then be started in the new location. Depending on the vendor, virtualization software may have quite a few capabilities and options to support service failover scenarios, although these often come at the cost of additional licensing. In case of a disaster, the virtual guests (which are the servers that deliver the business functionality) can be started or automatically fail over in the replicated data center.

Geographical Considerations

When maintaining multiple data centers as recovery sites, the **geographical distance** between data centers is an important consideration. In most cases, organizations should ensure that there is at least a 20 km distance between data centers so that disasters such as power outages, flooding, and earthquakes do not affect more than one data center. For earthquakes in particular, it is also a good idea to ensure that both data centers are not located on the same fault line. The exact parameters of this decision will depend on business needs, local geography, weather patterns, whether staff can be made available in the remote location, and available infrastructure, such as network connectivity and available resilience between sites.

Using the Cloud as a Recovery Site

In cloud-based infrastructure, such as **platform as a service (PaaS)** or IaaS, the desired resilience is usually decided at design time and implemented through one of the resilience mechanisms offered by the cloud provider. In general, for **infrastructure as a service (IaaS)** or PaaS, cloud providers will provide highly available, regionally stable infrastructure in which most of the specifics of failover are abstracted away from the designer, thus greatly simplifying the design and operation of disaster recovery.

As a general rule, the more available, robust, and resilient options carry a higher price tag, and as part of the disaster recovery planning, the business needs to decide how much to spend and where.

Multiple Processing Sites

One approach to resiliency and recovery is to process the data on multiple sites, for instance, one processing site in each geographic region. In **multiple-processing-site** scenarios, multiple sites are simultaneously responsible for delivering the necessary business services to the entire organization, and one site can take over for another site if a site suffers a failure. Multiple processing sites work best where a site serves relatively static content, such as news websites, or where a site reconciles transactions between multiple databases, such as two different sites serving different geographical regions. Multiple processing usually requires a private network between the processing sites to ensure that content and data can be synchronized between the two locations and to ensure that the two locations offer the same services in the same way.

From a security perspective, it is also necessary to assess the risk of these private connections and ensure that they offer acceptable levels of security. Processes that rely on transactions (such as online shopping or banking) will need fast and secure replication of transactions between the sites in order to facilitate multiple processing sites.

In addition to disaster recovery, a key component of availability is also the resilience of a system against the failure of its components. This brings you to the second half of this chapter, which will discuss system resilience.

Comparing Security and Resilience in Systems

System resilience is the ability of a system to withstand and survive adversary events such as natural disasters and cyberattacks. Key components of system resilience include **high availability** (**HA**), **quality of service** (**QoS**), and **fault tolerance**.

A resilient design involves different trade-offs between technical possibilities and costs than those necessary for security design. In security design, you consider a malicious adversary that is out to destroy, damage, or compromise the system. In designing for resilience, you have to include seemingly random events, such as earthquakes, storms, flooding, and outages caused by other parties.

Further, redundancy is a key element of designing for resilience, whereas in secure design, redundancy usually adds little if anything at all.

An example of all this is the concept of **redundant array of inexpensive Disks** (**RAID**), which is a system resilience concept related data disks on servers. In a single disk, such as in a laptop, phone or desktop, generally, the data is lost if the disk fails. In server operating systems, disks are busy and the chance of disk failure is increased. A **RAID** array is a method of protecting the data on the array by arranging the disks to work together in order to achieve better resilience. The key RAID levels are as follows:

- **RAID 0** is an array without redundancy or resilience. A RAID 0 array may consist of multiple disks and the total capacity of the array is the sum of the capacities of all the disks. Because RAID 0 still distributes the data across all the disks, the entire array will fail if a single disk fails.

- **RAID 1** is an array where data is always mirrored between two disks. That means if one disk fails, there is a second disk containing a copy of the data. The capacity of a **RAID 1** array is half the sum of the capacities of the individual disks.

- **RAID 5** is a more complicated system that needs a minimum of three disks and relies on maintaining a parity block in addition to the data. When a single disk in a RAID 5 array fails, the remaining data can be rebuilt from the parity data.

Understanding RAID in detail is less important when large parts of your infrastructure may be in the cloud, but it is still an important concept to understand.

There is a significant overlap between the properties of a resilient system and the properties of a secure system. A system that is not resilient and at risk of failure is also a threat to data integrity and perhaps data confidentiality, thus compromising all properties of a secure system (confidentiality, integrity, and availability).

Finally, failures of security and resilience can both lead to incidents and crises, which will need a response from a dedicated team. In the case of security incidents, a small and nimble team is required that can acquire additional resources (e.g., for deep-dive malware research or threat intelligence) as needed. On the other hand, in the case of an availability or reliability crisis, larger teams with more diverse points of view, such as network performance or storage capacity and capability, are required. To determine the source of a system outage during intermittent failures or non-responsive services, regardless of whether they are caused by an attacker or not, there is a lot of ground to be covered to quickly eliminate possible causes.

Security is often involved in resilience discussions and design because availability is one of the things security teams are responsible for, but be aware that adding fault tolerance and redundancy to systems also opens new and different avenues of attack. A redundant data center has a copy of your production data and must therefore also be secured using the same technology and processes that are in use on the main processing site.

Finally, you also need to consider the resilience of the security components of a system, such as authentication and authorization systems, firewalls, and VPNs, as well as monitoring and alerting infrastructure. The need for operational security must dictate whether these systems need to be engineered to similar standards of resilience as the systems they are monitoring or securing.

As a general rule, where systems are *in-line*, that is, on the critical network path of the process to be secured (as is the case with firewalls straddling the main network connections), they must be built to the same resilience standards as required by the solution. For *out-of-band* systems, that is, systems that get a copy of the relevant data, such as log collectors, detection systems, and security incident and event monitoring systems, the needs of the security team, alongside legal obligations, may determine the level of resilience to which these systems are engineered.

The remainder of this chapter will focus on some of the processes required to ensure that organizations have credible resilience and disaster recovery capabilities that do not fail when needed.

Implementing Disaster Recovery Processes

Apart from how disaster recovery is designed, it is also important to consider how it is operated during times of crisis and to plan accordingly. An organization needs to consider how best to respond in terms of judging the seriousness of the event and the best practices in handling it. You also need to know who the right people to respond are, how any issues will be communicated with stakeholders such as management and customers, and how to assess the extent of any damage. Finally, any plan will include details on the restoration process, and how to learn and implement learnings from any incident.

Again, it is quite hard to give specific recommendations on disaster recovery since the details of the process and plan to recover heavily depend on contextual factors of the organization. What constitutes a disaster for a facility that manufactures and stores dangerous chemicals is very different from what constitutes a disaster for a medical facility or a financial services firm. Plans will vary significantly across these organizations, and a large part of disaster recovery planning is ensuring that the plan meets the needs of the business and its key stakeholders. For some organizations, the latter may include the physical safety of the responders.

The following subsections will point out some common points that are found in all good plans.

Response

The first thing to consider when it comes to response is when the response process should start. What constitutes a disaster that will initiate failover procedures? While this may be clear in the case of a natural disaster or power outage, in other cases, the trigger point for declaring a disaster may be less clear. Is it an outage of 1 minute, 5 minutes, or 1 hour?

The next thing to consider is triaging the seriousness of the outage. Does it affect a single branch, a single person, a geographic region, or the entire organization?

It is important to check that the operative disaster recovery plans give a clear indication of which decision criteria to use to initiate the plan (declaring a disaster), as well as triaging the seriousness of the disaster.

All personnel involved in recovery should be aware of what the plan is and be trained in using its procedures.

Personnel

The second key decision point when developing a disaster recovery plan is the key personnel that should be involved in resolving a disaster and how and where they can be reached. Again, the operative disaster recovery plan should have an up-to-date list of people to involve and their contact details, as well as their *second-in-command* in case the primary person is unavailable. The business also needs to consider which personnel should have *after-hours* provisions in their job contracts and what these after-hours provisions are, to ensure that the key people needed in the case of a disaster are available when needed.

Communications

Communication with stakeholders in the business, such as customers, suppliers, staff, employees, and contractors, is very important during disasters. In general, organizations will suffer more during disasters if they do not have their communications under control and have processes to ensure that communications are both accurate and timely. Damage could be reputational, but also legal, for instance, where organizations have a duty to inform partners, customers, and regulators accurately and in a timely manner.

You need to consider a broad range of questions in order to determine a communication strategy as part of your disaster recovery plan. Among these questions are the following: How will the business communicate with its customers, staff, contractors, and providers during an outage? How public do they want an outage to be? But also, how will the internal team handling the crises communicate?

The plan for communications during incident response should include strategies for how the team will communicate if the internal systems in the business—such as email and chat—are considered unreliable. It is also important that the members of the crisis response team know how to communicate with each other in advance, and do not rely on contact information collected during the incident without independent verification. In small organizations, this can be as simple as making sure that each member of the response team has the other members in their phone contact list, for example.

For communications with the outside world, the communications department of an organization should be involved at this point in the plan, considering questions such as who can write and approve an interim holding statement? Who will talk to the press? Who will communicate with internal and external stakeholders such as customers and suppliers? Are special instructions to the call center required?

Assessment

The disaster recovery team needs to consider how they assess the disaster and the procedures to follow to verify what components of the infrastructure are functioning or not functioning. In case of a network outage, the assessment may be a simple ping command, but for a complex application, the assessment can get complex and involve functionality testing and the verification of content integrity, down to storage performance or errors on storage hardware controllers.

As outlined, an important part of the assessment is to determine the size of the disaster and what functionality in business applications is affected, in how many places (a branch, a region, or the entire organization), and for how long.

You can also refer to the assessment as a **triage**. The triage process originated in military hospitals during times of war and guided doctors and nurses to quickly assess which patients needed medical care first as well as the sort of medical care required. In the case of disaster recovery, triage often involves a process where you determine the size, location, and seriousness of the outage, as well as the team required to recover from the outage.

Service Restoration

In the **restoration** step, the team considers the procedures that will restore service to the business. This may involve the order in which data is restored, the order of which systems to bring online first, and any functionality testing required.

Rather than including guidelines for a full-service restoration in the disaster recovery plan, it is better to document service restoration procedures for individual components and refer to them during disaster recovery.

Training and Awareness

The disaster recovery plan needs to be communicated to the business and to the people involved in its execution, and these people need to be trained on how the plan is to be used when needed.

Users of critical business applications should also be made aware of any changes that may occur if the business changes over to the disaster recovery site. Such changes may involve changes in the locations of applications, changes in processes, and some data being lost during the disaster.

Lessons Learned

A blameless review after the completion of a disaster recovery process (or disaster recovery test) is an important step that allows the team to improve documentation and processes. For example, the review can include examining the steps in the plan that failed or information that was erroneous or incomplete. Based on the review results, improvements in the plan can be proposed.

The parameters of such a review are very similar to those followed in an incident response process. It is important that the review is blameless and treated as a *lessons learned* step and not an opportunity to assess and comment on the performance of individuals or teams. Trying to attribute blame not only demotivates individuals but can also discourage transparency in the process if individuals are reluctant to speak up for fear of repercussions. The focus needs to be on the plan and the system and how well it worked rather than on the strengths and failings of individuals.

It is also important that disaster recovery plans are regularly tested. The next section will outline some of the approaches that organizations may choose to test disaster recovery processes.

Testing Disaster Recovery Plans

In many poorly managed disasters, organizations often find that there was a disaster recovery plan, but that it was old, out of date, and incorrect. The purpose of testing disaster recovery plans is to ensure that plans reflect the current state of systems and processes, the business criticality of services, and accurate communication plans and call lists—in short, to ensure that the plan has the best chance of working when it is needed.

Disaster recovery plans can be tested at different levels, from a read-through, tabletop exercise to a full failover of an actual site.

Tabletop Exercise

A tabletop exercise is a scenario-based simulation of a business disaster aimed at testing the processes, plans, and team capability for handling an actual disaster. The scenario may start with a general prompt ("Your main business website is unavailable to customers") or develop several **injects**, that is, simulated events or simulated outcomes of further investigations ("You have looked at the web server. It is up and running correctly") to guide the team through various aspects of the plan.

In a tabletop exercise, the testing team examines an outage scenario and outlines what they would do in response to that scenario. They also refer back to the currently operative plan to make sure that it is complete and up to date and note any points for improvement of the plan.

An advantage of a tabletop exercise is that it can be done quickly and without any risk to the business. Usually, such an exercise takes place in a meeting room without making changes to production or disaster recovery environments. A disadvantage of a tabletop approach is that the team relies on written documentation about how failover is supposed to work, which may or may not be accurate in practice.

Walk-Through

In a **walk-through**, the testing team may also visit the data center and any other important sites in addition to following the plan and can verify at a physical level that the site is configured as described in the documentation. They can thus gain a certain level of assurance that the documented features of the site are described correctly and, if necessary, amend the plan by referring to photos of equipment and settings.

A walk-through may also involve checking that personnel lists are complete and up to date, and checking whether the plan recommends the right steps for the right scenarios.

Simulation of a Disaster

In a **simulation**, the team takes things a step further and simulates some services being offline, without affecting production.

The advantage of a simulation is that the team can run through the procedure in practice and verify that the recommended steps work. A good example of a simulation is a test restore of data from a backup and verifying that it can be restored to a working instance of the service. This way, teams can establish actual restore times (rather than estimates), which makes disaster recovery planning more reliable.

Parallel Testing

In a parallel scenario, the team brings up the disaster recovery site and verifies its operation without connecting it to the business operation network. A parallel recovery can give valuable insights into the intricacies of activating the recovery data center, the procedures to be followed, as well as the time the business can expect to elapse before the data center is fully operational.

Full Interruption

In a full interruption, the business brings up the recovery data center and fails over to it for a set period of time. This test helps the organization validate whether the disaster recovery site will function under the actual business load for a set period of time and whether it can be managed (and backed up in turn) to a satisfactory level. Further, at the end of a full interruption test, the business has to fail back to the main site, which will validate whether the documented failback and service restoration procedures work.

Summary

This chapter discussed the important aspects to consider when designing and implementing disaster recovery. Disaster recovery is usually run by the operations side of the business, not by security. However, availability is one of the components of the **Confidentiality, Integrity, and Availability (CIA)** triad, as well as the increasing prevalence of ransomware, leading to security being involved in the design and principles of disaster recovery planning in business.

You looked at the various approaches that organizations can take to implement disaster recovery, their advantages and disadvantages, and the operative principles that help an organization decide which approach is suitable for a given scenario and risk tolerance. You also saw how these approaches are carried out and documented, but it is also important that plans are tested to ensure that they will work as well as possible during an incident.

The topic of disaster recovery and resilience is large and depends critically on an interplay between processes and available technologies. There are also many technical options available to deliver resilience and disaster recovery capabilities. When studying for the exam, a good strategy is to consider which of these are currently in use in the place where you work, and how they work together to deliver an overall approach to resiliency and disaster recovery. A lot of this process depends on the context of the business that you are in, and what constitutes a *disaster* or *failure* in that context.

The next chapter will discuss business continuity, and some aspects of personnel and physical security that you need to be aware of.

Further Reading

It is possible to write an entire book on just the topic discussed in this chapter. For more details on disaster recovery and resilience, visit Google's site on site reliability engineering, `https://sre.google/`, where you can find very detailed information on how highly resilient services are engineered.

It is also worthwhile to consider what is currently being done in your place of work, for instance, what disaster recovery plans are available and how they work. Looking at this scenario from a perspective where you also know the context will help significantly with gaining a more in-depth understanding of these topics.

Exam Readiness Drill – Chapter Review Questions

Apart from a solid understanding of key concepts, being able to think quickly under time pressure is a skill that will help you ace your certification exam. That is why working on these skills early on in your learning journey is key.

Chapter review questions are designed to improve your test-taking skills progressively with each chapter you learn and review your understanding of key concepts in the chapter at the same time. You'll find these at the end of each chapter.

> **How to Access These Materials**
>
> To learn how to access these resources, head over to the chapter titled *Chapter 24, Accessing the Online Resources*.

To open the Chapter Review Questions for this chapter, perform the following steps:

1. Click the link – `https://packt.link/chapter18`.

    Alternatively, you can scan the following **QR code** (*Figure 18.1*):

Figure 18.1: QR code that opens Chapter Review Questions for logged-in users

2. Once you log in, you'll see a page similar to the one shown in *Figure 18.2*:

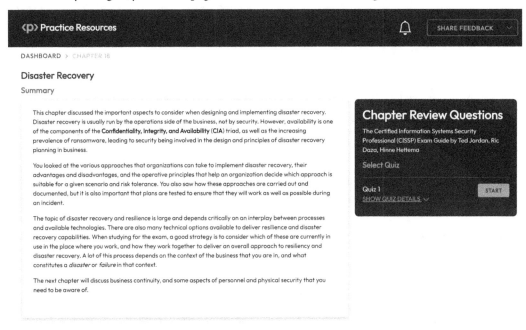

Figure 18.2: Chapter Review Questions for Chapter 18

3. Once ready, start the following practice drills, re-attempting the quiz multiple times.

Exam Readiness Drill

For the first three attempts, don't worry about the time limit.

ATTEMPT 1

The first time, aim for at least **40%**. Look at the answers you got wrong and read the relevant sections in the chapter again to fix your learning gaps.

ATTEMPT 2

The second time, aim for at least **60%**. Look at the answers you got wrong and read the relevant sections in the chapter again to fix any remaining learning gaps.

ATTEMPT 3

The third time, aim for at least **75%**. Once you score 75% or more, you start working on your timing.

> Tip
> You may take more than **three** attempts to reach 75%. That's okay. Just review the relevant sections in the chapter till you get there.

Working On Timing

Target: Your aim is to keep the score the same while trying to answer these questions as quickly as possible. Here's an example of how your next attempts should look like:

Attempt	Score	Time Taken
Attempt 5	77%	21 mins 30 seconds
Attempt 6	78%	18 mins 34 seconds
Attempt 7	76%	14 mins 44 seconds

Table 18.1: Sample timing practice drills on the online platform

> Note
> The time limits shown in the above table are just examples. Set your own time limits with each attempt based on the time limit of the quiz on the website.

With each new attempt, your score should stay above **75%** while your "time taken" to complete should "decrease". Repeat as many attempts as you want till you feel confident dealing with the time pressure.

Business Continuity, Personnel, and Physical Security

This chapter will discuss some auxiliary items that often come up for security teams: business continuity, personnel, and physical safety. They are "auxiliary" because they have little to do with information technology but affect the overall security posture and are often included in the activity scope of security teams. One characteristic that these items have in common is that there are hardly any generally accepted best practices that can be followed and implemented. Because these items deal with the specific physical and process aspects of the business, context matters a lot in all of them. However, security teams can focus on what questions need to be asked to resolve these items and what the answers subsequently mean for the business.

In terms of getting ready for the exam, this chapter will give you a general outline of the issues at play and the activities that security teams usually undertake when they get involved in them. But be aware that business continuity, physical security, and personnel security in most organizations also involve other teams, such as IT management, security guards, and HR. An important aspect of your study will be to ask questions of those departments in the place where you currently are to put them in context.

It is worthwhile considering how this situation came about. Because **availability** is part of the CIA triad, and because maintaining the baseline of this triad is traditionally the responsibility of security, many organizations also consider security a key player in business continuity and disaster recovery. However, this is only partially true: in modern IT infrastructure, ensuring business continuity and the capability to recover from a disaster fall well outside the scope of what most security teams are equipped to handle.

Hence, the first responsibility for teams when issues around business continuity come up is to ensure that the right teams are engaged in the discussion, including the software development, networking, and platforms teams, as well as the business itself. Without these teams taking an active part in the discussions and execution of business continuity and disaster recovery, security cannot be successful.

For many teams, the first thing they need to do when this issue comes up is to ensure that they are not made solely responsible for something they have limited control over. Business continuity involves almost all aspects of the IT organization and even involves many people outside of IT. Security is only one seat at a very large table. Having said that, security has an important role to play in both business continuity and disaster recovery, and in many organizations, security teams can get heavily involved in planning for disaster.

This chapter will briefly discuss the role of security in business continuity and disaster recovery, and we will discuss some aspects of business continuity that security teams should be particularly aware of.

By the end of this chapter, you will be able to answer questions on:

- Participating in planning and exercises
- Physical security
- Personnel safety

Planning and Exercises

Many organizations plan and exercise business continuity either by performing a tabletop exercise, which replays a scenario in a paper-based gamed scenario, often with accelerated time, or in live exercises, which fail over applications and data centers to another location during a time agreed on by the business.

The details of how a business fails over its infrastructure depend a lot on the context of the business applications. This context is made up of the business processes that the applications support, how these applications contribute to the bottom line, how people typically use them, and the continuity that the business requires to keep operating. It is therefore not generally possible to determine a set of best practices that will guarantee that every important aspect is considered. However, it is possible to consider common questions that are likely to come up, such as whether failover procedures are documented and tested, what the security posture is during and after failover, and how the services the application will depend on (such as authentication and file services) will fail over alongside it. Security teams should be asking these questions and evaluating the answers.

The role of an **exercise** (whether it is a **tabletop** or **real live-fire** exercise) is to bring to light the important aspects of the failover process and the simulation.

Security has a key role to play in these exercises since security is one of the key requirements of modern business, but also because availability is often seen as something that security teams are responsible for. Security teams need to specifically consider the following security-oriented items during planning:

- Whether the necessary security infrastructure will **fail over** securely along with the rest of the infrastructure

- Whether a defensible security posture is maintained during the failover
- Where all the secrets (passwords, tokens, API keys, and physical keys) necessary for the failover are available during and after the failover

The security architect has to consider what happens to the security infrastructure, such as network monitoring, firewalling, and log shipping during and after a failover. There are some security infrastructure engineering questions that we will discuss first, such as what parts of an infrastructure need failover and how security and key access are maintained during a failover.

Security Infrastructure Engineering and Business Continuity

It is a somewhat open question whether all security infrastructure must be engineered to fail over in the event of a failure. Security infrastructure needs to consider the **Recovery Time Objective (RTO)** and **Recovery Point Objective (RPO)**, like all other components of the IT infrastructure, and engineer high availability and failover procedures in line with what is required to secure the business.

As discussed in the previous chapter, RTO is the time taken to restore the functionality of a failed component. Note that the discussion on restore times must be had with the business and is not something that IT (or security) can just decide on its own. The amount of downtime on a component that is acceptable (or survivable) will depend on the business processes that depend on that component, and how long a business can operate (or survive) if these processes are not available.

RTO and RPO are also in play with security infrastructure. If we lose, say, 3 hours of network monitoring data, then there is no forensic investigation capability for that 3-hour time period. Is that acceptable? The answer in most of these cases is "it depends."

Another item to consider is the configuration of the security infrastructure. For **security infrastructure** that is in line—infrastructure that other business functionality directly depends on, such as firewalls— the RTOs and RPOs are determined by the business. For out-of-band security infrastructure, such as out-of-band authentication traffic collection, it is the security team itself that sets the RTOs and RPOs.

The two key questions are "How quickly must we have this system back?" and "How much data are we prepared to lose?".

Maintain Security Posture During Failover

It is also crucial for an organization to maintain its security posture while the infrastructure fails over to a secondary location. In practice, this implies that security teams need to review each stage of the failover process and determine whether an acceptable security posture is maintained.

In general, systems have dependencies. As an example, a business application will depend on other infrastructure, such as databases, authentication, container infrastructure, or file services. The security posture of these components determines the security posture of the overall system. During and after failover, what is the security posture of the system? Is there a risk, for instance, that systems are vulnerable to attack when they start up or shut down? What is the order in which things need to be brought online?

It is also necessary to investigate all the components that a system needs to start and run, and how they are made available in a disaster recovery scenario.

Key Management During Failover

Another important aspect to consider is the **storage** of all secrets, such as passwords, keys, and tokens, that are needed in the case of a failover. Specifically, can the relevant team access these keys when they are needed? Are the systems failing over in the right order?

It is important to consider these questions carefully in sequence; otherwise, it is possible to end up in a loop of dependencies in systems. For example, if the authentication and multi-factor systems depend on each other, and an administrator needs to log on with a second factor to enable the multi-factor system, you may effectively be locked out of your own system during recovery.

Some organizations create a **capsule** in which they store a set of *break glass keys* that will allow access into the infrastructure should all else fail. If that is the chosen solution, teams must pay careful attention to where this capsule is stored, who has access to it, and under what circumstances.

Implement and Manage Physical Security

Physical security is something that sometimes (but not often) falls under the scope of the security team. In practice, most organizations have a separate physical security department that is responsible for the issuing and maintenance of access cards, security cameras, guards, and anything else the organization needs.

Sometimes, security teams get involved in the technology of access cards or cameras and which ones the business should buy or are responsible for the security of the physical security system, which we will briefly review next.

In the next subsections, we will discuss how and where IT security may get involved in aspects of physical security, focusing especially on access card technology and IT security reviews of physical security systems.

Perimeter Security Controls

In this section, we will mainly discuss **access cards** and **access systems** and leave out a discussion of more significant physical security concerns that may exist, such as mobile phone coverage on a site, **Wi-Fi signal** leakage and **war-driving**, whether people are allowed to carry electronic devices while on site, and how to manage and regulate their use. While the latter considerations are important for high-security scenarios, they are not the most common ones to run into when securing an organization.

Access Cards

Access cards are physical cards that employees and visitors use to enter and exit the premises. They are usually purchased and installed by the people responsible for the physical security of the premises, which, in many organizations, is not the IT department or IT security team. Although modern access card and camera systems are critically dependent on computers, physical security teams often have a limited understanding of the problems that come up in cybersecurity, and in this area, security teams may be of value for reviewing and approving which systems to purchase.

Access cards come in a number of varieties with different technologies. Most of the time, security teams have little control over which technology is being used because, for instance, access card technology is already implemented in the building and is part of the lease of the premises. Larger organizations, such as universities, may want to mandate access card technologies because student ID cards are also access cards for the facilities.

Once these systems are in place, they are very hard to change. Changing all the access cards in a building often involves changing all the readers as well, and this can be a major undertaking for even a small or medium-sized building.

The main access card technologies to be aware of, along with their current security status, are as follows:

- **Prox cards** are easily copied, and card readers/copiers are readily available. A copied prox card will get access to a building, just like the original.

- **MIFARE** cards are capable of acting in a number of scenarios beyond building access (e.g., payments, transport ticketing, and parking) or as identification in scenarios such as access to meals in student halls. The MIFARE standard comes in a number of varieties with varying security properties. It is useful to read the overview on the standards site: `https://packt.link/WRcjo`

It is not easy for security teams to get involved in the process of selecting and security testing cards since, traditionally, these have been the purview of physical security, and any conversations need to be had carefully and with some sensitivity. The best approach is to have a conversation with the department responsible for physical security and see how and where IT security may be able to help with technology such as access cards, cameras, and monitoring.

Access Systems

Access systems refer to the IT components that make up the system that enables access cards. Access systems usually involve some dedicated hardware, such as card readers or (in the case of industrial control) a combination of card readers, programmable logic controllers, and a set of servers in the data center or the cloud maintaining the card system (for instance, keeping a record of which cards are authorized).

The IT components of physical security systems certainly fall under the scope of action of the **chief information security officer (CISO)** or the security team. This means that these systems need normal security management at a minimum, or, considering their importance to the business, even stricter security management.

This relationship between physical and IT security is one that security teams will need to build but build carefully. The teams responsible for physical security often work under different legal obligations and maintain close relationships with private security companies, the police force, and, for some types of sites, specific departments that govern the business conducted on that site (e.g., in ports or transportation sites, expect relationships with customs and the coast guard). Depending on your jurisdiction, there may be a lot of legislation here that governs how physical access systems are run and secured.

Security teams need to ensure that the IT components of physical security systems are managed following best practices and in accordance with the law. On many occasions, these systems collect information, which also falls under privacy regulations, and for this reason, compromises of physical security systems can be particularly devastating to organizations. From that perspective, security teams need to carefully design a management and auditing approach for these systems.

Security Auditing of Physical Access Systems

Similar to the checks on digital infrastructure, physical systems should also be checked in a regular and systematic way. Security auditing of these systems is a critical element of the overall security posture but carries significant complications. As we have already discussed, such systems can be subject to multiple legal and standards obligations (which not all need to say the same thing). Questions to ask prior to performing an audit of security infrastructure are the following:

- What legislation does your site fall under that may impact its physical security requirements?
- What technologies are used to secure access?
- Are these technologies managed in-house, by a security provider, or outsourced to a third party?
- What visibility do we have of how third parties manage the physical security posture?

It is hard to come up with a set of generic best practice answers in this scenario since there are significant variations between organizations, their sites, and their legal and standards obligations. Sites that carry out sensitive government operations, either directly or as a third party, might have minimum physical security expectations mandated. The same goes for sites that carry out biological research.

Electrical substations or transportation sites may have specific health and safety requirements related to extreme hazards, as well as legal security requirements due to being essential infrastructure.

As a result, the first step a security team may need to undertake is to get some of its members certified to access these sites (maybe while being accompanied) so that the team is capable of doing forensic data collection in the case of a security incident or carry out security reviews of installed infrastructure.

Best practice in these cases usually consists of opening a channel of communication to the physical security department, exploring what problems need to be addressed for the IT systems managing physical access, and then working out a plan of action.

Internal Security Controls

There are a number of process-based security controls, such as separation of duties and financial transactions, that are vital in keeping an organization secure. These are practices that sometimes fall under the acronym of **OpSec**, (**operational security**), and are designed to give an organization the assurance that security teams will follow a responsible and robust process when carrying out their duties.

Separation of Duties and Responsibilities

A good safeguard in process security is to ensure that the role of carrying out a transaction and the role of approving a transaction are separate. That means that one person initiates a transaction, and someone else approves it. This ensures that there are checks and balances in how funds get paid, certificates are issued, and processes operate.

Sometimes, organizations object to a separation of responsibilities because it can slow down the speed of a transaction, especially in an emergency. Another side of that coin is that sometimes this is exactly the purpose—to break the chain of urgency that attackers often skillfully create to get us to do things we later regret. A good example of such urgency that later proves to be misplaced is business email compromise and separation of duties in payment processes, as discussed in the next section.

Business Email Compromise and Processes for Payments

Business email compromise (BEC) is a process in which attackers forge emails pretending to be someone high up in the company and try to convince staff from finance departments that a bill needs to be paid in a non-standard way. BEC is still one of the most profitable cyber scams, leading to significant loss of funds in many organizations. Cybercriminals who specialize in BEC will do their homework by checking the social media pages of their targets and the person they are pretending to be.

BEC is technically simple because it involves crafting and sending an email. However, it is still one of the attacks through which organizations regularly lose considerable sums of money. The attack can take many forms:

- An attacker pretends to be the CEO or another C-level executive who is currently overseas and needs an urgent payment made in some non-standard way, such as gift cards or wireless transfers

- An attacker pretends to be a staff member who needs a change of bank account for their salary payment

- An attacker pretends to be a staff member who is overseas and in urgent need of funds to get out of a sticky situation

The dead giveaway with most BECs is that the *from* address of the email doesn't match the *reply to* address. The *from* address is the person the attacker aims to impersonate and the *reply to* address is the address of the perpetrator. Organizations could filter out such emails with the right tooling. However, such filtering may have unintended side effects, such as the blocking of legitimate emails. For this reason, many organizations do not explicitly implement filtering except reputation checks on the IPs and addresses of the sending server, where poor reputation addresses or servers (that is, addresses or servers sending lots of malicious emails) get blocked.

Since there are very few markers of a BEC email, the best protection that many organizations have is policies, staff training, and strict processes for payments and salaries that are adhered to at all times. It may be worthwhile noting that **large language models** such as ChatGPT are increasingly used to craft more convincing emails, which are harder to spot and harder to train people for.

To close this chapter, we will discuss some aspects of personnel safety and security concerns that security teams sometimes are involved in.

Personnel Safety and Security Concerns

When it comes to ensuring comprehensive security, organizations must also consider the personal safety of their employees, the security of these employees' data, and how these two interact. Personal and data risks may occur when employees travel or are in roles that involve them posting on social media. This section will discuss some of these aspects, focusing on some common scenarios with travel, security training and awareness, and social media.

Travel

Modern business **travel** mostly involves taking a laptop and other mobile devices to locations where they are hard to update, monitor, and secure. Crossing borders in some cases may involve handing these devices over to border guards or law enforcement for inspection and perhaps imaging of the data. Similarly, some locations may involve the traveler having to hand over their social media accounts to law enforcement.

Hence, it makes sense to think in advance about these scenarios and what they may mean for the business. It is sometimes a good idea to travel with **one-time-only (burner) devices** such as phones and laptops, which are set up and configured to use during travel but will not return to normal business use after travel, so anything implanted on them by a hostile foreign entity is unlikely to get much information.

For example, the following considerations may come into play:

- Should employees have burner phones and laptops for travel to some locations? These burner devices contain no business or personal data, and only the minimum to be functional, and will therefore yield little information to hostile foreign environments.

- If employees need to render the details of their personal social media accounts to enable business travel, there are human resource and privacy questions to consider. There are no good or best practice answers here, and organizations and people involved are likely to handle this in their own way.

- What is the approach an organization will choose for handing over laptops and other devices to law enforcement or customs in foreign locations?

Many organizations also engage with outside organizations that provide safety intelligence on foreign locations and provide location-specific tips to promote both physical security and cybersecurity.

Security Training and Awareness

Many organizations conduct regular security **training sessions**, which aim to educate the user base about security risks coming from spam, malicious web pages, and misinformation, and also inform them of their obligations with respect to privacy and financial transactions.

In the same vein, many organizations now run **friendly phishing** campaigns to train their employees to recognize malicious emails. All of these programs form part of an emerging practice of **security culture**.

Social Media Policy

Many attackers use social media to discover important facts about their intended victims, such as whether they are attending a conference, traveling, changing jobs, or looking somewhat disgruntled. Disgruntled employees can become a security risk; attackers can engage with them and maybe enroll their assistance in attacks on the organization.

Social media information can be weaponized, for instance, in a BEC scam, as we discussed earlier in this chapter.

It therefore makes sense to have policies that focus on social media behavior and what is deemed acceptable. Again, it is hard to specify best practice here in a generic sense, since context matters. In general, a good baseline policy to have is to demand that employees do not divulge confidential business information on social media pages, and in some particularly sensitive cases (such as in the case of takeovers), even avoid talking about travel.

Emergency Management

Most organizations have procedures in place that determine what to do if an employee gets in trouble or an emergency occurs during business travel. These procedures are commonly managed by the human resources or corporate travel department.

For security teams, it pays to find out what these procedures are, what they say about cybersecurity, and propose changes to these procedures if they are not comprehensive or do not meet the needs of the business. Usually, in most large organizations, these procedures are supplied by a third-party vendor that specializes in business travel insurance and they may not be open to amendment by cybersecurity professionals.

IT security teams should know, however, what these rules are, and supply additions where they think necessary.

Duress

An extreme situation exists with **duress**. When an employee is kidnapped or held for ransom, organizations need to work with law enforcement to ensure that they follow the proper procedure and do not jeopardize the safe return of their employees. This rarely involves the involvement of IT security, although in cases where employees are kidnapped or held to ransom with their devices, IT will usually be involved to determine what information may be at risk, or what the next steps are that need to be taken on the IT systems.

Summary

This chapter addressed some areas of physical and safety that affect the operation of cybersecurity teams such as business continuity, personnel safety, and the resilience of security monitoring systems, as well as the security aspects of systems used in protecting physical security.

Most of these topics involve developing and maintaining contact with various areas of the business outside of IT: finance, human resources, corporate travel, and physical security.

The topics of business continuity, physical safety, and travel often involve a very specific context, such as what the line of business of the organization is, what its legal obligations are, whether it owns or leases buildings, or how much data it is prepared to lose. For this reason, it is hard to come up with best-practice solutions to these items. Instead, the common recommendation is that teams develop and maintain relationships with the other departments in the business that are responsible for them and see what the common problems are, and how IT security may be able to assist.

Security leadership has to make a determined and genuine effort to foster and develop these relations and maintain fruitful conversations between the security team and the rest of the business. In the next chapter, you will read about the importance of security in the software development lifecycle.

Further Reading

ID Watchdog, *Cybersecurity for Employees while Travelling*, `https://packt.link/bNs3A`

Exam Readiness Drill – Chapter Review Questions

Apart from a solid understanding of key concepts, being able to think quickly under time pressure is a skill that will help you ace your certification exam. That is why working on these skills early on in your learning journey is key.

Chapter review questions are designed to improve your test-taking skills progressively with each chapter you learn and review your understanding of key concepts in the chapter at the same time. You'll find these at the end of each chapter.

> **How to Access These Materials**
>
> To learn how to access these resources, head over to the chapter titled *Chapter 24, Accessing the Online Resources*.

To open the Chapter Review Questions for this chapter, perform the following steps:

1. Click the link – `https://packt.link/chapter19`.

    Alternatively, you can scan the following **QR code** (*Figure 19.1*):

Figure 19.1: QR code that opens Chapter Review Questions for logged-in users

2. Once you log in, you'll see a page similar to the one shown in *Figure 19.2*:

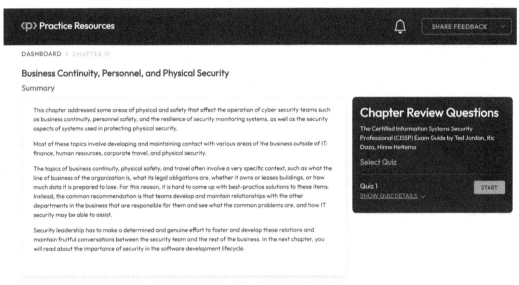

Figure 19.2: Chapter Review Questions for Chapter 19

3. Once ready, start the following practice drills, re-attempting the quiz multiple times.

Exam Readiness Drill

For the first three attempts, don't worry about the time limit.

ATTEMPT 1

The first time, aim for at least **40%**. Look at the answers you got wrong and read the relevant sections in the chapter again to fix your learning gaps.

ATTEMPT 2

The second time, aim for at least **60%**. Look at the answers you got wrong and read the relevant sections in the chapter again to fix any remaining learning gaps.

ATTEMPT 3

The third time, aim for at least **75%**. Once you score 75% or more, you start working on your timing.

> Tip
>
> You may take more than **three** attempts to reach 75%. That's okay. Just review the relevant sections in the chapter till you get there.

Working On Timing

Target: Your aim is to keep the score the same while trying to answer these questions as quickly as possible. Here's an example of how your next attempts should look like:

Attempt	Score	Time Taken
Attempt 5	77%	21 mins 30 seconds
Attempt 6	78%	18 mins 34 seconds
Attempt 7	76%	14 mins 44 seconds

Table 19.1: Sample timing practice drills on the online platform

> Note
>
> The time limits shown in the above table are just examples. Set your own time limits with each attempt based on the time limit of the quiz on the website.

With each new attempt, your score should stay above **75%** while your "time taken" to complete should "decrease". Repeat as many attempts as you want till you feel confident dealing with the time pressure.

20
Software Development Life Cycle Security

Applications should be designed and planned to be functional and secure right from the beginning. In this chapter, you will learn about software development methodologies used to ensure that applications meet minimum levels of security. Also, you will learn how to determine whether suppliers are meeting minimum levels of application security in their development processes.

Because development projects and the popularity of applications are growing rapidly, ensuring that they remain relevant is crucial. This is achieved through regular updates and patching.

This chapter will discuss how cross-functional or integrated product teams and change management keep applications functional and secure for the life of the product within their user communities.

This chapter will cover the following topics:

- Software development methodologies
- Maturity models
- Operations and maintenance
- Change management
- Integrated product team

Software Development Methodologies

To create a **secure software development project**, you must have a secure development process in place. This process includes proper planning and collection of software requirements, architecture and design, coding, testing, and release and maintenance. There are several software development frameworks in use today, including Waterfall, Agile, DevOps, and DevSecOps. These frameworks guide the process of developing secure software applications and provide a structured approach to releasing software projects.

The Agile model is an incremental development methodology that focuses on customer collaboration and feedback. Agile was developed to address the dissatisfaction that customers had when the final software project did not meet their expectations; it does so by involving them more in the process. For example, consider a project where a client requires an app to print two dots. Without involvement, the developer might complete the project and make the dots print horizontally. However, the customer expected the dots to be vertical. Had the client been involved in each stage of development, this issue would have been caught early enough in development and resolved.

The Waterfall model is a sequential approach to software development, where the project moves through several phases of gathering requirements, design, implementation, testing, and maintenance. Each phase must be completed before the next. This can lead to dissatisfied customers since they are not included in each step of the process.

DevOps is a set of practices that combines software **development (Dev)** and IT **operations (Ops)** to reduce the time taken in the **software development life cycle (SDLC)** and provide continuous delivery of high-quality software. By breaking down silos between development and operations teams, DevOps automates workflows and reduces bottlenecks to allow for faster, more efficient, and more reliable software releases. DevSecOps extends the DevOps approach by integrating security practices throughout the SDLC.

The following subsections cover the software development models that the CISSP exam focuses on, starting with the Waterfall model.

Waterfall

The Waterfall model follows a sequential approach to software development. It is characterized by a step-by-step approach, with each phase needing to be completed before the next phase can begin. It is very difficult to return to a completed phase, as shown in *Figure 20.1*.

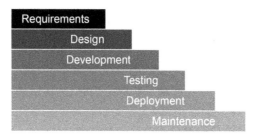

Figure 20.1: The Waterfall model

The Waterfall model was designed to improve the predictability of the pricing of software development projects. End users prefer knowing how much a project is going to cost so they can set a budget for it.

The Waterfall model is divided into six phases:

1. **Requirements**: In this phase, the project team gathers requirements from the customer, which include the customer's goals as well as constraints for the project.

2. **Design**: In this phase, the project team designs the overall architecture and the component interfaces of the system. This includes identifying system components, interactions, and the system's overall structure.

3. **Development or coding**: In the development/coding phase, the project team writes the code for the project.

4. **Testing**: In this phase, the project team tests the system to ensure that it meets the requirements detailed in the first phase. Testing includes unit testing, integration testing, and regression testing.

5. **Deployment**: Next, the team deploys the system to production. This includes installing, configuring, and training users on the system.

6. **Maintenance**: This phase deals with additional requirements. As clients use the application, they may discover they need more features added or problems that need to be resolved with a patch.

The Waterfall model is straightforward and well-suited for well-defined projects, but it can be inflexible and resistant to changes in requirements or technology as it makes going back to earlier phases of the project difficult. This means it is not well-suited for projects with changing requirements, such as if an app is undergoing customer testing and feedback is being used to make tweaks to the project. There is a testing phase in Waterfall, but this is to ensure that it works according to the plans, rather than testing it for customer satisfaction.

The Agile model mitigates some of the disadvantages of the Waterfall model and is discussed in this next section.

Agile

The Agile model is a software development methodology that emphasizes iterative development, frequent deliveries, and continuous improvement. It is based on the idea that software development is a complex process, so it is better to break the project down into smaller, manageable pieces, as shown in *Figure 20.2*.

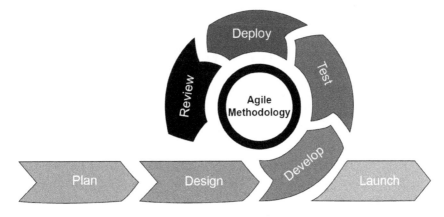

Figure 20.2: The Agile software development model

Agile has several advantages over Waterfall. First, the Agile model allows flexibility. As the project progresses, the Agile team can adjust the plan to accommodate changes in requirements or unforeseen circumstances. Second, the Agile model produces working software faster than Waterfall. This is because the Agile team is constantly delivering working software increments to the customer by developing the project in small, manageable cycles called **sprints**. Each sprint typically lasts a few weeks and culminates in a potentially shippable product increment. This approach allows for continuous feedback and iterative improvements throughout the development process. Finally, the Agile model is more likely to produce high-quality software because the software is constantly being tested and improved while it's being developed.

The Agile Manifesto emphasizes individuals and interactions over processes and tools, customer collaboration, and adapting to change. Agile methodologies such as Scrum and Kanban use short development cycles (called sprints) to prioritize and deliver small, functional components of software. This enables teams to rapidly respond to changing requirements and continuously improve the product.

> **Note**
>
> The original Agile Manifesto can be found at the Manifesto for Agile Software Development `https://packt.link/nYUG0`

Agile's disadvantages include that it can be more difficult to manage because the Agile team is constantly changing and adapting the plan. This leads to more difficulty managing the project timeline versus that of the Waterfall model.

- Here are some key features of the Agile model:
- **Iterative development**: Agile breaks down large projects into smaller, more manageable chunks, called sprints. Each sprint is typically 2–4 weeks long, and at the end of each sprint, the team delivers a working product increment. This allows the team to get feedback from stakeholders early and often, and to make changes to the product as needed.

- **Incremental development**: This means that the team delivers working software increments throughout the development process, rather than waiting until the end of the project to deliver a final product. This allows the end user to provide feedback early and often so that the team can make changes as needed.

- **Self-organizing teams**: Agile teams are self-organizing teams. This means that each team is responsible for setting its own goals, processes, and work. This approach gives a team a high degree of autonomy, which leads to increased productivity.

- **Collaboration**: Agile encourages collaboration. The team works closely with stakeholders to gather requirements, and team members work closely with each other to develop the product. This approach helps to ensure the product meets the needs of the customer, and that it is delivered on time and on budget.

- **Continuous improvement**: Agile enables continuous improvement. The team learns from its experiences and makes changes as needed. There are many different Agile frameworks available, each with its own strengths and weaknesses. These frameworks include **Scrum**, **Kanban**, and **Extreme Programming (XP)**.

The Agile model has become the preferred methodology for many software development projects. Agile is a good choice for projects that are subject to change or that require rapid development.

Though neither the Agile nor Waterfall methods are inherently insecure frameworks, neither are developed with security at the forefront. However, in the next section, you will see how DevSecOps integrates security into the development process.

DevOps and DevSecOps

While the Agile methodology is intended to create better products through constant testing and frequent code releases, bottlenecks can occur under traditional role separation, where development teams write code and operations teams deploy it. This separation can lead to delays and inefficiencies, as the hand-off between teams often slows down the overall process. The DevOps methodology seeks to address that by combining development and operations (hence DevOps) into one framework.

Using close collaboration and automation, DevOps aims to streamline the entire SDLC, from planning and coding to testing, deployment, and operations, with the result being **continuous integration and continuous delivery (CI/CD)**. Because a single person has complete control over processes, this violates the **separation of duties (SoD)** principle. Mitigate SoD violations with RBAC, code review, and change management.

The following are the core principles of DevOps:

- **Collaboration**: Developers, operators, customers, and other stakeholders collaborate throughout the entire SDLC

- **CI/CD**: Frequent integration of code changes into a central repository and automated deployment enable faster and more reliable updates

- **Automation**: By automating repetitive tasks, such as building, testing, and deployment, DevOps teams can reduce errors, save time, and improve efficiency

- **Infrastructure as code (IaC)**: Treating computer and network hardware as code through virtualization and containers allows for the automation and provisioning of these infrastructures

- **Monitoring and feedback**: Continuous monitoring of applications provides valuable feedback on performance and quality, allowing teams to identify issues and track metrics against baselines

- **Agile practices**: DevOps often aligns with Agile methodologies, embracing iterative and incremental development, frequent releases, and feedback loops

DevOps tools include version control systems (e.g., Git), continuous integration and deployment tools (e.g., Jenkins and Travis CI), configuration management tools (e.g., Ansible, Chef, and Puppet), containerization and orchestration platforms (e.g., Docker and Kubernetes), and monitoring and logging tools (e.g., Prometheus and the ELK stack).

Having combined development and operations, the next step is to introduce security into the process. **Development, security, and operations (DevSecOps)**, also known as shift-left security, is a security practice that integrates security into the entire SDLC. This means that security is not an afterthought but rather a core consideration from the very beginning of the development process.

Integrating security into the process means that security expands beyond centralized teams to be a concern of all product delivery functions. To deploy DevSecOps using a CI/CD pipeline, developers can integrate security scanning into the pipeline. This helps to prevent security vulnerabilities from making it into production. Methods include everything from reviewing source code in static analysis to testing code while it functions (dynamic analysis). Tools such as SonarQube (used for static application testing) or Burp Suite (for dynamic application testing) can be used to identify vulnerabilities in code before it is deployed. Common tools and techniques are discussed in the next chapter.

In addition to using security scanning tools, DevSecOps can also mean a cultural shift where software is designed more securely, and there is a great emphasis on education and awareness of security best practices across functions.

Maturity Models

Software development methodologies are crucial because software development is a complex process with competing requirements such as efficiency, security, and market fit. If the SDLC emphasizes security excessively without considering the end user, the end product might be highly secure but suffer from a slow and difficult user experience. Conversely, a process that balances user experience and security without considering shipping time might create an excellent product that takes so long to reach the market that the company loses its competitive edge.

An effective SDLC helps an organization balance all necessary factors appropriately, but achieving this balance is not as simple as applying a model from day one. Teams must continuously tweak and improve their processes based on the specific requirements of the business, function, or project. In the early stages of software development, a new team might experience chaos as they discover unexpected requirements and dynamics. They may need to react and make significant process changes to address unforeseen issues. Over time, as the team progresses, their processes will become more stable, and the number of unexpected issues will decrease. Eventually, the team will develop robust working practices with efficient, predictable outcomes.

This progression illustrates the process of a team maturing, which can be effectively measured using maturity models. Maturity models provide a framework to assess and guide the development of processes, helping teams to identify areas for improvement and track their progress toward achieving efficient and reliable practices.

There are two popular models that focus on software development security and software security overall: the **Capability Maturity Model Integration (CMMI)**, which has replaced the earlier **Capability Maturity Model (CMM)**, and the **Software Assurance Maturity Model (SAMM)**. CMMI is a more general framework that can be used to improve other aspects of software development, while SAMM is a more specific framework that focuses on software security.

Let's start by discussing CMMI.

CMMI

CMMI is a process improvement approach that helps teams increase their ability to deliver secure applications. This is done by using the CMMI framework to evaluate processes and determine the current maturity level of the team. In doing this, organizations can identify areas for improvement and the next steps for development. This structured assessment enables teams to systematically advance their capabilities, leading to more effective and secure software delivery.

CMMI defines five levels of process maturity:

1. **Initial**: Organizations at this level have no defined processes.
2. **Repeatable**: Here, organizations have defined processes and can repeat them consistently.
3. **Defined**: Organizations have defined processes and can measure their performance.
4. **Managed**: These organizations have defined processes and can control their performance.
5. **Optimized**: Optimized organizations at this level have defined processes and are continuously improving their performance.

CMMI is not a prescriptive model that tells organizations what processes to use but, rather, provides a framework for organizations to define their own processes that improve their security over time. However, CMMI does offer roadmaps and best practices in areas such as project development, and it is used by software developers to develop higher-quality and more secure applications.

For example, imagine a team developing a new suite of virtual audio effects. The initial idea, maybe from a late-night brainstorm from a group of friends who had studied audio technology at university, might have shown some real promise when sandboxing. Much of the development process could have been done late at night after day jobs had ended, and thus the emphasis was on creativity and doing things cheaply with no formal team, record keeping, and documentation. This would be the *Initial* phase. However, the friends want to create an initial product that they can test with potential consumers, and so want to reach level 2, *Repeatable*, where they start to properly use an SDLC, with systematic coding practices. The CMMI model suggests that they implement processes such as requirements management, project planning, and software tracking. They start to reuse code and do proper testing. The more formal process helps them to notice problem areas sooner and create more efficient code.

The next stage in the process might be to define the process formally, taking them to level 3, *Defined*. Companies value firms that adopt CMMI. For example, when an organization issues a **request for proposal (RFP)**, it might state that the software developer is certified for CMMI level 3 or higher. Organizations want to be certain that a minimum level of security development processes is being followed.

SAMM

The **Open Worldwide Application Security Project (OWASP)** is an open source organization that created SAMM to help organizations improve their software security by offering a framework for SDLC implementation. It can be used by firms of all sizes and industries, and can even be applied when using third-party vendors for development. SAMM's philosophy is that software security is everyone's responsibility, and it provides guidance on how to involve all stakeholders in the security process.

SAMM is divided into five levels:

- **Level 1, Awareness**: Level 1 focuses on raising awareness of software security within the organization.

- **Level 2, Basic protection**: Basic protection involves implementing basic security controls to protect software assets.

- **Level 3, Managed processes**: This level focuses on establishing and managing security processes throughout the SDLC.

- **Level 4, Continuous improvement**: Continuous improvement advances the organization's software security posture.

- **Level 5, Advanced practices**: Level 5 focuses on implementing advanced security practices, such as threat modeling and security architecture.

Through SAMM, OWASP provides a suite of tools to help organizations implement the framework. These include a self-assessment tool that helps organizations assess their current software security posture, guidance on how to improve the organization's software security posture, and a set of security controls and processes that can be implemented to protect software assets and security posture.

Operations and Maintenance

The **operations and maintenance (O&M)** phase of a software development project occurs after the software is released to production. During this phase, software maintenance and support ensure that the application continues to meet the needs of the users. This includes bug fixes, performance improvements, and adapting the software to any new requirements or environments.

The O&M phase typically includes the following activities:

- **Monitoring**: The application is monitored to ensure that it is performing as expected. This includes auditing software performance and security.

- **Troubleshooting**: When problems occur, they must be resolved. This involves root cause analysis, development, and implementation of the solution.

- **Patching**: Security vulnerabilities are patched as soon as possible so that the application maintains minimal levels of security.

- **Upgrades**: The software is upgraded to deploy new features and improve performance.

- **Training**: Users are trained on how to use the software via books, training sessions, or online tutorials.

- **Support**: Users are provided email or phone support for problems encountered with the application.

Some key challenges of O&M include the following:

- The software environment is constantly changing, so it can be difficult to keep up with modifications.

- O&M can be costly due to software monitoring, troubleshooting, patching, upgrading, training, and support.

- Security solutions make software more difficult to use, and usability improvements often create vulnerabilities. This makes striking a balance between the two difficult.

Despite these challenges, O&M is an important part of the SDLC. It ensures that the software continues to meet the needs of its users and that it is protected from security vulnerabilities.

To manage O&M effectively, the organization needs to plan ahead to ensure the necessary resources are available and use automation to reduce the cost and complexity of O&M. Next, a communication plan is needed – it should include contact information for support staff, as well as instructions on how to report problems. Finally, the organization should continuously improve processes by identifying areas for improvement and implementing changes.

Change Management

Organizations are always changing. New computers, new staff, and new applications all cause a company to function differently. In *Chapter 17*, *Security Operations*, you read about general changes in IT architecture, In software development projects, change management is also essential for ensuring that changes to the project are implemented efficiently because once one thing changes, everything changes. For example, implementing a new feature might require changes to the database schema, updates to the user interface, modifications to the API, and additional testing.

The change management process typically includes the following stages:

1. **Identifying the need for change**: This can be done by stakeholders or project managers and is initiated because the client desires some new function in the application or even a patch.

2. **Assessing the impact of change**: The changes might improve the product but cause a vulnerability, or break another functionality.

3. **Developing a change management plan**: The plan should include the purpose, scope, impact, affected users, communications, training, testing, and implementation planning.

4. **Communicating the change**: Make sure the correspondence is clear, concise, and informative.

5. **Training**: Training should ensure that stakeholders understand the change and its impacts.

6. **Testing the change**: It's very important to make sure the change has no negative impacts on the project and provides the intended result.

7. **Implementing the change**: This implementation should be done in a controlled and orderly manner. Create a "rollback" plan in case the change causes an unexpected result.

8. **Monitoring the change**: Ensure the change is working as intended and that there are no negative impacts on the project.

By following the preceding steps, project managers can confirm that changes to their projects are successful.

Change management, along with SDLC methodologies such as Agile and DevOps and maturity models with integrated security assessments, extends beyond the traditional functions of a developer. Efficiently planning, creating, and shipping high-quality digital products requires more than just coding; it involves a variety of roles and functions that are often integrated within a single team, as will be discussed in the next section.

Integrated Product Team

An **integrated product team (IPT)** is a multidisciplinary group of people who are collectively responsible for delivering a defined software project. IPTs are used in complex development programs for review and decision-making. In a large-scale software development project, an IPT might include developers, testers, UX designers, and product managers who collaborate to ensure that a new feature is seamlessly integrated into the system. When facing a critical decision on how to address a major security vulnerability, the IPT reviews various solutions, evaluates their impact, and decides on the best approach to mitigate the risk while maintaining project timelines.

The emphasis of the IPT is on the involvement of all stakeholders, including users, customers, management, developers, and contractors in a collaborative forum. IPTs may be addressed at the program level, but there may also be **oversight IPTs (OIPTs)** or **working-level IPTs (WIPTs)**.

IPTs are created most often as part of structured system engineering methodologies, focusing attention on understanding the needs of each stakeholder. Because IPTs bring job roles of different disciplines together, communication is smoother and collaboration is easier. This leads to better decision-making and faster problem-solving. The use of IPTs assures that the needs of the customer are met throughout the development process, resulting in customer loyalty, cost reductions, and innovation.

However, IPTs require a change in the way that people work. The cross-functional collaboration will impact everything from meetings to requirement documentation and sign-offs at critical stages. This can be frustrating, as it can create the perception of delays and barriers due to differing concerns among job roles, particularly in organizations accustomed to traditional, hierarchical structures. With training and good management, these issues can be mitigated. Some strategies for creating and managing an effective IPT include having a clear vision, selecting the right team members, continuous communication, and above all, celebrating success.

Summary

In this chapter, you learned about the several frameworks used for creating applications, including Waterfall, Agile, and DevOps. Each of these has a suite of tools that work along with them to help teams create secure operating systems, programs, and websites.

You also learned about the importance of ensuring that suppliers meet minimal security levels and have application security policies in place. These can be measured with maturity models such as CMMI and SAMM.

Operations and maintenance are essential considerations. Once an application is released, it's important to determine whether patches will need to be created or whether the application will be updated with new features. You learned that applications are not static and must be changed for various reasons. Changes are planned and designed as part of the SDLC.

Finally, you learned how changes need to be tracked and managed, and that as you improve applications, you should make sure that changes are secure. You learned how IPTs help you by making sure applications meet the needs of the customer, vendors, suppliers, and other working teams in the organization. The next chapter will cover software development security controls.

Further Reading

Here are some good references on the SDLC, change management, and integrated product teams:

- NIST Special Publication 800-218, *Secure Software Development Framework (SSDF) version 1.1: Recommendations for Mitigating the Risk of Software Vulnerabilities*, `https://packt.link/YMv9D`, February 2022.

- *Change Management Best Practices Guide*, `https://packt.link/mABie`, United States Agency International Development, May 8, 2015.

- *Unifying R&D with Integrated Product Teams*, Dr. R. Brothers, `https://packt.link/haAIh`, October 26, 2015.

Exam Readiness Drill – Chapter Review Questions

Apart from a solid understanding of key concepts, being able to think quickly under time pressure is a skill that will help you ace your certification exam. That is why working on these skills early on in your learning journey is key.

Chapter review questions are designed to improve your test-taking skills progressively with each chapter you learn and review your understanding of key concepts in the chapter at the same time. You'll find these at the end of each chapter.

> **How to Access These Materials**
>
> To learn how to access these resources, head over to the chapter titled *Chapter 24, Accessing the Online Resources*.

To open the Chapter Review Questions for this chapter, perform the following steps:

1. Click the link – `https://packt.link/chapter20`.

    Alternatively, you can scan the following **QR code** (*Figure 20.3*):

Figure 20.3: QR code that opens Chapter Review Questions for logged-in users

2. Once you log in, you'll see a page similar to the one shown in *Figure 20.4*:

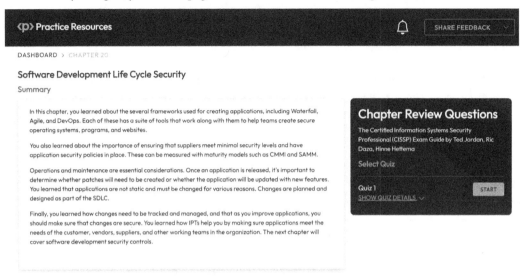

‹ρ› Practice Resources

DASHBOARD > CHAPTER 20

Software Development Life Cycle Security

Summary

In this chapter, you learned about the several frameworks used for creating applications, including Waterfall, Agile, and DevOps. Each of these has a suite of tools that work along with them to help teams create secure operating systems, programs, and websites.

You also learned about the importance of ensuring that suppliers meet minimal security levels and have application security policies in place. These can be measured with maturity models such as CMMI and SAMM.

Operations and maintenance are essential considerations. Once an application is released, it's important to determine whether patches will need to be created or whether the application will be updated with new features. You learned that applications are not static and must be changed for various reasons. Changes are planned and designed as part of the SDLC.

Finally, you learned how changes need to be tracked and managed, and that as you improve applications, you should make sure that changes are secure. You learned how IPTs help you by making sure applications meet the needs of the customer, vendors, suppliers, and other working teams in the organization. The next chapter will cover software development security controls.

Chapter Review Questions

The Certified Information Systems Security Professional (CISSP) Exam Guide by Ted Jordan, Ric Daza, Hinne Hettema

Select Quiz

Quiz 1   START
SHOW QUIZ DETAILS ⌄

Figure 20.4: Chapter Review Questions for Chapter 20

3. Once ready, start the following practice drills, re-attempting the quiz multiple times.

Exam Readiness Drill

For the first three attempts, don't worry about the time limit.

ATTEMPT 1

The first time, aim for at least **40%**. Look at the answers you got wrong and read the relevant sections in the chapter again to fix your learning gaps.

ATTEMPT 2

The second time, aim for at least **60%**. Look at the answers you got wrong and read the relevant sections in the chapter again to fix any remaining learning gaps.

ATTEMPT 3

The third time, aim for at least **75%**. Once you score 75% or more, you start working on your timing.

> Tip
>
> You may take more than **three** attempts to reach 75%. That's okay. Just review the relevant sections in the chapter till you get there.

Working On Timing

Target: Your aim is to keep the score the same while trying to answer these questions as quickly as possible. Here's an example of how your next attempts should look like:

Attempt	Score	Time Taken
Attempt 5	77%	21 mins 30 seconds
Attempt 6	78%	18 mins 34 seconds
Attempt 7	76%	14 mins 44 seconds

Table 20.1: Sample timing practice drills on the online platform

> Note
>
> The time limits shown in the above table are just examples. Set your own time limits with each attempt based on the time limit of the quiz on the website.

With each new attempt, your score should stay above **75%** while your "time taken" to complete should "decrease". Repeat as many attempts as you want till you feel confident dealing with the time pressure.

21
Software Development Security Controls

Software development security controls are measures implemented throughout the **software development lifecycle** (**SDLC**) to mitigate application vulnerabilities. Such controls focus on secure coding practices and secure coding standards. Secure coding practices help developers write code that mitigates vulnerabilities. Secure coding standards are rules that developers must follow to ensure code meets baseline levels of security. Both coding practices and secure coding standards cover the entire lifecycle of development.

Computer programming, **continuous integration and continuous delivery** (**CI/CD**), and software security automation help implement effective software security controls and reduce the risk of vulnerabilities being introduced into applications. You will first learn about security around computer programming.

In the context of CI/CD, security controls focus on integrating security checks into the pipeline. This includes using **static application security testing** (**SAST**) tools to scan code for vulnerabilities, using **dynamic application security testing** (**DAST**) tools to scan running applications for vulnerabilities, and using penetration testing to identify exploitable vulnerabilities.

Software security automation is the use of tools and techniques to automate security tasks throughout the SDLC. This can include automating the deployment of security controls, automating the execution of security tests, and automating the remediation of vulnerabilities.

By the end of this chapter, you will be able to answer questions on the following:

- Computer programming
- Continuous integration and continuous delivery
- Software security automation

You will start with a review of computer programming.

Computer Programming

Computer programming is the process of developing a set of instructions to direct a computer to perform a specific task such as updating an electronic address book or recording a video. These instructions (known as code) are written in a programming language that defines a syntax of instructions that the computer can understand. It is important to consider security during the writing of these instructions. Malicious, or even poorly written, code can introduce security flaws into a system. In this section, you will learn about the essential components of the development environment, which are **programming languages**, **libraries**, **tool sets**, **integrated development environments** (IDEs), and **runtime**, and their implications for security.

There are a multitude of programming languages available, including **Python**, **Java**, **C/C++**, **JavaScript**, and **PHP**. Libraries provide programmers with access to commonly used functionality such as access to a keyboard and screen. Toolsets assist programmer developers by providing security solutions. IDEs provide programmers with tools for writing and debugging code. Runtimes provides support for securely executing code. This section will start with a look at programming languages.

Programming Languages

A **programming language** contains a set of instructions that tell a computer what to do, such as running your operating system or processing information in an app. There are hundreds of programming languages available, each with its own purpose, such as web development, data analysis, or mobile apps. The language used for a project is determined by the requirements and the customer. **Processing chips** work solely in binary code, 1s and 0s, so all programming languages are essentially translated into binary code, although they do so through various levels of abstraction. Understanding these levels of abstraction can help in identifying potential security vulnerabilities in code, as different languages and their abstractions may impact how security flaws are introduced and addressed.

Computer languages are categorized into five different levels or generations, listed next. Security is managed somewhat differently depending on the generation:

- **First-generation (1GL, machine language)**: 1GL is the lowest-level language and consists of only binary code, 0s and 1s. Coders seldom write at this level as it is long and complex, so it is almost impossible to implement any security measures. This also makes it more difficult for malicious actors to work at this level, though any attacks or other bugs are difficult to detect by simply reading 1s and 0s.

- **Second-generation (2GL, assembly language)**: Assembly is a low-level symbolic representation of machine language. Symbolic addresses and mnemonics represent operations, such as `mul` for multiply, and `sub` to subtract for an x86 processor. An assembler converts assembly language into machine code. Assembly language provides low-level access to system resources, which can be risky if not managed properly. Although symbolic, 2GL is still very close to the hardware and can be prone to errors and vulnerabilities that are hard to debug. It is often used in contexts where buffer overflow vulnerabilities can be exploited because memory is managed manually, rather than automatically.

- **Third-generation (3GL, high-level language)**: 3GLs are the languages most programmers use today, and are either compiled or interpreted. Compiled languages include **Fortran, C, C++,** and **C#**. In these languages, source code is converted into machine code by a compiler before being distributed to end users. This compiled code is often in a binary format, which makes it harder for end users to modify or understand directly. This makes it harder to inject malicious code, but also harder to check for it. While compiled code is less readable and thus harder to alter, it can still be reverse-engineered or decompiled to inspect for malicious code or vulnerabilities.

  Interpreted languages include **Python, JavaScript,** and **Ruby**. For these languages, the source code is executed by an interpreter at runtime rather than being compiled into machine code beforehand. Because the code is often distributed in a more readable form, it can be easier for an attacker to insert malicious code. On the other hand, since the code remains accessible in a more human-readable format, it can also be easier to detect such malicious code.

  The distinction between compiled and interpreted languages is blurred. Java uses source code compiled to bytecode, which is interpreted by the **Java Virtual Machine (JVM)**.

- **Fourth generation (4GL, very high-level language)**: 4GLs are designed to be closer to human language. They perform specific tasks, such as SQL database manipulation or report generation. 4GLs are declarative and aim to accomplish more with fewer lines of code compared to 3GLs. This simplicity also makes it easier to manipulate the language for malicious purposes. A common example is SQL injection, where a user can use an input field on a webform to crack into a database. This is covered in *Chapter 8, Architecture Vulnerabilities and Cryptography*.

- **Fifth generation – (5GL, natural language)**: 5GLs aim to simplify human-computer interaction in the most natural way possible. Examples include Prolog and Lisp, which are associated with artificial intelligence and expert systems. Interaction with large data models and natural language has opened up a whole new area of security, and there are ongoing conversations about how AI should be used securely and ethically. Working with large data models means there is a risk of data poisoning in an attempt to manipulate models with false data or trying to fool language models into producing privileged or dangerous information.

Libraries

A library contains reusable code that is used to perform common tasks. Libraries are typically organized by **functions**, such as **networking, databases, graphics**, and more. These save programmers time by providing pre-written code used in their own programs. Other benefits include improved code quality because you use a set of well-tested, well-documented, and well-secured code. However, because you are not the creator of the library, you should always be aware of what the library is and who developed it, so that you can keep track of any reported security vulnerabilities.

Because libraries are widely used by many developers, they often benefit from extensive testing and feedback. This leads to quicker identification and resolution of bugs, which can enhance the overall security of applications by ensuring that patches and updates are implemented more swiftly. Additionally, using well-established libraries can make your application more readable and maintainable, as these libraries are typically well-documented and familiar to other developers.

Libraries include standard, third-party, and in-house programs. The programming language provides a set of standard libraries to support keyboards and mice, for example. Third-party libraries are made by outside organizations, and in-house libraries are made by your organization. All libraries must be evaluated for security to mitigate backdoor and buffer overflow attacks, as discussed in *Chapter 22*.

Tool Sets

Code can be written in **rich text** and checked manually before the next stage, but this can be time-consuming and prone to errors, especially given the precise syntax programming languages require. Just like a writer might use a word processing program with grammar checks, word prediction, and autoformatting, developers have access to development tools.

A programming toolset helps programmers save time and improve code quality by making processes easier. It includes a collection of tools, namely editors, compilers, and debuggers. Because tool sets are themselves applications, you should be aware that they have the same vulnerabilities as any other code-based asset. Toolsets should be run in secure environments, and you should pass all the checks that any other application needs. Make sure the programming tool set supports your development language and platform, whether it is Linux, Windows, Mac, or another. Only use toolsets provided by firms that follow security protocols, otherwise your applications could be laced with malware.

Integrated Development Environments

IDEs are applications where developers can write code efficiently. They boost productivity for software developers by providing a unified environment for developing, debugging, and testing software. IDEs include a text editor, **compiler**, **debugger**, **syntax checking**, and **version control**.

Common IDEs include **Visual Studio**, developed by **Microsoft** for **Windows**, **Eclipse** for Windows, Mac, and Linux, and **NetBeans**, developed by Oracle for Windows, Mac, and Linux. IDEs support several programming languages, including C++, C#, and Java. The Eclipse IDE is shown in *Figure 21.1*.

Figure 21.1: The Eclipse integrated development environment

In addition to the standard checks mentioned above, IDEs can include integrated security checking software such as Veracode. Because IDEs have direct access to your code, you should ensure you trust them, and that they're run in a secure environment. You should always use IDEs provided by firms that follow security protocols.

Runtime

The runtime environment is a type of virtual machine that provides an execution environment for an application to run within. Runtimes allow a developer to create an application that runs in any environment, such as a website. Key concerns include vulnerabilities within the runtime itself, improper sandboxing and isolation, code injection risks, insecure configurations, dependency management issues, memory management flaws, inadequate access control, interpreted code security, logging and error handling exposures, and weak enforcement of security policies.

To mitigate security issues, it is important to keep runtime updated, check security certificates during auditing, and ensure configurations are set properly. If runtime environments accept dynamic inputting, ensure input validation.

Examples of runtime environments include the **Java Virtual Machine (JVM)**, **Microsoft .NET Framework (.NET)**, and the **Python Virtual Environment (PVM)**. These virtual machines are available for many platforms, including Windows, Mac, and Linux.

Continuous Integration and Continuous Delivery

Continuous Integration and Continuous Delivery (CI/CD), as covered in *Chapter 20, Software Development Life Cycle Security*, is a **development, security, and operations (DevSecOps)** process that integrates frequent code changes into a central repository. Deployments are automated and enable faster and more reliable updates, as shown in *Figure 21.2*. This agile practice includes collaboration with stakeholders, monitoring quality, and continuous feedback.

CI/CD can introduce automated scanning tools into the pipeline, uncovering vulnerabilities before applications make it into production. However, these integrated systems can introduce their own security concerns.

DevOps and **DevSecOps** make use of code repositories such as **GitHub** that open code to exploitation if not secured properly with techniques such as multi-factor authentication. Automated build scripts should also be properly managed to ensure no misconfigurations that open up vulnerabilities in applications. Pipelines make use of dependencies such as libraries, which need to be properly assessed, or even third-party services for record-keeping that also need to be vetted. Finally, a CI/CD pipeline can be the direct target of an attack, so access controls, audit logs, and scanning tools can be used to check for malicious actors.

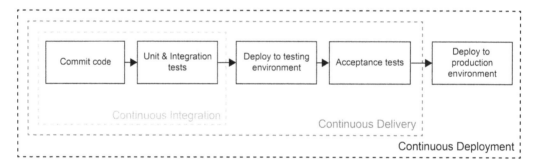

Figure 21.2: CI/CD pipeline workflow

Application Security Testing

When it comes to securing applications within the CI/CD pipeline, there are two primary methods. The first method is visually reviewing source code for security vulnerabilities, known as **static application security testing (SAST)**. Each developer reviews another developer's code, without executing it. There are also automated tools designed to assist in **SAST**. SAST tools include **SonarQube**, Veracode, and **Checkmarx**, which will analyze code for known vulnerabilities.

The second method is known as **dynamic application security testing (DAST)**, which tests for security vulnerabilities in the running state. Rather than inspecting the code, the application is tested by some kind of proxy user, such as a tester or software, checking the app does what it's intended to do, and logging any errors. Black box testing means the tester does not have knowledge of the inner workings of the software or source code. DAST tools include OWASP ZAP and Nessus, which find vulnerabilities in running applications.

Interactive Application Security Testing (IAST) combines elements of SAST and DAST to identify security vulnerabilities while the application is running, for example, testing how well the application responds to fuzzy inputs.

The effectiveness of these methods depends on the capabilities of the tools integrated into your CI/CD pipeline and the experience of the development team.

Software Security Automation

Organizations can improve the security of their software development processes with tools that automate security tasks, track software changes, and store source code in such a manner that multiple developers can access the latest version. This section introduces the following frameworks and tools that address these requirements:

- **Security orchestration**, **automation**, and **response**
- **Software configuration management**
- **Code repositories**

Benefits of these systems include increased efficiency, improved collaboration, and, most importantly, better security.

Security Orchestration, Automation, and Response

Security orchestration, automation, and response (**SOAR**) is a security framework that integrates and automates security tools and processes. SOAR solutions help security teams automate repetitive tasks and respond to incidents faster.

Orchestration integrates security tools such as a **SIEM**, **IDS**, **IPS**, and firewalls to efficiently respond to incidents. Automation robotizes repetitive tasks, such as threat hunting, incident response, and compliance reporting via playbooks that respond to specific attacks, such as moving phishing attacks to a spam folder. Finally, response systematizes the response to security incidents, such as triaging alerts and remediating vulnerabilities.

Software Configuration Management

Software configuration management (**SCM**) is the process of tracking and controlling changes in software. SCM helps to ensure that changes are made in a coordinated manner, which is especially important in a continuous development pipeline that includes multiple iterations of code. SCM includes audit trails to track what changes to code were made, by whom, and when; fine-grained access controls so people can only change the sections they need to; and configuration management to reduce the risks of configuration issues across platforms introducing vulnerabilities. SCM should also include strict update policies to maintain security best practices and rollback to make it easy to revert to earlier versions of code if bugs or attacks occur.

SCM can be implemented using a variety of tools and techniques, such as **Ansible**, **Chef**, **Puppet**, and **SaltStack**, which enable **infrastructure-as-code** (**IaC**). IaC systems allow administrators to build hundreds of preconfigured computer environments in minutes in the cloud.

Code Repositories

Code repositories are essential tools for software development teams because they allow teams to store, manage, and share code securely. Several are available, and one of the largest is github.com. However, code repositories can also be a target for **cyberattacks**. Common threats to code repository security include **data breaches** and **denial-of-service** attacks. **GitHub** fell victim to one of the largest DDoS attacks in history, at 1.3Tbps.

Steps to secure code repositories include strong passwords, two-factor authentication, and code scanning tools mentioned previously to identify security vulnerabilities before being committed to the application. Code scanning tools are often built into code repositories such as **GitHub**, **GitLab**, and Bitbucket.

Summary

In this chapter, you explored various aspects of software development. Understanding the different levels of programming languages helps in analyzing the foundational principles of how code interacts with computer systems. Earlier-generation languages, which are closer to machine language and pure binary, provide more direct access to system resources, while later-generation languages are easier to read and more portable. The differing characteristics of these programming languages lead to different security concerns when considering secure coding practices. Additionally, you examined how the compilation process and runtime environments can introduce new security challenges.

You saw CI/CD security controls focus on integrating security checks into each stage of the software development life cycle. This includes using SAST tools to scan code for vulnerabilities, and DAST tools to scan running applications for weaknesses. Security analysts also use penetration testing to identify flaws.

Software security automation is the use of tools that automate security tasks throughout the SDLC. This includes automating the deployment of security controls, security tests, and remediation of vulnerabilities. The next chapter will cover securing software development.

Further Reading

Below are some good references for designing and developing secure applications using the SDLC and CI/CD:

- National Center for Education Statistics. "Safeguarding Your Technology," US Department of Education. `https://packt.link/ioWdg`. 2012. Web.

- "Defending Continuous Integration/Continuous Delivery (CI/CD) Environments," Cybersecurity and Infrastructure Security Agency, `https://packt.link/UJKG1`, June, 2023.

- "Automating Risk Analysis of Software Design Models," M Frydman, et. al. National Library of Medicine. `https://packt.link/y2Mm1`. 2014.

Exam Readiness Drill – Chapter Review Questions

Apart from a solid understanding of key concepts, being able to think quickly under time pressure is a skill that will help you ace your certification exam. That is why working on these skills early on in your learning journey is key.

Chapter review questions are designed to improve your test-taking skills progressively with each chapter you learn and review your understanding of key concepts in the chapter at the same time. You'll find these at the end of each chapter.

> **How to Access These Materials**
>
> To learn how to access these resources, head over to the chapter titled *Chapter 24, Accessing the Online Resources*.

To open the Chapter Review Questions for this chapter, perform the following steps:

1. Click the link – https://packt.link/chapter21.

    Alternatively, you can scan the following **QR code** (*Figure 21.3*):

Figure 21.3: QR code that opens Chapter Review Questions for logged-in users

2. Once you log in, you'll see a page similar to the one shown in *Figure 21.4*:

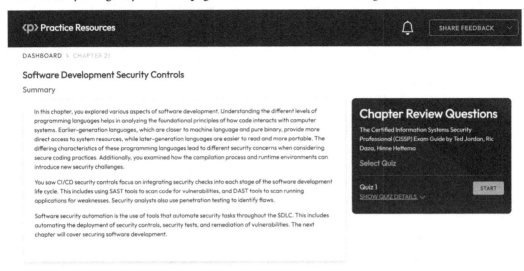

Figure 21.4: Chapter Review Questions for Chapter 21

3. Once ready, start the following practice drills, re-attempting the quiz multiple times.

Exam Readiness Drill

For the first three attempts, don't worry about the time limit.

ATTEMPT 1

The first time, aim for at least **40%**. Look at the answers you got wrong and read the relevant sections in the chapter again to fix your learning gaps.

ATTEMPT 2

The second time, aim for at least **60%**. Look at the answers you got wrong and read the relevant sections in the chapter again to fix any remaining learning gaps.

ATTEMPT 3

The third time, aim for at least **75%**. Once you score 75% or more, you start working on your timing.

> Tip
>
> You may take more than **three** attempts to reach 75%. That's okay. Just review the relevant sections in the chapter till you get there.

Working On Timing

Target: Your aim is to keep the score the same while trying to answer these questions as quickly as possible. Here's an example of how your next attempts should look like:

Attempt	Score	Time Taken
Attempt 5	77%	21 mins 30 seconds
Attempt 6	78%	18 mins 34 seconds
Attempt 7	76%	14 mins 44 seconds

Table 21.1: Sample timing practice drills on the online platform

> Note
>
> The time limits shown in the above table are just examples. Set your own time limits with each attempt based on the time limit of the quiz on the website.

With each new attempt, your score should stay above **75%** while your "time taken" to complete should "decrease". Repeat as many attempts as you want till you feel confident dealing with the time pressure.

22
Securing Software Development

In the previous chapter, you saw the different types of programming languages and how modern software development impacts how we consider security. Software is everywhere, including inside phones, cars, and medical devices. But even the best programmers make mistakes, and as software gets more complex, so does the error count. Also, as software complexity grows, the more likely life can be endangered. For example, self-driving cars now use software to control the engine and brakes. A software glitch could cause a crash.

By the end of this chapter, you will be able to answer questions on the following:

- Assessing the effectiveness of software security
- Assessing the security impact of acquired software

Let's start by learning about the effectiveness of software security so that once you obtain your CISSP certification, you will be prepared to examine security policies for software development.

Assessing the Effectiveness of Software Security

After the initial coding phase, any software needs a rigorous testing regime. This involves ensuring that the system, application, and processed data remain confidential and accessible. Identifying potential vulnerabilities that could expose data or compromise the system is critical. Organizations need to evaluate the software development process, assessing its effectiveness and pinpointing areas for improvement. This can be done through two methods:

- Auditing and logging of changes
- Risk analysis and mitigation

The following subsections explain these methods in detail.

Auditing and Logging of Changes

Auditing and logging are essential components of the **software development lifecycle** (**SDLC**). Through tracking changes, organizations can enhance software quality and security. Comprehensive auditing provides a detailed historical record of the software's evolution, enabling effective troubleshooting.

Effective logging begins in the early stages of development. Utilizing version control systems such as Git to track code changes allows developers to revert to previous code versions if bugs appear in later versions. For example, if *TicTacToe v.2.1* is released and found to have critical bugs, it's relatively easy to roll back to *TicTacToe v2.0* until the bugs are fixed.

Issue tracking tools document bug reports, feature requests, and other development tasks. These practices establish a solid foundation for further analysis. As development progresses, it's crucial to log events such as code builds, deployments, and test results to identify vulnerabilities and bottlenecks.

To ensure the integrity of the software, security-focused auditing, such as tracking user activities, system access, and configuration changes, is paramount. Monitoring these areas allows organizations to detect unauthorized access and compliance violations promptly. Finally, detailed logs aid in forensic investigations of security incidents.

As software evolves, it becomes more complex. Consider Debian Linux, with its 50 million lines of code; the error potential for this would be very high. Formal frameworks can help. Following frameworks such as the SEI **Capability Maturity Model Integration** (**CMMI**), which is discussed in *Chapter 20, Software Development Life Cycle Security*) can lower the error rate to just 1 per 1,000 lines, versus the average error rate of about 30 errors per 1,000 lines, by setting out the best coding practices, such as systematic testing. This minimizes risks and makes complex projects such as LibreOffice more manageable. *Figure 22.1* shows the five maturity levels with a short description.

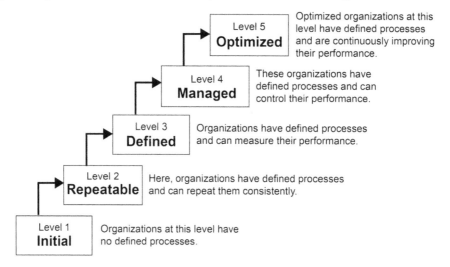

Figure 22.1: Characteristics of CMMI maturity levels

When purchasing software from a third-party vendor, you can use this as a guide for how skilled you want the developer to be. For example, a start-up might be operating at **Level 1**, which means their process is ad hoc and without set processes. This could potentially mean that testing is not systematic and, therefore, their products are less trustworthy. However, their solution might be really important. By understanding the maturity level, an organization can weigh up the risks, and mitigations of using the product against the cost and benefit. For instance, the company could purchase the product but run its own intensive checks on it.

Testing is crucial. Robust testing methodologies with tools such as **SonarQube** for static testing and **Fortify** for dynamic and interactive testing are essential for large software projects (see *Chapter 21, Software Development Security Controls*). They can help catch errors before they become major security vulnerabilities. To understand the kinds of errors that could be present, it's worth looking at the types of software vulnerabilities.

Risk Analysis and Mitigation

Understanding the types of risk you face is step to ensuring that your software is being developed in a secure way. This section provides an overview of common application vulnerabilities and tracks research provided by the **Open Worldwide Application Security Project (OWASP)**. OWASP publishes a reference standard for the most critical application risks, the OWASP Top Ten. The IT security industry is essentially an arms race, with attacks evolving and new vulnerabilities emerging all the time. It is the security professional's job to stay abreast of changes. The following are currently the most common, and are worth looking at in more detail:

- Injection attacks
- Buffer overflows
- Race conditions
- Cross-site scripting
- Cross-site request forgery
- Path traversal

Let's learn about these vulnerabilities and their proper mitigations.

Injection Attacks

An **injection attack** occurs when an attacker provides unexpected input. For example, the application might request the user to enter a postal code. In the United States, this is a 5-digit number such as **48231**; however, the attacker instead enters something like \ ! * (n in the hopes of either creating a denial of service by shutting the system down or hacking into the system to obtain credit card numbers.

One way to mitigate these attacks is the use of **input validation** or **data validation**. By rejecting certain characters or character strings, an application can prevent malicious input. In *Figure 22.2*, the left image shows how an attacker can reach the PII database because there is no validation. The attacker is now able to steal credit card numbers, addresses, phone numbers, and so on.

The right image shows the input validation stage. Now, data is confirmed to be valid before it's processed.

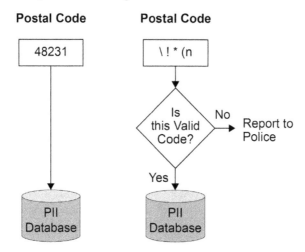

Figure 22.2: SQL injection example

SQL Injection Attacks

SQL injection is a special class of injection in which a hacker launches database commands from an input. For example, from a login screen where a user might simply use their login name and password such as ted@jordanteam.com and MyP@ssw0rd, the attacker simply enters login name montrie' OR 1=1-- *without* a password (the single quote that follows montrie is important for the attack to work). If the system is vulnerable to SQL injection, the attacker will instantly be logged in as the first user. This works because the attacker logs in with a SQL command that is always true; 1 will always equal 1. Again, input validation can mitigate these attacks by not allowing certain characters (such as -) or certain strings (such as a number followed by an equals sign). This is explained in more detail in *Chapter 8, Architecture Vulnerabilities and Cryptography*.

Software errors are inevitable. Even the best programmers make mistakes, and even one error can have serious security implications. In 2014, a bug was discovered in the OpenSSL cryptography library that allowed attackers to read sensitive data directly from the memory of the server. This enabled them to steal keys and passwords. Dubbed the Heartbleed bug, it was caused by a single line of code but impacted websites including Tumblr, Google, and Dropbox. You can read more about it at the HeartBleed website https://packt.link/k2CZg.

Buffer Overflows

A buffer overflow occurs when a program writes more data to a memory buffer than the latter can accommodate. This excess data spills into adjacent private or raw memory areas causing the computer to fault in unexpected manners. The overflow could overwrite critical data, causing a system to crash and create a **denial of service (DoS)**.

For example, an application might have a field for a username. Instead of the attacker placing a normal entry such as gwcarver, they enter doctor-george-washington-carver-the-peanut-researcher because the hacker knows usernames should be 16 characters or less. The application attempts to read the long login name, and the memory buffer overflows, causing the application or server to fault, as shown in *Figure 22.3*.

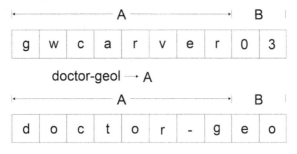

Figure 22.3: Buffer overflow example

A skilled attacker could even force an overwrite of a flag in memory indicating users and administrators, then gain administrator rights by inserting their own credentials.

Certain programming languages (notably **C** and **C++**) are more susceptible to buffer overflow risks. These languages do not inherently block the access or overwriting of data in unallocated memory areas. Additionally, they don't automatically ensure that data written to arrays, which act as built-in buffers, stays within the defined boundaries of those arrays.

One approach to counteract buffer overflows is bounds checking, which involves extra coding and processing to ensure that data stays within its designated array. Despite its effectiveness, this method adds to the overall complexity and processing requirements of the program. The Heartbleed bug mentioned earlier in the chapter was caused by such a boundary error.

In response to these vulnerabilities, modern operating systems have developed several defensive strategies.

These include randomizing the memory layout to make it less predictable and using **canaries**, which are strategically placed bits of data, normally integers, adjacent to buffers. The term "canaries" derives from when miners worked in coal mines. If a caged canary died, that signaled miners to leave the cave immediately because there was odorless yet poisonous gas. Similarly, canary values can be checked, and if they have been overwritten, they indicate a buffer overflow and data in that section of memory will not be trusted.

Race Conditions

One example of a race condition is when two people are purchasing an online ticket for an event. If they press "Buy Now" at the exact same time, the system sells two tickets even though there is only one available. This occurs because there is a slight delay between the system checking for ticket availability, finding that there is only one, and then updating that value to zero. During this window, both purchase requests can be validated as successful, even though only one ticket is available. Exploiting the time it takes for a system to check and update data can enable attackers to carry out activities such as financial fraud, causing a DoS, or even gaining elevated permissions, such as switching accounts while a check is underway. This type of vulnerability is also known as a **time of check to time of use** (**TOCTOU**) attack.

One mitigation to this type of attack is to validate each transaction so that if two entries come at the exact same time, only one is allowed. Most databases use some form of the ACID algorithm (**ACID** stands for **atomicity, consistency, isolation, and durability**). This algorithm is based on the principle that when two entries attempt to make an identical purchase, drop both entries and force the user to try again. In the earlier example, one customer will eventually be able to purchase a ticket. This is explained further in *Chapter 23*.

Cross-Site Scripting

Cross-site scripting (**XSS**) is a security vulnerability that allows attackers to inject malicious code into emails and social networking sites. The attack commonly uses JavaScript to launch malware that can steal PII, perform session hijacking, or install ransomware.

One common attack is to hide an XSS attack within an email. For example, a reader may receive an email to click a link to see pretty cats:

`To see pretty cats, CLICK HERE!`

However, underneath the CLICK HERE link, there is HTML code that will launch some malicious code, such as this example:

`To see pretty cats,<script href="malware.js">CLICK HERE!</script>`

In the preceding example, the victim will be fooled into installing malware because they do not see the underlying attack, only the CLICK HERE button. The best mitigation is to deploy input validation. Proper input validation via a host-based firewall will deny the execution of any scripts by either noticing the < and > symbols or by blocking all JavaScript.

Cross-Site Request Forgery

Cross-site request forgery (**XSRF** or **CSRF**) is an exploit that tricks a victim into running unwanted commands on a website where they are authenticated. For example, suppose an individual is logged in to their bank account online; an attacker could send an email with a link designed to execute an action in the bank, such as transferring money. The attacker takes advantage of the fact that the user is already logged in.

Here is an example of how a CSRF attack might be carried out:

```
<a href = "https://bankingwebsite.com/
transfer?amount=1000&toAccount=attackerAccount">Click here!</a>
```

The best mitigations are to use input validation on HTML POST requests so that certain requests or character strings are blocked, or to always require tokens in requests that are verified by the server.

> **Note**
>
> See this video and watch an XSRF attack in action where Linus Tech Tips temporarily lost his major YouTube channel due to the attack: `https://packt.link/CeHFF`.

Path Traversal

Path traversal (also known as directory traversal) occurs when an insecure web application allows attackers to access all files on a server, including private data. This allows attackers to steal PII and execute website defacement. For example, imagine a server that makes you $1 million per month now earning you about $5 per month because malicious actors now control the website.In a secure system, the customer is restricted to the website area. However, if the system is vulnerable to path traversal, an attacker could manipulate the URL by using `../` (dot-dot-slash), which means "go up one directory." So, an attacker might run `http://example.com/download?file=../../` `etc/passwd`. This exploits the vulnerability and allows the attacker to access the `etc/passwd` file on a Linux system.

Mitigations for this attack include input validation and using secure websites allowing only HTTPS access instead of HTTP access. **HTTP Strict Transport Security** tells web browsers to always connect using HTTPS, which mitigates path traversal and other cyberattacks.

To uncover vulnerabilities, conduct assessments, such as penetration testing and code reviews. Also, prioritize vulnerabilities based on their severity, impact, and exploitability so that you can focus resources on the most important threats, as discussed in *Chapter 2*.

Next, apply mitigations, such as patching, input validation, access control, and intrusion detection. Patching known vulnerabilities remains the most effective defense, so having a robust patch management system in place ensures swift deployment of security updates. Mitigation is not a one-time fix, so continuous monitoring is critical. Mitigation combined with a culture of security awareness can significantly reduce software risks.

Assess the Security Impact of Acquired Software

Ideally, as technical professionals, you would prefer to trust what software vendors say about their applications. However, as security professionals, you must verify and validate all applications' capabilities and security. The first step is understanding your business requirements of the application and then making sure it fulfills those requirements in a secure manner.

In this section, you will learn how to assess the security of different types of acquired software such as the following:

- **Commercial off-the-shelf (COTS)**
- Open source
- Third-party
- Managed services

Let's first cover secure approaches for COTS software.

COTS

When selecting COTS software, thoroughly evaluate vendor claims to ensure they align with the product's actual capabilities and your requirements. Conducting a bake-off, in which multiple COTS products that meet your requirements are tested and validated against each other, is a valuable method. Avoid relying on product roadmaps or what some call *vaporware*, as these represent future ideals rather than guaranteed features.

Security integration must also be considered. Software should align with your organization's existing security measures, such as firewalls, intrusion detection systems, identity and access management, and encryption protocols. Assessing whether the software can integrate with these tools will prevent security gaps. For example, if your organization uses a specific authentication method, the COTS product should support this method to ensure secure access control. When you evaluate the **total cost of ownership** (**TCO**), which includes both capital expenses and annual maintenance costs to make informed financial decisions, consider extra security measures, the cost of security maintenance, and the cost of changing any existing infrastructure.

To gain valuable insights into selected COTS solutions, consider utilizing third-party research organizations. This can save you time and money from running thorough in-house testing. Research organizations provide assessments, comparisons, and a starting point for evaluation.

Engage with existing customers of the software candidates to gain real-world perspectives. However, exercise caution when relying on contacts provided by the vendor, as they may not be objective. Independent feedback from users through personal networks or online forums can be more beneficial.

All software will need routine security patch updates, so when evaluating COTS solutions, consider potential alternate vendors in case your primary vendor goes out of business. Investigate the availability of support via third-party organizations to ensure ongoing patch updates.

Open Source Software

Open source software, or **copyleft** (a play on copyright) software, provides the blueprints to an application – in other words, the instructions to make the application run are visible. Open source software is governed by various licensing agreements, including the **GNU Public License (GPL)**, **Lesser GPL (LGPL)**, **Massachusetts Institute of Technology (MIT/X)** license, **Berkeley Software Distribution (BSD)** license, and the **Apache** license. Under some open source licenses, you are permitted to use, modify, and distribute the source code in your own applications, but you must provide proper attribution to the original authors of that code.

GPL is the most commonly used license and champions the concept of free software (where "free" means the liberty of users to utilize, modify, and distribute software), although this is not unique among open source software.

The GPL's copyleft provision mandates that any alterations to GPL-licensed software must also remain free, prohibiting the transformation of GPL code into proprietary software. In contrast, other free licenses such as BSD and LGPL permit the transition of licensed code into proprietary applications.

From a security point of view, a significant benefit of open source software is that it's available for everyone to see. Since there are more eyes on it, flaws can be found quicker. Malware is discovered very quickly in popular projects because vulnerabilities are spotted in the source code. This is part of the reason why Linux, in its many versions, has grown in popularity. Before deciding on an open source application, you should search newsgroups, such as Reddit, and CVE for known vulnerabilities.

Unlike proprietary software, open source projects might not offer formal support or maintenance contracts. This means users need to be more proactive in managing the software. It might also lack the organizational capabilities to ensure timely communication of updates. You should regularly check for security patches and other important information in open source communities as well as the same systematic testing you do for other software.

Third-Party

Custom-developed software is an alternative to COTS, providing bespoke solutions tailored to specific needs. However, this approach has risks and benefits. When engaging with third-party development partners, clear contract language, verification and validation, NDAs, and **service-level agreements** (**SLAs**) are essential for all aspects of the project, including security.

Never assume that security will be built in by default unless explicitly mandated in the contract. Clearly define security requirements both during contract negotiations and SLA drafting. To ensure the vendor's commitment, incorporate these security requirements into the SLA and verify that they align with the vendor's security policy. You may wish to agree on minimum security standards and penalties for their not being met.

As with COTS, it's crucial to address critical questions related to vendor sustainability because of the impact on security updates, feature gaps, and support. What happens if the third-party development company ceases operations? Do they offer a code escrow solution? How will missing critical features be addressed? How accessible is support for the custom-developed application?

Managed Services

Managed services are security and information services provided by third-party providers known as **managed service providers** (**MSPs**). MSPs allow clients to focus on their core business while they manage IT-related issues such as operations and security. These services can include **infrastructure as a service** (**IaaS**), **platform as a service** (**PaaS**), or **software as a service** (**SaaS**), as discussed in *Chapter 8*. As with any other supplier, it's essential to ensure that the MSP has documented security policies in place. One way to verify this is by reviewing their **service organization control** (**SOC**) reports, which provide detailed information on their internal controls and security practices. SOC reports are discussed in detail next.

SOC Reports

SOC is a reporting framework that demonstrates a firm's commitment to security. SOC reports are generated by independent auditors to ensure an organization has installed security controls and can demonstrate their effectiveness in protecting PII, PHI, IoT, desktops, servers, and other assets.

There are three types of SOC reports:

- **SOC 1** focuses on internal controls over financial reporting. SOC 1 reports are used by financial services organizations such as banks, mortgage, investment, and other firms that must comply with financial regulations.

- **SOC 2** borrows techniques from the financial industry but focuses on security and privacy. If you are using an MSP, you can ask them to provide SOC 2 reports, which verify that the firm has controls in place (**Type I**), or validate that the controls are effective (**Type II**).

- **SOC 3** reports redact private details from the SOC 2 reports and are public-facing reports posted on an MSP's websites and other marketing channels. SOC 3 reports demonstrate the MSP's commitment to security.

After reviewing the MSP's SOC 3 report, you may be required to sign an NDA to review either type of SOC 2 report. It's important to choose between Type I, which provides a snapshot of controls at a specific time, and Type II, which evaluates the effectiveness of those controls over a longer period. Key areas to assess include the MSP's security measures, availability, processing integrity, confidentiality, and privacy controls to ensure they align with your organization's needs. Conduct a detailed analysis of the control objectives, testing methods, and results, paying close attention to any deficiencies and the MSP's action plans for addressing them. You should ensure continuous monitoring by regularly receiving updated SOC 2 reports and maintaining open communication with the MSP to discuss any ongoing or emerging security concerns.

Further Securing Managed Services

- In addition to the SOC reports, there are several other ways to secure managed services. Best practices include the following:Establish a clear SLA. The SLA should clearly define the security responsibilities of both the client and the MSP, and detail who is responsible for areas of overlap such as APIs, scripts, patching, and data security.

- Implement strong authentication and access controls by deploying **multi-factor authentication (MFA)** and **role-based access control (RBAC)**. This is discussed in *Chapter 12, Identity, Access Management, and Federation.*

- MSPs will assist you with monitoring, logging, and disaster recovery activities to detect suspicious activity and investigate incidents quickly and effectively. MSPs work with you during an incident. Therefore, in case of a security incident, it is critical to notify the MSP as soon as possible so that they can start their portion of the incident management process.

By following these best practices, you can secure managed services and protect private data and computing technologies.

Summary

This chapter started with a discussion about the effectiveness of software security. You examined how to audit software changes and the importance of logging software changes. When users write thousands of lines of code, errors are more likely.

One system that helps to measure the software development process for developers is the CMMI model. The closer the development team gets to the top of the scale, or 5, the better that developer's processes are. Then you learned about software vulnerabilities; the top 3 have to do with poor authentication, broken encryption, and injection attacks. Other vulnerabilities are SQL injection and buffer overflow attacks.

Next, you learned about working with software developers and COTS software. In most cases, you're not going to be able to see the source code, so it's important to make sure that these developers have strong software development processes and secure policies; again, the CMMI model can help find the best developer.

Finally, the chapter concluded with an overview of managed software providers and a review of the SOC 3 or SOC 2 report to make sure they follow software security policies. To confirm that developers have security controls in place and that they are effective, you need to view the SOC 2 Type II report. The SOC 2 Type I report only confirms that the vendor has security controls in place.

You will read about secure coding practices and issues around artificial intelligence and security in the next chapter.

Further Reading

Here are some good resources for learning more about secure software development and secure software policies:

- *A Quick Guide to GPLv3*: `https://packt.link/7w5ld`
- SQL injection: `https://packt.link/5XrHL`

Exam Readiness Drill – Chapter Review Questions

Apart from a solid understanding of key concepts, being able to think quickly under time pressure is a skill that will help you ace your certification exam. That is why working on these skills early on in your learning journey is key.

Chapter review questions are designed to improve your test-taking skills progressively with each chapter you learn and review your understanding of key concepts in the chapter at the same time. You'll find these at the end of each chapter.

> **How to Access These Materials**
>
> To learn how to access these resources, head over to the chapter titled *Chapter 24, Accessing the Online Resources*.

To open the Chapter Review Questions for this chapter, perform the following steps:

1. Click the link – `https://packt.link/chapter22`.

   Alternatively, you can scan the following **QR code** (*Figure 22.4*):

Figure 22.4: QR code that opens Chapter Review Questions for logged-in users

2. Once you log in, you'll see a page similar to the one shown in *Figure 22.5*:

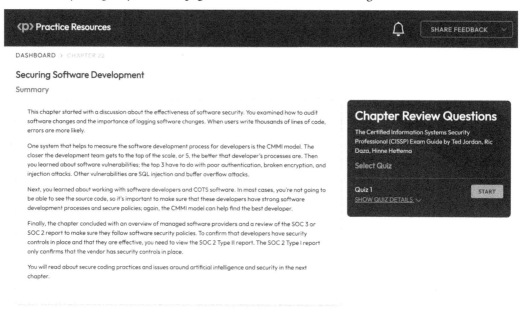

Figure 22.5: Chapter Review Questions for Chapter 22

3. Once ready, start the following practice drills, re-attempting the quiz multiple times.

Exam Readiness Drill

For the first three attempts, don't worry about the time limit.

ATTEMPT 1

The first time, aim for at least **40%**. Look at the answers you got wrong and read the relevant sections in the chapter again to fix your learning gaps.

ATTEMPT 2

The second time, aim for at least **60%**. Look at the answers you got wrong and read the relevant sections in the chapter again to fix any remaining learning gaps.

ATTEMPT 3

The third time, aim for at least **75%**. Once you score 75% or more, you start working on your timing.

> Tip
>
> You may take more than **three** attempts to reach 75%. That's okay. Just review the relevant sections in the chapter till you get there.

Working On Timing

Target: Your aim is to keep the score the same while trying to answer these questions as quickly as possible. Here's an example of how your next attempts should look like:

Attempt	Score	Time Taken
Attempt 5	77%	21 mins 30 seconds
Attempt 6	78%	18 mins 34 seconds
Attempt 7	76%	14 mins 44 seconds

Table 22.1: Sample timing practice drills on the online platform

> Note
>
> The time limits shown in the above table are just examples. Set your own time limits with each attempt based on the time limit of the quiz on the website.

With each new attempt, your score should stay above **75%** while your "time taken" to complete should "decrease". Repeat as many attempts as you want till you feel confident dealing with the time pressure.

23
Secure Coding Guidelines, Third-Party Software, and Databases

Keeping architectures secure is not simply about ensuring malicious actors cannot enter your system with viruses or false identities. The applications that underpin the everyday use of information technologies can also be a source of vulnerabilities. Malicious code can be placed in seemingly useful but benign software you might purchase to improve your workflow. Programming errors might create tiny logical errors that can cascade into large systematic issues, and code that has not been updated for a few years can create vulnerabilities that the original developers did not anticipate.

To ensure applications are safe, security must be designed into the software development process at the beginning of any software project. Both developers and suppliers must follow security standards, and users must understand the risks and know how to test for and mitigate them.

For many organizations, one of the central applications is a database. While database security often aligns with overall system security, it is crucial to differentiate between various types of databases and to analyze key concepts such as concurrency, SQL, and polyinstantiation.By the end of this chapter, you will be able to answer questions on the following:

- Security vulnerabilities at the source code level
- Security of **Application Programming Interfaces (APIs)**
- Secure coding practices
- Establishing secure databases
- Artificial intelligence

This chapter details secure coding practices, the protection of supplier codes, and issues around artificial intelligence and security. Let's first discuss security vulnerabilities at the source code level.

Security Vulnerabilities at the Source Code Level

Software powers everything from critical infrastructure to personal devices, so the security of source code is of utmost importance. Not only can attackers compromise applications, as discussed in *Chapter 22, Securing Software Development*, but rogue developers can also write malware into "trusted" applications. One positive way this is done is by developers programming "cheat codes" into a video game. These are used for video game testing but are later leaked so that users can enjoy the game more. However, malicious developers can write malware that establishes a communication link, known as beaconing, allowing the hacker to compromise your computer or even a nuclear facility (see the article on Stuxnet in *Further Reading*). In this section of the chapter, we will first look at different application testing approaches and then discuss some root causes of security weaknesses.

Testing

Computer applications can be tested with different approaches. Three popular methods include **white-box**, **black-box**, and **gray-box** testing. With white-box testing, the examiner has access to all of the source code and knows all the internals of the application. The goal is to fully evaluate code for logical errors and other vulnerabilities, testing its internal workings rather than user-facing output. It often includes unit testing, integration testing, and line-by-line security assessments. A good example of code that can be white-box tested for security threats is an open source application, such as Linux or LibreOffice, where the source code is available to all users.

With black-box testing, the examiner only knows what outputs are created from certain inputs and knows nothing about the internals of the application. In black-box testing, the source code is not provided, so the functionality of the application is assessed, mimicking the experience of an end user. It also mimics how an external attacker might approach the application so it can reveal security flaws that might not have been apparent by focusing directly on the code. This can be done with external penetration testing or even bug bounties (see *Chapter 14, Designing and Conducting Security Assessments*). Applications such as macOS and Lotus Notes are closed source and examiners will not have access to the source code, so security testing is done using black-box testing.

In gray-box testing, the examiner has access to some of the source code—for example, an API they may have developed—but they don't have access to the source code that the API connects to. For example, a shopping website might offer APIs to developers to calculate taxes, but the shopping cart will not allow developers to view how authentication is handled.

Systematic testing will uncover weaknesses, but it is also important to understand the root cause of these weaknesses, which will be covered in the next section.

Security Weaknesses

Software security weaknesses often arise from unintentional errors, logic loopholes, legacy burdens—such as reliance on outdated operating systems or other dependencies—third-party trust, which involves relying on external code such as libraries, and code escrow, where third-party code is stored as a safeguard in case a company goes out of business. The next section will look at these in more detail.

Unintentional Errors

At the heart of many vulnerabilities lie simple coding mistakes, such as those discussed in *Chapter 21, Software Development Security Controls*. These **unintentional errors** could include unvalidated user inputs, insecure storage, and weak authentication. Such mistakes can allow attackers to inject malicious code, access sensitive information, or bypass security controls.

Logic Loopholes

Logic loopholes are not necessarily mistakes on their own. A block of code in isolation might have no issues, but as software development projects grow from hundreds to thousands to millions of lines of code, they are more likely to harbor hidden vulnerabilities. These loopholes arise when the combined behavior of different parts of the software introduces unintended consequences. As discussed in *Chapter 22, Securing Software Development*, race conditions, unexpected program interactions, buffer overflows, and poorly programmed algorithms can all be exploited to manipulate programs and gain unauthorized access.

Because these vulnerabilities are part of the normal application, they can be very difficult to identify, so thorough analysis and testing must be done.

Legacy Burdens

Legacy burdens are when you are dependent on outdated software that may not have been maintained with proper security patching. Users depending on application and operating system upgrades can be at risk of security vulnerabilities when a software company goes out of business, or when software is no longer supported. For example, thousands of users still use Windows 98 even though it is no longer supported by Microsoft. This is usually because of some software program the client was using that didn't work when they tried to upgrade to the latest version of Microsoft Windows.

The best recommendation you can give a client like this is to make sure the system never accesses the internet. This is because new malware is written every day, and unsupported systems will not have the updates to mitigate it.

Third-Party Trust

Many applications rely on **third-party libraries** and **APIs**. These offer great convenience to users, but such dependencies introduce additional risks because they must be verified and validated for security. Vulnerabilities in external code can be just as damaging as those in your own, creating a chain reaction of potential exploits. Careful assessment and management of third-party dependencies are crucial for maintaining overall security.

Code escrow acts like insurance for the software source code that your organization needs. If you license a software application but the developer goes out of business, stops supporting it, or simply disappears, you are stuck with an application that can no longer be updated or patched; code escrow mitigates this risk.

With code escrow, the developer entrusts the program's source code to a third-party escrow organization. This code remains locked away, like money in an escrow account. However, under specified circumstances, such as the developer going bankrupt, the escrow agent can release the source code to you. Now, you can maintain the application yourself, hire another developer, or even sell the rights.

Security of APIs

Modern websites employ detailed controls, often interacting with multiple web services such as online payment processors, social media platforms, and shipping providers. To smooth these communications, allow apps to directly talk to each other via function calls.

Think of an API as a special phone line directly connected to a service's "brain." By using the right codes (called **function calls**), you can make the service do things such as post updates on social media or check product availability. For example, you might have a system automating orders from a third party. The third party will set up an API for their inventory system and provide the documentation telling you how to code requests, as well as a key or token to use when making the request. The following is an example. GET is the method used to make a request. 12345 is the product code and /availability is the request to check whether the product is in stock. YOUR_ACCESS_TOKEN is the authentication key supplied by the third party:

```
GET https://api.onlinestore.com/v1/products/12345/availability
Authorization: Bearer YOUR_ACCESS_TOKEN
```

API keys are simple identifiers used to track and control access to an API, often embedded directly in your code. Access tokens, on the other hand, are more secure, short-lived credentials used to access protected resources, often on behalf of a user.

Security practice recommendations for APIs include deploying **API authentication**, such as passwords to prevent unauthorized access, or the use of special keys for sensitive actions, such as placing a high-value order. Developers must obtain these keys or tokens from the service provider and implement them in their applications. They are responsible for verifying that the keys or tokens are valid and properly configured before allowing their application to interact with the API and access its resources.

> **Note**
> Treat API keys like passwords, including sending them over encrypted communication channels only.

APIs need thorough security testing to identify vulnerabilities. Software tools such as **curl**, **links**, or **telnet** can be used to simulate API requests and check for weaknesses in API communication. *Figure 23.1* is an example of using telnet to test a connection. The `Unable to connect` response tells the tester that the connection is closed.

```
bosko@pnap:~$ telnet google.com 22
Trying 142.251.39.78...
Trying 2a00:1450:400d:80e::200e...
telnet: Unable to connect to remote host: Network is unreachable
```

Figure 23.1: Using telnet on the command-line interface

The following figure, *Figure 23.2*, tells the tester the connection is open because the response is `Connected to google.com`.

```
bosko@pnap:~$ telnet google.com 80
Trying 142.251.39.78...
Connected to google.com.
Escape character is '^]'.
```

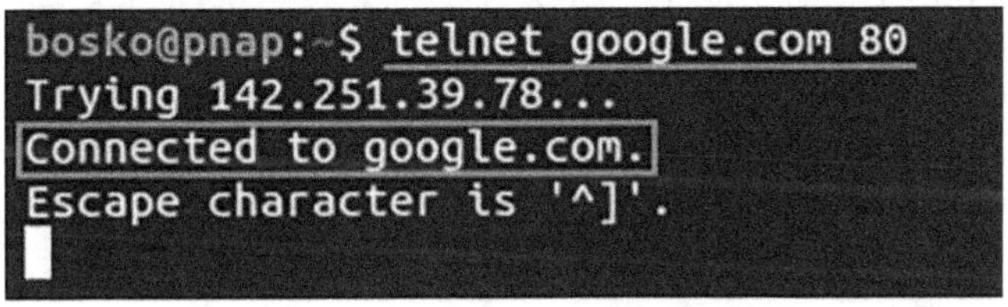

Figure 23.2: Response for an open connection

The use of APIs is growing, as they allow smoother, automated interaction with third-party services, social media sites, and even AI models such as ChatGPT. But as with all third-party software interaction, careful consideration of security concerns needs to be paramount, and tokens or keys should be protected with the same vigilance as passwords. Typically, APIs exist within larger blocks of code, and thus the standard secure coding practices should be applied. These are discussed in the next section.

Secure Coding Practices

In addition to the secure coding practices defined by OWASP, as discussed in *Chapter 22*, another consideration around software applications includes how a software program responds to failures. For example, if you are depositing $1,000 into your bank account and the ATM malfunctions during this process, you don't want the ATM to lose your money. In this case, you would want the ATM to fail in such a way that your money is still secure. For example, the drawer closes tight so that only a trusted bank employee can access your deposited money.

How systems behave in failure scenarios is crucial for building reliable applications. There are four choices when planning for software failure:

- **Fail-secure**: When an application is designed with fail-secure features, it prioritizes preventing damage in case of failure. In the event of an error, the system transitions to a safe state, even if it means sacrificing some functionality. For example, an e-commerce website that fails during checkout cancels the transaction to prevent unauthorized charges, even if it means losing a potential sale.

- **Fail-open**: A fail-open system prioritizes availability, even if it means leaving that system vulnerable. For example, if a news website encounters a database issue, it can still display cached content, even if users won't see the latest updates.

- **Fail-closed**: The opposite of fail-open, fail-closed systems prioritize data security by denying access upon failure. This is useful for sensitive systems where threats pose a significant risk. For example, if an online banking application encounters a security breach, it automatically locks down accounts to prevent any access until resolution.

- **Fail-over**: Finally, fail-over applications switch to redundant systems upon failure to minimize downtime. For example, a web server experiencing a hardware failure will automatically route to a backup server, ensuring continued service when designed with a fail-over system.

Design your applications as needed and weigh several factors, such as the type of application, its criticality, risks, and user expectations. Carefully consider the trade-offs between safety, availability, integrity, and continuity to choose the best fail method for your software.

Software-Defined Security

With the increase of cloud-based architectures, such as infrastructure as a service, more security operations that used to be handled with hardware are now done by software. **Software-defined security (SDSec)** automates security management by replacing dedicated hardware appliances such as firewalls, IDSs, and IPSs with software replacements such as SDNs, as discussed in *Chapter 10*, and SOARs, as discussed in *Chapter 21*. SDSec also uses features of artificial intelligence and machine learning to analyze security data, identify threats, and predict potential attacks. SDSec centralizes an organization's security, making it more efficient and adaptable.

Overall, an SDSec approach secures organizations by streamlining security operations, improving defenses, and staying ahead of emerging threats. However, because SDSec is software, the same secure practices and software testing should be done as any other software, especially if it is provided by a third-party vendor.

Secure software practices involve systematic testing throughout the development, deployment, and operational phases of your code. This also includes ensuring that the application handles failures in a manner that mitigates potential damage and maintains security. To effectively apply these practices, it is essential to analyze the types of applications and architectures being used, understand their usage contexts, and assess the trustworthiness of third-party developers and their components. One of the largest and most important, as well as most common, applications you will come across is a database. This will be discussed in the next section.

Establishing Secure Databases

Organizations rely on databases to store crucial information, including customer details, order tracking, credit card numbers, and employee data. Securing databases is paramount to mitigate unauthorized access, data manipulation, and destruction.

This section details the different types of **database management systems (DBMSs)** and discusses how they are used and their key features. Then, you will look at **open Database Connectivity (ODBC)**, attacks and databases, mitigations, the role of machine learning, and artificial intelligence. First, we will start with database models.

Database Models

Databases play an important part in most organizational structures, as they are where the data is kept. There are various database models, with different types offering different ways of managing and organizing data effectively. The hierarchical database model, with its tree-like structure, illustrates parent-child relationships and is suitable for hierarchical data. Conversely, centralized databases centralize storage in a single location, simplifying management but posing scalability challenges, while distributed databases address these challenges by spreading data across multiple locations. The relational database model, which structures data into tables and uses keys to define relationships, offers robust data management and querying capabilities, supported by SQL for precise control and transaction integrity.

Hierarchical Database Model

A **hierarchical database model** represents a tree-like structure, which highlights parent-child relationships. An example of a hierarchical database model is shown in *Figure 23.3*, where each record can have several or no children.

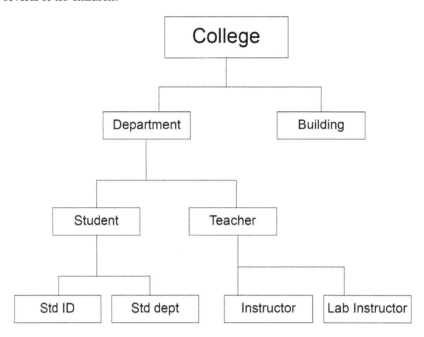

Figure 23.3: Hierarchical database model

The topmost record is known as the **root**. Any record that contains information and can have child records is known as the **parent** record. **Child** records contain information about their parent and can also have child records. A record will only have one parent, while a parent can have numerous child records. For instance, in the preceding diagram, a department might have many teachers, but a teacher will only be a member of one department. Every record, except those at the top and the bottom of the many records, will be the child of one record and the parent of one or more records.

Hierarchical database models work best where data naturally falls into a treelike structure, with clear single relationships between records—such as a company organization chart, where everyone has one line manager but a line manager has many reports, or a product catalog, with set categories such as electronics and subcategories such as computers, mobile devices, printers, and so on. Filesystems also commonly use a hierarchical structure with folders and subfolders. However, as the tree grows and more complex relationships, such as horizontal or multiple parent connections, are introduced, hierarchical databases can become inefficient—a relational database model is better.

Relational Database Model

The **relational database model** structures data into **tables**, where relationships are defined by columns called **attributes** and rows called **tuples**. *Table 23.1* shows an example of a table where the columns define the characteristics of an entry, and a row represents a single record or tuple.

Customer ID	Name	Address	Phone
1	Jai Lei	123 Birch St.	987-1234
2	Neha Sharma	456 Walnut St.	877-4242
3	Mike Lee	789 Acorn Ave.	393-1441

Table 23.1: A database table

Think of data in tables like entries in a phonebook, in which, to find a person, you would use their name. In databases, special sets of fields called **keys** act like the name. In the preceding table, the key is the customer ID.

Keys are unique identifiers for each record or tuple and can connect different tables. For example, if you want to see which phone numbers belong to people in a specific city, you can join the Phonebook table with a City table, using the City field as a key. You should be familiar with the following keys for the CISSP exam:

- **Primary key**: The primary key is a unique identifier for each record within a table. It ensures that no duplicate records exist. In the Customer table shown in *Figure 23.4*, Customer ID acts as the primary key.

- **Foreign keys**: Tables are linked together through foreign keys. A foreign key in one table references the primary key of the other, establishing relationships between related records. For example, an Order table might have the Customer ID column from *Figure 23.4* as a foreign key, connecting it to this Customer table.

Relational databases use a language called **Search Query Language** (**SQL**), which allows you to store, retrieve, and manage the database using simple commands. Different database companies such as **Oracle**, **Xano**, and **Immuta** have slightly different versions of SQL, but they share a core set of features, including how to define, query, and manipulate your database. A major strength of SQL is that it uses precision to offer great security. Think of SQL as the Swiss Army knife for databases in that it has a range of different tools that allow administrators and users to manage and access information.

Centralized/Decentralized Models

In addition to the structure of your database, where you store your database is also important as it will impact the security, efficiency, and reliability of your data. A **centralized database model** stores the entire database in a single location. The location is usually a server that can be accessed through a network. Although the database is at a single location, which simplifies management, it does not scale well and acts as a **single point of failure (SPoF)**.

To improve the shortcomings of the centralized database, many organizations use a **distributed database model**, which places data in several locations within an organization or across different regions. Each database location is called a node and provides benefits of improved scalability, availability, and performance over the centralized database model.

Processing Database Transactions

Relational databases use **transactions** to ensure **data integrity** and **consistency**. They are groups of SQL instructions that make sure everything goes smoothly. Each transaction is like a single action in that it either works entirely or not at all. If any part of the transaction fails, the entire transaction is rolled back, meaning that none of the changes are applied. This ensures that the data stays consistent and accurate.

For example, if two people attempt to purchase the same flight from Detroit to Milwaukee but there's only one seat left, the transaction attempts to execute a COMMIT command and finds that the transaction cannot be completed for both. The transaction is then aborted using the ROLLBACK feature, and both customers must attempt to purchase the ticket again until one of them gets it. This ensures that the integrity of the database is maintained.

Relational databases have four important characteristics: **atomicity**, **consistency**, **isolation**, and **durability**, or **ACID**, which help to maintain database integrity. Atomicity means all database activities are all-or-nothing transactions. Consistency means that the environment remains consistent with the database rules (for example, that each tuple should have a unique primary key). Isolation requires that transactions operate separately from each other. This mitigates race conditions by forcing one transaction to complete before another transaction is allowed to act upon the same data.

Finally, durability means once transactions are committed to the database, they are preserved. In a durable database, transactions are saved to log files and backed up by system administrators.

Overall, the relational database model is a versatile and robust approach for organizing and managing structured data, particularly in situations where data integrity, ease of querying, and standardized practices are crucial. However, it's essential to consider its potential limitations regarding complexity and performance for large-scale data management needs.

Regardless of how it is structured, databases tend to hold some of the most valuable and sensitive information of an organization. They can hold personal details, such as customer email addresses and credit card details, or personnel files. They are also used to hold passwords, keys, financial details, and patents. Even if your database is set up with ACID properties to ensure transaction integrity, it remains a prime target for malicious actors. Therefore, security must always be a top priority, as will be discussed in the next section.

Database Security

The strategies and techniques taught in this book that protect your architecture as a whole will logically also protect your database. However, there are also specific considerations to take into account when designing and building a database. For instance, you might have a situation where you are combining secret-level data with unclassified data. One approach is to use labels or tags to define the clearance and control access based on the user's security level. However, maintaining such a system is complex, and there is a risk of unauthorized access if the system fails or is improperly configured. For instance, a user with lower-level access might accidentally gain visibility into sensitive or classified information. The best solution is to have separate databases so the classified and unclassified information is handled separately.

The complexity and sophistication of, and reliance upon, modern databases opens them up to various forms of attacks, such as aggregation, inference, and concurrency. Strategies such as polyinstantiation and different database models can help protect sensitive data. In this section, you will look at key types of database attacks, how they can compromise information security, and effective mitigations to safeguard against vulnerabilities.

Aggregation Attacks

The process of combining data from different databases to reveal valuable insights is known as an **aggregation attack**. For example, a threat actor decides to target Liara because they believe she has a lot of money to steal. The attacker engages with her on Facebook and is able to learn her birth date. From public records, they learn her address. From a breach site, the attacker learns her tax identification number. Now the actor has all the information needed to create a fake credit card in Lisa's name.

On a larger scale, attackers who gain access to different databases can combine them using SQL techniques to reveal more information. For instance, a customer database might have names and order IDs. A product database might have order IDs, shipping addresses, and contact numbers. Combining the two could reveal customer names, addresses, and contact numbers, opening them up to phishing attacks, and harming the reputation of the business the data was stolen from.

A combination of security mitigations such as **defense in depth**, **need-to-know**, and **least privilege** help prevent such attacks. For example, deploying firewalls, restricting tax ID numbers to the purchasing department only, and allowing only managers to view tax ID numbers are great first steps. These are discussed in more detail in *Chapter 12, Identity, Access Management, and Federation*.

Inference Attacks

We saw how attackers can exploit seemingly harmless data to gain valuable insights through aggregation. **Inference attacks** employ a similar tactic. An attacker pieces together non-sensitive information to unlock classified data. Unlike aggregation's reliance on connecting a few databases, inference attacks exploit human deduction.

Imagine a student who has access to the names of all other students but not individual grades. The student notices the instructor posted the grades in alphabetical order of the students' names, even though all of the names have been redacted. Simply using the alphabet, they can infer the grades of every single student in the class. Just like with aggregation attacks, the best defense against inference attacks is careful permission control, defense in depth, and least privilege.

To combat aggregation and inference attacks, database partitioning can be used. This involves splitting the database into sections, each with its own security level and data type. This makes it harder for attackers to gather sensitive information by piecing together data from different parts.

This is also discussed in *Chapter 8, Architecture Vulnerabilities and Cryptography*.

Concurrency Attacks

Concurrency, also known as edit control, is the ability of a system, such as a database, to handle multiple processes or users accessing and modifying data at the same time. Properly implementing concurrency, such as with the methods mentioned earlier on in the chapter, in the section titled *Processing Database Transactions*, prevents database mishaps.

Ensuring only one process updates data at a time protects the database from the following issues:

- **Lost updates**: Imagine 2 shipping trucks, each with 5 packages, making a total of 10 packages. One truck delivers a single package, but due to an error, both trucks seem to have delivered a package. As a result, instead of the database showing nine packages remaining, it incorrectly shows eight packages. This issue occurs because the system doesn't properly coordinate the updates, leading to a scenario where one truck's update overwrites the others. Without proper concurrency control, the database might mistakenly show that both trucks have four packages left, even though one still has five.

- **Dirty reads**: If one truck's update is interrupted (e.g., due to a system crash) while it is adjusting the package count, another process might read the partially updated data. This could cause the database to display incomplete or inaccurate information until the original update is either completed or rolled back properly.

Concurrency Control

Concurrency control mechanisms, such as locks, are used to manage access to data. When one truck (or process) is updating a record, it locks that record, preventing other processes from reading or changing it until the update is finished. Once the update is complete, the lock is released, allowing others to access the now-consistent and accurate data.

In some cases, concurrency can act as a detective. By tracking changes with **logging** or **auditing**, administrators can see who modified which data and when. This helps them identify and repair errors that might have been missed. By keeping the data traffic organized, concurrency helps ensure the accuracy, integrity, and availability of information in the database.

Polyinstantiation

With polyinstantiation, you store multiple instances of the same data object within a database under different security levels; this allows people with higher privileges to see the record's PII, and those with lower privileges only to see the record's name—for example, a secured database tracking the locations of soldiers. Normally, these positions are classified, but what about a special division on a top-secret mission? It might be the case that only top officials know about the existence of this special force; revealing their actual location could raise suspicion, and even marking a database or a file as TOP SECRET could alert those with lower security clearances to the existence of the division.

This is where polyinstantiation comes in because it can create a secret identity for data. Instead of one record, you create two:

- **Top secret truth**: One record, classified "top secret," holds the team's actual location, accessible only to those with the highest clearance
- **Secret cover story**: The other record contains misinformation, so those viewers will not know about the true mission

Polyinstantiation relies on the ability to have granular control over a database. This is a bit like locking individual drawers in a filing cabinet instead of the entire cabinet. Polyinstantiation has two main approaches:

- **Content-dependent access control**: Security based on the actual content of an object. This offers detailed control but requires more processing power.
- **Cell suppression**: Hiding specific data cells or applying stricter security restrictions.

Similar to content-dependent control, **context-dependent access control** takes the bigger picture into account, by analyzing how each piece of data, such as location, time, user's role, and so on, relates to the entire operation. A seemingly harmless data point might become suspicious when combined with other information.

Another method similar to polyinstantiation, called **noise and perturbation**, involves adding misleading or false data to hide personal details. For example, in a medical trial, the ages and other statistical details such as the weight or height of subjects might be slightly altered by randomly adding or subtracting small amounts to the data points. These changes are designed to be minimal enough to preserve the overall data integrity for analysis but significant enough to prevent the identification of individual subjects.

So far, you have seen the most common database structures, attacks, and key security concerns. The way the world uses data is large and complex, and as data becomes valuable, so do methods of storing and processing data. In the final sections of this chapter, we will look at how different databases can connect to each other, ways of using databases without SQL, and the key data considerations around machine learning and artificial intelligence.

Open Database Connectivity

There are many different database systems out there, each with its own language. **ODBC**) acts as a translator, allowing your application to speak to any database, regardless of type, by providing a common language.

If you create an application that communicates via ODBC, the latter acts as a connector to translate your application's requests into the language that a particular database understands. When the database responds, ODBC translates the response back into the format that your application understands, as shown in *Figure 23.4*.

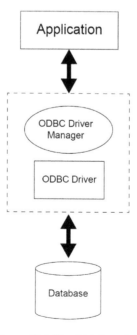

Figure 23.4: The ODBC model

This frees you from having to learn the specific language of every database system you want to use. ODBC lets you focus on building your application and handles the database communication.

NoSQL

The traditional relational model does not always fit the needs of the user in every circumstance—for example, in the case of unstructured data such as emails. Also, when data is being generated and processed at high speeds, it does not fit neatly into rows and columns. One example is vehicle temperatures and pressures that come from a sensor and are constantly changing. **NoSQL** (which stands for **Not Only SQL**) databases offer an alternative.

Common NoSQL types include the following:

- **Key-value stores**: Think of these as digital sticky notes that store data in simple pairs using a unique key, or label, and a value, called information or data—for example, *Car:Cadillac, Model:XT5*. This simple structure makes them lightning-fast for recording and processing large datasets.

- **Document stores**: Similar to key-value stores, these use keys for access, but the stored information is more complex. Documents, often in formats such as JSON or XML, hold the data. This is useful for situations where data has a more flexible structure, such as product information with varying details.

- **Graph databases**: Graph databases work using nodes for objects and connections (edges) to show relationships. This makes them ideal for anything network-related, such as social media, maps, or anything with interconnected data.

NoSQL databases differ significantly from relational ones, so security professionals must understand the specific features of the NoSQL solution and design security control solutions with database teams.

Knowledge-Based AI

Computers are great number crunchers, but what about tasks that require reasoning and expertise? **Artificial intelligence** (**AI**) is making strides in this arena and knowledge-based systems are at the forefront. They are a type of AI that attempts to emulate human expertise and decision-making processes. Knowledge-based systems draw on vast amounts of data, rules, and models to make decisions to mimic human expertise and decision-making, rather than following simple logic like traditional computing systems.

This section explores two key types of knowledge-based AI: expert systems and neural networks. Expert systems capture the knowledge of a human expert by tracking their work activity and use this stored knowledge to solve problems and make decisions. This is often used in medical contexts, where machines can learn from the diagnosis of expert doctors to build their own diagnostic capabilities. These results, used as a tool by a human doctor, allow for better patient care.

On the other hand, the creation of neural networks is inspired by the human brain; neural networks are complex AI systems that learn and improve through experience. Computer vision is an example of this. The next subsections will discuss how these systems can be powerful tools that aid in computer security.

Expert Systems

Expert systems employ AI that collects the knowledge of a human specialist in a specific area and then creates logic trees. They then analyze situations and provide solutions accordingly using the stored knowledge.

Expert systems function via two components:

- **Knowledge base**: This acts as the brain of the system and stores the expert's knowledge in a collection of rules, typically written as `if-then` statements.

- **Inference engine**: The inference engine is the problem solver. It receives information about a situation (for example, a car that is braking ahead) and uses knowledge-based rules to reach a conclusion, such as stop your car or turn and go past the stopping car.

Expert systems excel at consistent decision-making, especially in high-pressure situations where emotions can cloud judgment. For instance, when evaluating new hires, a business might use an expert system to remove potential bias from the process.

However, inaccurate or incomplete information in the knowledge base can lead to flawed decisions and systems struggle the more complex the requirements get. Despite these limitations, expert systems remain valuable tools in specific fields, including computer security—such systems can create mitigations for common cyberattacks.

Machine Learning

Instead of relying on explicit programming, as in expert systems, machine learning allows computers to analyze data and learn from it. Imagine a technical support engineer who builds expertise by studying and practicing tons of examples—this is the core idea behind machine learning. The data can then be used to predict and/or identify solutions in case of security events. There are two important ways machine learning approaches data. These are **supervised learning** and **unsupervised learning**.

Supervised learning uses labeled data, where each data segment is categorized or has a known outcome. The algorithm analyzes the labeled data, learns the underlying patterns and relationships, and then applies the knowledge.

For example, a security professional might train a machine learning model to detect network-based DoS attempts. The analyst provides a dataset of network traffic labeled as **malicious** or **normal**. The model analyzes the data, learns the characteristics, and uses the results to determine whether any network traffic is malicious.

Unsupervised learning is like giving a student a giant pile of unsorted data and asking them to make sense of it. It works with unlabeled data, such as seemingly random actions on an e-commerce website from millions of users. Such data has no predefined categories or outcomes. The algorithm's job is to find hidden patterns.

Imagine the same network traffic but without labels indicating which traffic is malicious. The unsupervised learning model might group network traffic based on factors such as time of day, location, or volume. Engineers can then review the data and discover patterns of suspicious activity.

By using supervised and unsupervised learning machine-learning techniques, computers become more sophisticated, making them valuable tools in cybersecurity.

Neural Networks

Neural networks are a complex type of machine-learning system inspired by the structure and function of the brain. Unlike expert systems with pre-programmed rules, neural networks consist of interconnected processing units that work together to solve problems.

Neural networks are becoming more advanced every day. Through training, they learn from vast amounts of data and can identify complex patterns. They can adjust their internal connections (called **weights**) based on new information, which allows them to continuously improve.

Neural networks are well-suited for voice recognition, face recognition, and weather prediction. For example, neural networks can analyze sound waves and distinguish between different voices. They can identify faces in images with impressive accuracy and analyze complex weather datasets to predict future weather patterns.

Neural networks are not pre-programmed with all the answers but learn through a process called **training** where they are exposed to data with known outcomes. The network adjusts the connections between its processing units based on how well it performs on the training data. This allows it to learn the underlying patterns and relationships within the data.

While neural networks don't yet replicate the full reasoning power of the human brain, they represent a significant leap forward in AI. As research continues, they have the potential to revolutionize cybersecurity because their ability to detect anomalies is unparalleled.

Summary

This chapter covered secure coding guidelines around working with suppliers and other third-party vendors. Whether you are developing applications on your own or have outsourced the work, it is important to design applications with security built in from the requirements stage of the SDLC. There are some risks of having third parties develop applications for you. One important risk is the possibility of the firm going out of business. Processes such as third-party trust and code escrow can mitigate the risks of you losing your entire project should the worst happen to your supplier. Most applications work with some type of database, so securing these database systems is important. You learned how ACID secures database transactions and data. Finally, you learned about the differences between expert systems, machine learning, and neural networks, and how these can secure applications.

Security is an ongoing process. It is the recommended best practice to integrate security measures throughout the entire SDLC, from the planning and development phases to the deployment and maintenance ones. By prioritizing application security, you can safeguard your organization's most valuable asset.

Further Reading

Here are some good articles on developing secure applications and managing third-party developers:

- *Your Guide to Source Code Escrow*, `https://packt.link/ngmIw`, Tech UK, 29 Mar 2023.

- *Certified Secure Software Lifecycle Professional* `https://packt.link/J9NFd`, International Information System Security Certification Consortium, 2024.

Exam Readiness Drill – Chapter Review Questions

Apart from a solid understanding of key concepts, being able to think quickly under time pressure is a skill that will help you ace your certification exam. That is why working on these skills early on in your learning journey is key.

Chapter review questions are designed to improve your test-taking skills progressively with each chapter you learn and review your understanding of key concepts in the chapter at the same time. You'll find these at the end of each chapter.

> **How to Access These Materials**
>
> To learn how to access these resources, head over to the chapter titled *Chapter 24, Accessing the Online Resources*.

To open the Chapter Review Questions for this chapter, perform the following steps:

1. Click the link – `https://packt.link/chapter23`.

    Alternatively, you can scan the following **QR code** (*Figure 23.5*):

Figure 23.5: QR code that opens Chapter Review Questions for logged-in users

2. Once you log in, you'll see a page similar to the one shown in *Figure 23.6*:

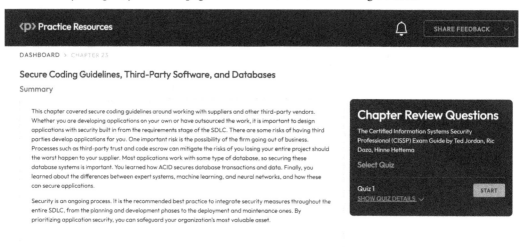

Figure 23.6: Chapter Review Questions for Chapter 23

3. Once ready, start the following practice drills, re-attempting the quiz multiple times.

Exam Readiness Drill

For the first three attempts, don't worry about the time limit.

ATTEMPT 1

The first time, aim for at least **40%**. Look at the answers you got wrong and read the relevant sections in the chapter again to fix your learning gaps.

ATTEMPT 2

The second time, aim for at least **60%**. Look at the answers you got wrong and read the relevant sections in the chapter again to fix any remaining learning gaps.

ATTEMPT 3

The third time, aim for at least **75%**. Once you score 75% or more, you start working on your timing.

> **Tip**
>
> You may take more than **three** attempts to reach 75%. That's okay. Just review the relevant sections in the chapter till you get there.

Working On Timing

Target: Your aim is to keep the score the same while trying to answer these questions as quickly as possible. Here's an example of how your next attempts should look like:

Attempt	Score	Time Taken
Attempt 5	77%	21 mins 30 seconds
Attempt 6	78%	18 mins 34 seconds
Attempt 7	76%	14 mins 44 seconds

Table 23.2: Sample timing practice drills on the online platform

> **Note**
>
> The time limits shown in the above table are just examples. Set your own time limits with each attempt based on the time limit of the quiz on the website.

With each new attempt, your score should stay above **75%** while your "time taken" to complete should "decrease". Repeat as many attempts as you want till you feel confident dealing with the time pressure.

24
Accessing the Online Practice Resources

Your copy of *Certified Information Systems Security Professional (CISSP) Exam Guide* comes with free online practice resources. Use these to hone your exam readiness even further by attempting practice questions on the companion website. The website is user-friendly and can be accessed from mobile, desktop, and tablet devices. It also includes interactive timers for an exam-like experience.

How to Access These Materials

Here's how you can start accessing these resources depending on your source of purchase.

Purchased from Packt Store (packtpub.com)

If you've bought the book from the Packt store (`packtpub.com`) eBook or Print, head to `https://packt.link/cisspunlock`. There, log in using the same Packt account you created or used to purchase the book.

Packt+ Subscription

If you're a *Packt+ subscriber*, you can head over to the same link (`https://packt.link/cissppractice`), log in with your `Packt ID`, and start using the resources. You will have access to them as long as your subscription is active.

If you face any issues accessing your free resources, contact us at `customercare@packt.com`.

Purchased from Amazon and Other Sources

If you've purchased from sources other than the ones mentioned above (like *Amazon*), you'll need to unlock the resources first by entering your unique sign-up code provided in this section. **Unlocking takes less than 10 minutes, can be done from any device, and needs to be done only once**. Follow these five easy steps to complete the process:

STEP 1

Open the link `https://packt.link/cisspunlock` OR scan the following **QR code** (*Figure 24.1*):

Figure 24.1: QR code for the page that lets you unlock this book's free online content

Either of those links will lead to the following page as shown in *Figure 24.2*:

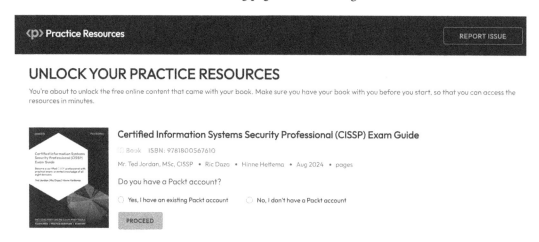

Figure 24.2: Unlock page for the online practice resources

STEP 2

If you already have a Packt account, select the option `Yes, I have an existing Packt account`. If not, select the option `No, I don't have a Packt account`.

If you don't have a Packt account, you'll be prompted to create a new account on the next page. It's free and only takes a minute to create.

Click `Proceed` after selecting one of those options.

STEP 3

After you've created your account or logged in to an existing one, you'll be directed to the following page as shown in *Figure 24.3*.

Make a note of your unique unlock code:

EUV9707

Type in or copy this code into the text box labeled 'Enter Unique Code':

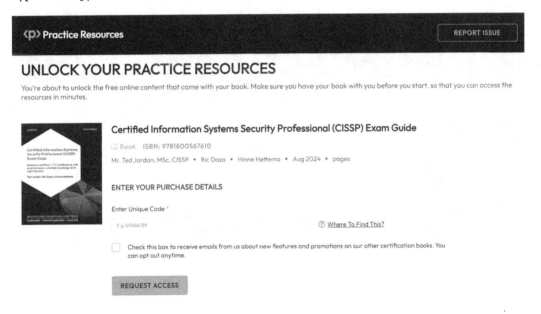

Figure 24.3: Enter your unique sign-up code to unlock the resources

> **Troubleshooting tip**
>
> After creating an account, if your connection drops off or you accidentally close the page, you can reopen the page shown in *Figure 24.2* and select `Yes, I have an existing account`. Then, sign in with the account you had created before you closed the page. You'll be redirected to the screen shown in *Figure 24.3*.

STEP 4

> **Note**
>
> You may choose to opt into emails regarding feature updates and offers on our other certification books. We don't spam, and it's easy to opt out at any time.

Click `Request Access`.

STEP 5

If the code you entered is correct, you'll see a button that says, OPEN PRACTICE RESOURCES, as shown in *Figure 24.4*:

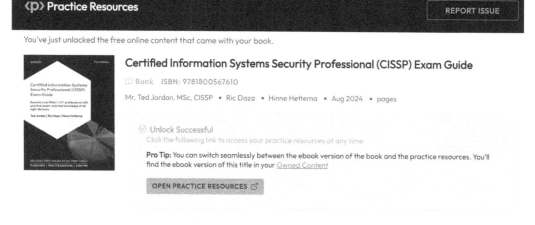

Figure 24.4: Page that shows up after a successful unlock

Click the OPEN PRACTICE RESOURCES link to start using your free online content. You'll be redirected to the Dashboard shown in *Figure 24.5*:

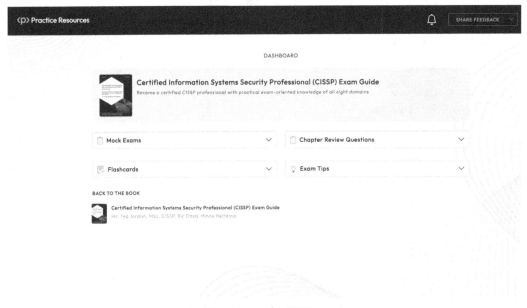

Figure 24.5: Dashboard page for CISSP practice resources

Bookmark this link

Now that you've unlocked the resources, you can come back to them anytime by visiting `https://packt.link/cisspractice` or scanning the following QR code provided in *Figure 24.6*:

Figure 24.6: QR code to bookmark practice resources website

Troubleshooting Tips

If you're facing issues unlocking, here are three things you can do:

- Double-check your unique code. All unique codes in our books are case-sensitive and your code needs to match exactly as it is shown in *STEP 3*.

- If that doesn't work, use the `Report Issue` button located at the top-right corner of the page.

- If you're not able to open the unlock page at all, write to `customercare@packt.com` and mention the name of the book.

Share Feedback

If you find any issues with the platform, the book, or any of the practice materials, you can click the `Share Feedback` button from any page and reach out to us. If you have any suggestions for improvement, you can share those as well.

Back to the Book

To make switching between the book and practice resources easy, we've added a link that takes you back to the book (*Figure 24.7*). Click it to open your book in Packt's online reader. Your reading position is synced so you can jump right back to where you left off when you last opened the book.

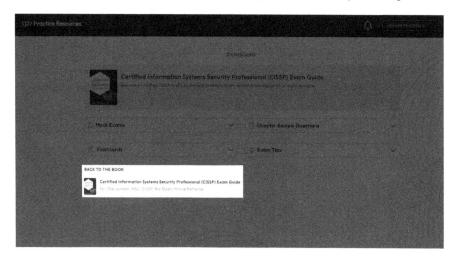

Figure 24.7: Dashboard page for CISSP practice resources

> **Note**
> Certain elements of the website might change over time and thus may end up looking different from how they are represented in the screenshots of this book.

Index

T

U

www.packtpub.com

Subscribe to our online digital library for full access to over 7,000 books and videos, as well as industry leading tools to help you plan your personal development and advance your career. For more information, please visit our website.

Why subscribe?

- Spend less time learning and more time coding with practical eBooks and Videos from over 4,000 industry professionals

- Improve your learning with Skill Plans built especially for you

- Get a free eBook or video every month

- Fully searchable for easy access to vital information

- Copy and paste, print, and bookmark content

At www.packtpub.com, you can also read a collection of free technical articles, sign up for a range of free newsletters, and receive exclusive discounts and offers on Packt books and eBooks.

Other Books You May Enjoy

If you enjoyed this book, you may be interested in these other books by Packt:

CISSP ISC2 Certification Practice Exams and Tests

Ted Jordan

ISBN: 978-1-80056-137-3

- Understand key principles of security, risk management, and asset security
- Become well-versed with topics focused on the security architecture and engineering domain
- Test your knowledge of IAM and communication using practice questions
- Study the concepts of security assessment, testing, and operations
- Find out which security controls are applied in software development security
- Find out how you can advance your career by acquiring this gold-standard certification

CCSP ISC2 Certified Cloud Security Professional Exam Guide

Omar A. Turner and Navya Lakshmana

ISBN: 978-1-83898-766-4

- Gain insights into the scope of the CCSP exam and why it is important for your security career
- Familiarize yourself with core cloud security concepts, architecture, and design principles
- Analyze cloud risks and prepare for worst-case scenarios
- Delve into application security, mastering assurance, validation, and verification
- Explore privacy, legal considerations, and other aspects of the cloud infrastructure
- Understand the exam registration process, along with valuable practice tests and learning tips

Share Your Thoughts

Now you've finished *Certified Information Systems Security Professional (CISSP) Exam Guide*, we'd love to hear your thoughts! Scan the QR code below to go straight to the Amazon review page for this book and share your feedback or leave a review on the site that you purchased it from.

https://packt.link/r/1800567618

Your review is important to us and the tech community and will help us make sure we're delivering excellent quality content.

www.ingramcontent.com/pod-product-compliance
Lightning Source LLC
LaVergne TN
LVHW081508050326
832903LV00025B/1416